Work-Based Learning

PRINCIPLES OF LIGHT VEHICLE TECHNOLOGY

Graham Stoakes

Heinemann is an imprint of Pearson Education Limited, Edinburgh Gate, Harlow, Essex, CM20 2JE.

www.pearsonschoolsandfecolleges.co.uk

Heinemann is a registered trademark of Pearson Education Limited

Text © Graham Stoakes 2012
Edited by Melanie Birdsall, Caroline Low and Liz Evans
Designed by Pearson Education Limited
Typeset by Phoenix Photosetting, Chatham, Kent
Original illustrations © Pearson Education Limited 2012
Illustrated by KJA Artists and Phoenix Photosetting
Cover design by Wooden Ark Ltd
Cover photo/illustration © Alamy/Jon Mikel Duralde

The rights of Graham Stoakes to be identified as author of this work have been asserted by them in accordance with the Copyright, Designs and Patents Act 1988.

First published 2012

ARP impression 98

British Library Cataloguing in Publication Data
A catalogue record for this book is available from the British Library

ISBN 978 0 435 07564 4

Copyright notice
All rights reserved. No part of this publication may be reproduced in any form or by any means (including photocopying or storing it in any medium by electronic means and whether or not transiently or incidentally to some other use of this publication) without the written permission of the copyright owner, except in accordance with the provisions of the Copyright, Designs and Patents Act 1988 or under the terms of a licence issued by the Copyright Licensing Agency, Saffron House, 6–10 Kirby Street, London EC1N 8TS (www.cla.co.uk). Applications for the copyright owner's written permission should be addressed to the publisher.

Printed in Great Britain by Ashford Colour Press Ltd

Websites
There are links to relevant websites in this book. In order to ensure that the links are up to date and that the links work we have made the links available on our website at www.pearsonhotlinks.co.uk. Search for this title Level 3 Diploma in Light Vehicle Technology or ISBN 9780435075644.

Acknowledgements

Pearson Education Limited would like to thank the following people for providing technical feedback: Ian Gillgrass and Beverley Lilley of the Institute of the Motor Industry (IMI), Eric Sykes of Babcock Training, Richard Chambers of Boston College and Chris Morris of Stephenson College for their invaluable help in the development of this title.

Graham Stoakes would like to thank Stella Mbubaegbu CBE, Highbury College Principal and Chief Executive, for the use of the college workshops during the photo shoot. Graham Stoakes would also like to thank Holly Stoakes and Andrew "Swede" Wickham for modelling during the photo shoot and Anita Stoakes for her support during this project.

The author and publisher would like to thank the following individuals and organisations for permission to reproduce photographs:

(Key: b-bottom; c-centre; l-left; r-right; t-top)

Alamy Images: Blend Images 305c, David J. Green (Laser thermometer) 34, 84, 99c, 178, 254, Douglas Pulsipher 157, Izmostock 224, James Schutte (Logic probe) 33, 83, 177, 182, 263, Matthew Richardson (Rolling road0, 34, 84, Montgomery Martin 325, stu49 112, Yang Yu 116, Zoonar GmbH 234; **Getty Images:** Comstock Images 23, PhotoDisc 316; **Glow Images:** Blend RF 311; **Graham Stoakes:** 45tl; **Pearson Education Ltd:** Gareth Boden 15t, 15cl, 15cr, 15bl, 17, Naki Photography 7, 16; **Plainpicture Ltd:** hasengold 305bl; **Sally Farndon:** 227, 229tr; **Shutterstock.com:** AJC Designs 193, ben bryant 219, Chungking 249, Deek 135, Karen Roach 229cr, Lucasz Kwiatowski 188bc, MaxStockPhoto 173, Nomad_soul 308, Olinchuk 1, Pan Xunbin 79, Rob Wilson 69, Thor Jorgen Udvang 29, Tyler Olsen 303, Vallefrias 241, 244, Vereshchagin Dmimtry 236, Woodsy 324; **SuperStock:** Exotica IM 305cl; **www.imagesource.com:** Richard Lewisohn 305tl

Cover images: *Front:* **Alamy Images:** Jon Mikel Duralde

All other images © Pearson Education LTD / Clark Wiseman, Studio 8

The author and publisher would like to thank the following individuals and organisations for permission to reproduce material:

p.326 British Standards Institution (BSI) Kitemark®

Every effort has been made to trace the copyright holders and we apologise in advance for any unintentional omissions. We would be pleased to insert the appropriate acknowledgement in any subsequent edition of this publication.

Contents

	Introduction	iv
	Features of this book	x
1	Introduction to light vehicle technology diagnosis & rectification	1
2	Diagnosis & rectification of light vehicle chassis system faults	29
3	Diagnosis & rectification of light vehicle engine faults	79
4	Diagnosis & rectification of light vehicle auxiliary electrical faults	173
5	Diagnosis & rectification of light vehicle transmission & driveline faults	249
6	Identifying & agreeing motor vehicle customer service needs	303
	Common acronyms/abbreviations	344
	Index	348

Introduction

Welcome to Principles of Light Vehicle Technology.

Working in the automotive industry is going to give you a great opportunity to work here in the UK or overseas: it's a challenging, stimulating and fulfilling career.

Working in this sector combines many different practical skills with a knowledge of specialised diagnostic techniques. It also requires good people skills: customers are as much a part of the sector as the light vehicles. This book will develop your abilities to work with the light vehicle systems, engines, chassis, transmission and electrics using systematic advanced diagnostic routines.

About this book

This book has been produced to help you build a sound knowledge and understanding of all aspects of the Diploma and NVQ requirements associated with the automotive industry.

The topics in this book cover all the information you will need to attain your Level 3 qualification in Light Vehicle Maintenance and Repair Principles. Chapter 1 of the book provides you with an introduction to the course and a brief health and safety legislation overview. Chapters 2–5 of the book relates to particular units of the Diploma and provides the information needed to form the required knowledge and understanding of that area. Chapter 6 covers the topic of customer service. It can be used for any awarding organisations qualification including IMIAL and City & Guilds.

Although there are many routes into a career in the automotive industry, most will involve undertaking an apprenticeship. An apprenticeship means you will probably be employed by a garage and study at a college one to two days a week. At college you will learn the theory needed to enable you to gain a 'technical certificate' which includes the knowledge and skills requirements of your qualification. At work you will practice the skills and gather evidence to show that you are able to undertake the required practical tasks competently. This will enable you to gain the NVQ or SVQ part of your apprenticeship qualification.

This book has been written by an experienced trainer who has many years of experience within the sector. He believes in providing you with all the necessary information you need to support your studies and ensuring it is presented in a style which is both manageable and relevant.

This book will also be a useful reference tool for you in your professional life once you have gained your qualifications and are working in the sector.

Introduction

About the Automotive industry

75 per cent of households in the UK have access to at least one car and 30 per cent own two or more cars. This means that there are more than 33 million cars licensed for use in the UK.

Around two million cars are sold in the UK each year and many of these will have advanced drive systems and electronics. These advanced systems require the industry to have highly skilled technicians trained to diagnose and rectify faults in these modern cars. During your Level 3 Diploma you will develop the highly complex technical skills required.

The training that you are now undertaking will enable you to advance your career in the motor industry. Your career doesn't have to end in the UK either – what about taking the skills and experience you are developing abroad? Having these automotive skills will give you a vocation you can take with you wherever you go. There's always going to be a car that needs your attention.

Qualifications for the automotive industry

There are many ways of progressing in the automotive industry, but the most common method is as an apprenticeship.

Apprenticeships

You can become an apprentice by being employed, usually be working for a garage or a vehicle manufacturer.

The Institute of the Motor Industry (IMI) operates the sector skills council which is responsible for setting the standards of automotive training in the UK.

The framework of an apprenticeship is based around an NVQ (or SVQ in Scotland). These qualifications are developed and approved by industry experts and will measure your practical skills and job knowledge on-site.

You will also need to achieve:

- a technical certificate
- the appropriate level of Functional skills assessment
- an Employees Rights and Responsibilities briefing.

Diploma

The Level 3 Diploma in Light Vehicle Maintenance and Repair Principles VRQ (Vocational Related Qualification) provides the knowledge requirements for its related VCQ (Vocational Competence Qualification) and forms the knowledge component of the IMI SSC Maintenance and Repair Apprenticeship framework (for Light Vehicle). The assessment of this VRQ is made up of two components:

- practical tasks
- online testing.

The Level 3 Diploma meets the requirements of the new Qualifications and Credit Framework (QCF) which bases a qualification on the number of credits.

In order to pass the qualification, you must achieve a minimum of 73 credits derived from mandatory and optional units. A minimum of 47 credits must be achieved at Level 3 or above.

As part of the Level 3 Diploma you will gain the skills needed for the NVQ as well as the Functional skills knowledge you will need to complete your qualification.

National Vocational Qualifications (NVQs)

NVQs are available to anyone, with no restrictions on age, length or type of training. There are different levels of NVQ (for example 1, 2, 3), which in turn are broken down into units of competence. NVQs are not like traditional examinations in which someone sits an exam paper. An NVQ is a 'doing' qualification, which means it lets the industry know that you have the knowledge, skills and ability to actually 'do' something.

NVQs are made up of both mandatory and optional units and the number of units that you need to complete for an NVQ depends on the level and vehicle area.

Author

Graham Stoakes MIMI is a lecturer in automotive engineering for light vehicles and motorcycles at a large college of further education. With his background as a qualified master technician, senior automotive manager and specialist diagnostic trainer, he brings 29 years of technical industry experience to this title.

About the Institute of the Motor Industry (IMI)

The IMI is here to help you progress throughout your career and give you the recognition you deserve.

If you're looking to start a career in the sector, **1st Gear** is here to help you do just that. With free advice from industry experts, 1st Gear aims to help point you in the right direction to get started. To join for free, visit www.1stgear.org.uk. For those of you just starting your career in the sector, **Accelerate** is the IMI's fast paced online community for students in training who are at the beginning of their career journey into the automotive sector. It's where, from the start of your career, you can get help and information to assist you in your formal training. Accelerate is here to plug you into up to date information and guidance. It hooks you up with fellow industry apprentices and trainees. It provides the support that will help you achieve your goals and make your career a success.

Not only does Accelerate help your career, it also gives your toolbox and social life a real boost. To take a look at the benefits and join, visit www.motor.org.uk/accelerate.

For more information on the work of the IMI, visit: www.motor.org.uk.

Introduction

Qualification mapping grid

The table below maps the knowledge covered in this candidate handbook against the IMI National Occupational Standards (NOS) for Light Vehicle Maintenance and Repair Principles at Level 3 (same references as IMIAL Level 3 Diploma), against the corresponding QCF unit references, City & Guilds units and Edexcel BTEC Level 3 Subsidiary Diploma, Diploma and Extended Diploma in Light Vehicle Maintenance and Repair.

*Relevant VCQ units list where the BTEC units provides partial coverage of the underpinning knowledge and understanding.

Chapter	QCF unit reference	IMIAL	City & Guilds	BTEC*
1 Introduction to light vehicle technology diagnosis & rectification	D/601/6171 Y/601/7254 T/601/6175 J/601/6262 K/601/6237 Y/601/6279	Introduction to the course and brief Health and Safety legislation overview G102, G3, G4 covered by Level 2 Diploma in Light Vehicle Maintenance & Repair Candidate Handbook (ISBN 9780435048167)	Unit 001 Unit 003 Unit 004 Unit 051 Unit 053 Unit 054	Unit 24 Level 2 Extended Certificate and Diploma in Vehicle Technology awaiting approval Unit 1
2 Diagnosis & rectification of light vehicle chassis system faults	L/601/3735 R/601/3879 D/601/3738 H/601/3885	LV08 K and S LV11.3 K and S	Unit 158 Unit 108 Unit 181 Unit 131	Unit 1 Unit 3 Unit 7 Unit 14
3 Diagnosis & rectification of light vehicle engine faults	F/601/3733 J/601/3877 R/601/3736 L/601/3881	LV07 K and S LV11.1 K and S	Unit 157 Unit 107 Unit 161 Unit 111	Unit 2 Unit 3 Unit 7 Unit 8 Unit 9 Unit 10 Unit 11
4 Diagnosis & rectification of light vehicle auxiliary electrical faults	A/601/3746 H/601/3868	AE06	Unit 456 Unit 406	Covered in Level 3 Auto-Electrical and Mobile Electrical Installations Unit 3 Unit 6 Unit 7 Unit 13
5 Diagnosis & rectification of light vehicle transmission & driveline faults	D/601/3741 T/601/3888 Y/601/3737 D/601/3884	LV13 K LV13 S LV11.2 K and S	Unit 163 Unit 113 Unit 171 Unit 121	Unit 1 Unit 3 Unit 7 Unit 12
6 Identifying & agreeing motor vehicle customer service needs	R/601/6247 M/601/6286	G8 K and S	Unit 008	N/A

ix

Level 3 Light Vehicle Technology

Features of this book

This book has been fully illustrated with technical artworks and photographs. There are also a large number of oscilloscope graphs to help you develop your diagnostic skills. These will help to give you more information about a concept or a procedure, as well as helping you to follow a step-by-step technical skill procedure or identify a particular tool or material.

This book also contains a number of different features to help your learning and development.

Working practice pages

These pages in Chapter 1 highlight the key health and safety areas you need to be aware of as you work on particular light vehicle systems.

Introduction

Before you start and before you finish

These sections can be found at the beginning and end of Chapters 2–5 and cover a number of themes relating to the practice of diagnosis and rectification of light vehicles.

Before you start

These sections include information on:

- safe working when carrying out light vehicle diagnostic and rectification activities
- electronic and electrical safety procedures
- information sources
- operation of electrical and electronic systems and components
- tooling.

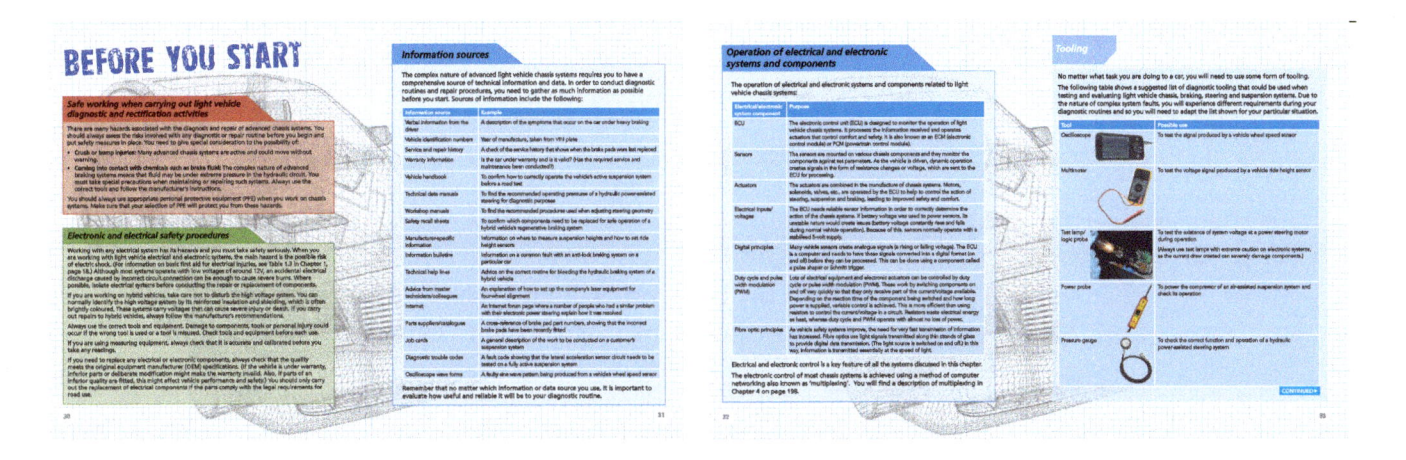

(This features is also included in Chapter 6 where it covers information on skills for work.)

Before you finish

These sections include information on recording and making suitable recommendations.

Technical skills

Throughout the book you will find step-by-step procedures to help you practice the technical skills you need to complete be successful in your studies.

Introduction

Safety tips

These four features give you guidance for working safely on cars and in the workshop.

Emergency
Green safety tips provide useful information about SAFE conditions in emergency situations and your personal safety.

Safe working
Red safety tips indicate a PROHIBITION (something you **must not** do).

Safe working
Blue safety tips indicate a MANDATORY instruction (something that you **must do**).

Safe working
Yellow safety tips indicate a WARNING (hazard or danger).

Other features

NEW TECH
This feature provides you with relevant knowledge and information about new or emerging technologies.

Key term
These are new or difficult words. They are picked out in **bold** in the text and then defined in the margin.

Did you know?
This feature gives you interesting facts about the automotive industry.

Action
These are short activities and research opportunities, designed to help you gain further information about, and understanding of, a topic area.

Case study

This feature provides real-life diagnostic and rectification scenarios along with a checklist of things that should be done to deal with these situations in order to help you develop your working practices.

Skills for work
This feature provides real-life customer service scenarios that will help you to reflect on your own abilities and plan your development.

CHECK YOUR PROGRESS

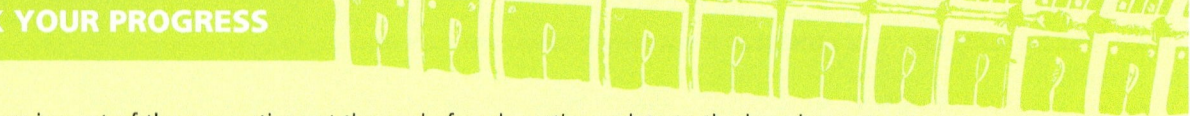

This feature is a set of three questions at the end of each section, relate to the learning outcome covered in that section and test you on the knowledge you have just learnt.

FINAL CHECK

This feature is a series of multiple choice questions at the end of each chapter, in the style of the end of unit tests.

GETTING READY FOR ASSESSMENT

This feature provides guidance for preparing for the practical assessment. It will give you advice on using the theory you have learnt about in a practical way.

1 Introduction to light vehicle technology diagnosis & rectification

This chapter gives you a brief overview of the health and safety legislation that applies to working in the motor industry. You will be expected to work safely at all times and observe the relevant health and safety regulations. The chapter also introduces you to the subjects covered in the light vehicle advanced technology chapters and gives details of methods that will help you during your diagnosis of complex vehicle system faults. It supports you by providing knowledge that will aid you when undertaking both theory and practical assessments. It will help you develop routines that can reduce the risks involved when working in a light vehicle workshop environment.

This chapter covers:

- Working practice:
  - Appropriate personal and vehicle protective equipment
  - Systematic maintenance and repair routines
- Health and safety
- Health and safety legislation
- Accident prevention
- Environmental protection
- Basic first aid
- Advanced technology topics
- Routes to diagnosis

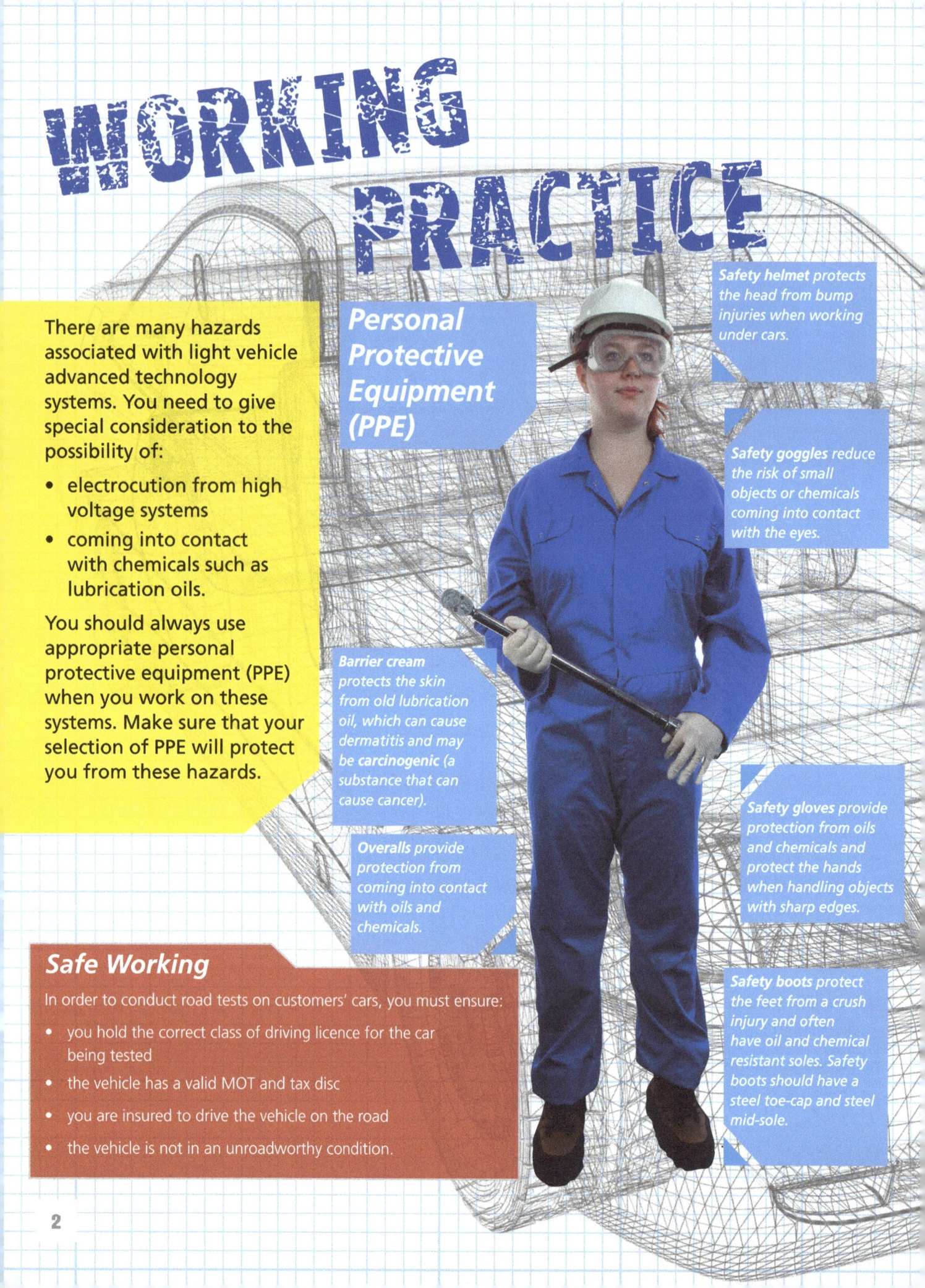

WORKING PRACTICE

There are many hazards associated with light vehicle advanced technology systems. You need to give special consideration to the possibility of:

- electrocution from high voltage systems
- coming into contact with chemicals such as lubrication oils.

You should always use appropriate personal protective equipment (PPE) when you work on these systems. Make sure that your selection of PPE will protect you from these hazards.

Personal Protective Equipment (PPE)

Safety helmet protects the head from bump injuries when working under cars.

Safety goggles reduce the risk of small objects or chemicals coming into contact with the eyes.

Barrier cream protects the skin from old lubrication oil, which can cause dermatitis and may be *carcinogenic* (a substance that can cause cancer).

Overalls provide protection from coming into contact with oils and chemicals.

Safety gloves provide protection from oils and chemicals and protect the hands when handling objects with sharp edges.

Safety boots protect the feet from a crush injury and often have oil and chemical resistant soles. Safety boots should have a steel toe-cap and steel mid-sole.

Safe Working

In order to conduct road tests on customers' cars, you must ensure:

- you hold the correct class of driving licence for the car being tested
- the vehicle has a valid MOT and tax disc
- you are insured to drive the vehicle on the road
- the vehicle is not in an unroadworthy condition.

To reduce the possibility of damage to the car, always use the appropriate vehicle protective equipment (VPE):

Wing covers

Steering wheel covers

Seat covers

Floor mats

If appropriate, safely remove and store the owner's property before you work on the vehicle. Before returning the vehicle to the customer, always check the interior and exterior to make sure that it hasn't become dirty or damaged during the repair operations. This will help promote good customer relations and maintain a professional company image.

Vehicle Protective Equipment (VPE)

Safe Environment

During the diagnosis and repair of light vehicle systems, you may need to drain lubrication oil. Under the Environmental Protection Act 1990 (EPA), you must dispose of oils in the correct manner. They should be safely stored in a clearly marked container until they are collected by a licensed recycling company. This company should give you a waste transfer note as the receipt of collection.

To further reduce the risks involved with hazards, always use safe working practices, including:

1. Immobilise the vehicle by removing the ignition key. Where possible, allow the engine to cool before starting work.

2. Prevent the vehicle moving during maintenance by applying the handbrake or chocking the wheels.

3. Follow a logical sequence when working. This reduces the possibility of missing things out and of accidents occurring. Work safely at all times.

4. Always use the correct tools and equipment to avoid damage to components and tools or personal injury. Check tools and equipment before each use.
   - Inspect any mechanical lifting equipment for correct operation, damage and hydraulic leaks.
   - Never exceed safe working loads (SWL).
   - Check that measuring equipment is accurate and calibrated before you take any readings.

5. If components need replacing, always check that the quality meets the original equipment manufacturer (OEM) specifications. (If the vehicle is under warranty, inferior parts or deliberate modification might make the warranty invalid. Also, if parts of an inferior quality are fitted, this might affect vehicle performance and safety.)

6. Following the replacement of any vehicle components, thoroughly road test the vehicle to ensure safe and correct operation.

Preparing the car

Health and safety

Health and safety at work is a key element of every activity that you undertake. At level 2 you will have learned about the methods and regulations relating to health and safety. As you progress into the role of an advanced diagnostic technician it is possible that your responsibilities in the workplace will change and this will have an effect on how you approach health and safety. While conducting your day-to-day work you may have to manage and supervise others and as a result you will become responsible for the assessment and control of risks.

The information in this chapter is intended to act as a reminder of the practices and procedures that you learned at level 2. However you should now approach your role in the workplace with regard to health and safety from a level 3 perspective. This will involve you looking again at safety from a position of greater responsibility where your role will directly affect those working under you and you are therefore reliant on your knowledge to protect them from harm.

Health and safety legislation

Your workshop should have in place health and safety policies. These will have been developed by your employer to make sure that government **legislation** is observed. It is important that you are aware of the legislation and your rights and responsibilities, as well as those of your employer. It is your right to expect your employer to fulfil their responsibilities and it is your employer's right to expect you to fulfil yours. Legislation is the law and, if you do not observe it, you are committing an offence.

The Health and Safety Executive (HSE)

The Health and Safety Executive (HSE) is the national independent watchdog for work-related health, safety and illness. They are an independent regulator and act in the public interest to reduce work-related death and serious injury across all workplaces in the UK. HSE inspectors have powers to issue **improvement notices** and **prohibition notices** if they believe that there are any poor health and safety practices in a workplace they are inspecting.

If an employer does not comply with an improvement or prohibition notice, they can be fined or even imprisoned. If employers do not properly look after the safety of their employees, they can be prosecuted. This could result in:

- charges of **corporate manslaughter**
- personal fines
- corporate fines
- imprisonment
- loss of production
- loss of income
- bad publicity in the media
- being served with a HSE prohibition notice.

> **Key terms**
>
> **Legislation** – laws that have been passed by government and which are enforced by the police and other bodies.
>
> **Improvement notice** – notification that the employer must eliminate a risk, for example a bad working practice such as draining petrol into open tanks. The improvement notice states a specific period of time by which the employer must eliminate the risk.
>
> **Prohibition notice** – notification that the employer has to immediately stop all work until the safety risk is eliminated. A prohibition notice is only issued for serious safety risks, such as a damaged building that may collapse.
>
> **Corporate manslaughter** – if an employee dies while at work, and the death was the result of the company's failure to provide a safe working environment, then the employers can be charged with the crime of corporate manslaughter. This could result in the company being fined and the staff responsible receiving a prison sentence.

Some of the laws that are relevant to the automotive industry are described below.

The Health and Safety at Work Act 1974 (HASAWA)

The health and safety of everyone in your workplace is protected by the Health and Safety at Work Act 1974 (HASAWA). This law protects you, your employer and all employees while at work. It also protects your customers and the general public when they are visiting your workplace. Table 1.1 lists what you and your employer are responsible for under this law.

Table 1.1 Who's responsible for what under the Health and Safety at Work Act 1974 (HASAWA)?

You are responsible for …	Your employer is responsible for …
• taking care that you do not endanger yourself or others who may be affected by your work • following your workplace's health and safety policies and procedures • not damaging any machinery or PPE provided for your safety • checking machine guards and reporting any problems.	• providing you with safe equipment and safe ways of carrying out work tasks • making sure that the equipment you use when handling, storing and transporting materials and substances is safe and without health risks • providing information, instruction, training and supervision to guarantee health and safety • maintaining the workplace, including means of entry and exit, in a safe condition • providing a safe and healthy working environment with adequate facilities and arrangements for employees' welfare • providing for employees a written statement of policy, organisation and arrangements for health and safety (employers with less than five employees do not have to do this but it is advisable to do so).

Other health and safety regulations

In addition to the Health and Safety at Work Act 1974, the following regulations apply across the full range of workplaces:

1 **The Management of Health and Safety at Work Regulations 1999:** require employers to carry out risk assessments, put in place measures to minimise risks, appoint competent people and arrange for appropriate information and training for their staff.

2 **Workplace (Health, Safety and Welfare) Regulations 1992:** cover a wide range of basic health, safety and welfare issues such as ventilation, heating, lighting, workstations, seating and welfare facilities.

3 **Health and Safety (Display Screen Equipment) Regulations 1992:** set out requirements for work with computers and visual display units (VDUs).

4 **Personal Protective Equipment (PPE) at Work Regulations 1992:** require employers to provide appropriate protective clothing and equipment for their employees.

5 **Provision and Use of Work Equipment Regulations 1998 (PUWER):** require that the equipment provided for use at work, including machinery, is safe.

Figure 1.1 Health and safety poster (required under the Health and Safety Information for Employees Regulations)

6 **Manual Handling Operations Regulations 1992:** cover the moving of objects by hand or physical force.

7 **Health and Safety (First Aid) Regulations 1981:** cover the requirements for first aid, including the number of trained first-aiders required in the workplace.

8 **The Health and Safety Information for Employees Regulations 1989:** require employers to display posters telling employees what they need to know about health and safety.

9 **Employers' Liability (Compulsory Insurance) Act 1969:** requires employers to take out insurance against work-related accidents and ill health involving employees and visitors to the premises.

10 **Reporting of Injuries, Diseases and Dangerous Occurrences Regulations 1995 (RIDDOR):** require employers to report accidents, near misses and ill health to the HSE and to keep records of these events. See Figure 1.2 on page 8 for an example of an accident record sheet.

CHECK YOUR PROGRESS

1 Name four pieces of health and safety legislation.

2 List four things that your employer is responsible for under the Health and Safety at Work Act 1974.

3 List four things that the employee is responsible for under the Health and Safety at Work Act 1974.

Level 3 Light Vehicle Technology

Report of an Accident, Dangerous Occurrence or Near Miss

Date of incident _14 February 2012_   Time of incident _10.45 AM_
Location of incident _St Oaks Motors (vehicle workshop)_

Details of person involved in incident
Name _Colin Mopp_   Date of birth _16 August 1991_   Sex _Male_
Address _Flat 22, Pineway House, Ingrham Crescent, East Harbour, Portsmouth, Hampshire, PO98 3QQ_   Occupation _Light vehicle technician_
Date off work (if applicable) _14-02-2012_   Date returning to work _20-02-2012_
Nature of injury _Bump to head causing temporary unconsciousness_

Management of injury
☐ First Aid only   ☐ Advised to see doctor
☒ Sent to casualty   ☒ Admitted to hospital

Account of accident, dangerous occurrence or near miss
(Continued on separate sheet if necessary)

While working under a vehicle that had been correctly raised and supported on a vehicle hoist, a large section of corroded exhaust system became dislodged. As most of the exhaust mountings had been removed to conduct a repair, a heavy silencer fell down and struck Colin in the forehead just below the edge of his bump cap, rendering him temporarily unconscious. Colin was sent to casualty and was admitted to hospital for observation.

Witnesses to the incident
(Names, addresses and occupations)

Simon Wing (apprentice technician)
24 Gordon Road
Portsmouth
Hampshire
PO99 5QR

Was the injured person wearing PPE? If yes, what PPE? _Steel-toe-cap safety boots, overalls, bump cap, rough service gloves, safety glasses._
Signature of person completing form _Bryan Kerr_
Occupation _Service manager_   Date _15 February 2012_

Figure 1.2 An accident record sheet (required under RIDDOR)

Dust | Toxic | Flammable | Irritant | Corrosive | Oxidising agent

Figure 1.3 Warning labels that you will find on containers of hazardous chemicals and substances. Labels such as these are required under the COSHH regulations

The safe use of chemicals and other substances

Company Name:

The findings should be recorded and the staff concerned informed as to their outcome and of the safe method of work. If you decide there are no risks to health, or risk is minimal this does not need to be recorded.

STEP 1 – HAZARD	YES	NO	Comments/Action	Date Completed
Have all hazardous substances (marked with orange square with black symbol) been listed?			List substances:	
Have safety data sheets (supplier legally obliged to provide) being obtained?			Where are the data sheets located?	
Has the method of use of the substance been determined?				
Have the number of people using the substance been determined?				
Has the effect of the substance on employees been assessed?			List effects:	
Has the risk been assessed using the above information?			List risks:	
STEP 2 – DECIDE ON PRECAUTIONS	YES	NO	Comments/Action	Date Completed
Has information from manufacturer and safety data sheets been used to determine suitable precautions?			What precautions needed?	
Has good working practices and standards recommended by trade association been used to determine suitable precautions?			What precautions needed?	
STEP 3 – PREVENT OR CONTROL EXPOSURE by making use of one or more of the following hierarchy of measures	YES	NO	Comments/Actions	Date Completed
Have you changed the process or activity (so that the subtance isn't needed)?			What was used? What is now used?	
Have you replaced the substance with a safer alternative (eg. water-based paints rather than solvent-based)?				
Have you made use of the substance in safer form (eg. pellets rather than loose powder)?				
Have you totally enclosed the process (removing all change of contact)?				
Have you partially enclosed the process and used local exhaust ventilation (eg. fume cabinets with extraction)?				
Have you provided adequate general ventilation (eg. in beauty salons, plenty of fresh air to avoid build-up of fumes from acetone etc)?				
Have you used methods of work which minimise handling or chance of spillage?				
Have you reduced the number of persons exposed or minimised time of exposure?				
Have you used personal protective equipment (eg. disposable gloves, overalls, safety goggles etc)?				
STEP 4 – CHECK YOUR ASSESSMENT AND PROCEDURES to assure the assessment's effectiveness	YES	NO	Comments/Actions	Date Completed
Have you checked that all staff are following procedures outlined in Step 2 and 3?				
Is the equipment used for controlling substances maintained and tested for effectiveness?				
Is air monitoring carried out if the assessment concludes there would be a serious risk to health if the controls should fail?				
Has health surveillance been carried out where specific health problems are associated with the substance (information on safety data sheet)? In most cases, this would be in the form of a simple question and answer or in cases such as potential dermatitis, regular examination of hands and other exposed areas.				

Figure 1.4 COSHH risk assessment checklist

11 **Noise at Work Regulations 1989:** require employers to take action to protect employees from hearing damage.

12 **Electricity at Work Regulations 1989:** require people in control of electrical systems to ensure they are safe to use and maintained in a proper working condition.

13 **Control of Substances Hazardous to Health Regulations 2002 (COSHH):** require employers to assess the risks from hazardous substances and take appropriate precautions.

Accident prevention

An automotive workshop is an extremely hazardous environment. There are many dangers that could result in an accident causing injury or even death. The management of these **hazards** is key to reducing the **risks** involved while working in the motor vehicle repair industry. Not all hazards can be removed, but they can be identified and measures can be put in place to reduce the dangers that they pose; this is the purpose of a risk assessment.

Risk assessment

A risk assessment is an important step in protecting workers and businesses, and is necessary in order to comply with the law. It is designed to focus on the risks that really matter in the workplace – the ones with the potential to cause real harm. In many cases, straightforward measures can control risks, for example ensuring that:

- spillages are cleaned up promptly so that slip hazards are avoided
- walkways are kept clean and clear of tools, etc., to avoid trip hazards
- adequate lighting is provided and inspection pits are covered to avoid falling hazards.

The law does not expect you to eliminate all risk, but you are required to protect people as far as is **reasonably practicable**.

A risk assessment is simply a careful examination of what, in your work, could cause harm to people. It allows you to weigh up whether you have taken enough precautions or should do more to prevent harm.

There are five main steps to risk assessment:

Step 1

Identify the hazards – conduct an inspection of your workplace and make a list of all of the hazards you find.

Step 2

Decide who might be harmed and how – remember to include all those who may be at risk, for example:

- staff/colleagues
- customers
- people with disabilities.
- contractors
- the general public
- delivery operatives
- young people

> **Key terms**
>
> **Hazard** – something that has the potential to cause harm or damage.
>
> **Risk** – the likelihood of the harm or damage actually happening.
>
> **Reasonably practicable** – can be carried out without incurring excessive effort or expense.

Step 3

Evaluate the risks and decide on precautions – for example:

- Can you get rid of the hazard completely? If not, what needs to be done to control the risk of harm?
- Is there a less risky option?
- Can the hazard be guarded or access prevented?
- Is PPE required? (Note that PPE should only be used when other methods of reducing the risk are not practical.)

Step 4

Record your findings and implement them – keep a written record of your risk assessment and what you have done to control the hazards.

Step 5

Review your assessment and update if necessary – working situations change, so make sure that you regularly check that your assessment still covers all hazards.

STEP 1

Hazard

Look only for hazards which you could reasonably expect to result in significant harm under the conditions in your workplace. Use the following examples as a guide.

- ☐ slipping/tripping hazards (e.g. poorly maintained floors or stairs)
- ☐ fire (e.g. from flammable materials)
- ☐ chemicals (e.g. battery acid)
- ☐ moving parts of machinery (e.g. blades)
- ☐ work at height (e.g. from mezzanine floors)
- ☐ ejection of material (e.g. from plastic moulding)
- ☐ pressure systems (e.g. steam boilers)
- ☐ vehicles (e.g. fork-lift trucks)
- ☐ electricity (poor wiring)
- ☐ dust (e.g. from grinding)
- ☐ fumes (e.g. welding)
- ☐ manual handling
- ☐ noise
- ☐ poor lighting
- ☐ low temperature

STEP 2

Who might be harmed?

There is no need to list individuals by name – just think about groups of people doing similar work or who may be affected, for example:

- ☐ office staff
- ☐ maintenance personnel
- ☐ contractors
- ☐ people sharing your workplace
- ☐ operators
- ☐ cleaners
- ☐ members of the public

Pay particular attention to:

- ☐ staff with disabilities
- ☐ visitors
- ☐ inexperienced staff
- ☐ lone workers

They may be more vulnerable.

STEP 3

Is more needed to control the risk?

For the hazards listed, do the precautions already taken:

- ☐ meet the standards set by a legal requirement?
- ☐ comply with a recognised industry standard?
- ☐ represent good practice?
- ☐ reduce risk as far as reasonably practicable?

Have you provided:

- ☐ adequate information instruction or training?
- ☐ adequate systems or procedures?

If so, then the risks are adequately controlled, but you need to indicate the precautions you have in place. (You may refer to procedures, company rules, etc.) Where the risk is not adequately controlled, indicate what more you need to do (the 'action' list).

STEPS 4 and 5

Record, review and revision

Set a date for review of the assessment.

On review check that the precautions for each hazard still adequately control the risk. If not indicate the action needed. Note the outcome. If necessary complete a new page for your risk assessment.

Making changes in your workplace, for example when bringing in new machines, substances, or procedures may introduce significant new hazards. Look for them and follow the 5 steps.

Figure 1.5 The five steps involved in carrying out a risk assessment

Key terms

Controlled waste – any waste which cannot be disposed of to landfill, including liquids, asbestos, tyres and waste that has been decontaminated. There are three types of controlled waste listed under the Environmental Protection (Controlled Waste) Regulations 2004: household, industrial and commercial waste. Damage to the environment can be caused by contaminating the atmosphere, water supply or drainage system.

Ecotoxic – describes a substance that is harmful to the environment.

Figure 1.6 Symbol for ecotoxic substances

Environmental protection

During the diagnosis and repair of light vehicle systems, you may need to dispose of controlled waste. Under the Environmental Protection Act 1990 (EPA) and The Waste (England and Wales) Regulations 2011, you must dispose of controlled waste in the correct manner. It is an offence to treat, keep or dispose of **controlled waste** in a way that is likely to pollute the environment or harm people. Any waste should be safely stored in a clearly marked container until it is collected by a licensed recycling company. You should give this company a signed waste transfer note, describing the type of waste collected and copies of this should be kept for a minimum of two years.

Your employer should have procedures in place for working with and disposing of any **ecotoxic** material that has potential to harm the environment. Examples of controlled waste from an automotive workshop include:

- old engine oil
- oil filters
- batteries
- tyres
- antifreeze
- airbags
- components known to contain mercury (SRS, ECUs or HID bulbs for example).

Environmental Protection (Duty of Care) Regulations 1991

These regulations describe the actions which anyone who produces, imports, keeps, stores, transports, treats, recycles or disposes of controlled waste must take. These people must:

- store the waste safely so that it does not cause pollution or harm anyone
- transfer it only to someone who is authorised to take it (such as someone who holds a waste management licence or is a registered waste carrier)
- when passing it on to someone else, provide a written description of the waste and fill in a transfer note
- keep these records for two years and provide a copy to the Environment Agency if they ask for one.

Safety signs

To assist with health and safety in the workplace, signs are often used to communicate instructions or to give warnings. These signs will normally contain images to convey their meaning. They use the following formats:

1 Introduction to light vehicle technology diagnosis & rectification

Red signs

Red signs indicate a PROHIBITION (something you **must not** do). They often have a red diagonal line over the symbol on the sign.

Do not oil or clean this machine whilst in motion
Warning to protect against damage to machine and operator

Do not drink the water

No smoking

Figure 1.7 Red prohibition signs tell you what you **must not** do

Look for boxes with this symbol throughout the book for PROHIBITIONS you should be aware of:

> **Safe working**
> Do not measure the voltages of a hybrid drive system unless you have been specifically trained. Hybrid drives operate with high voltages that can cause electric shock and death.

Blue signs

Blue signs indicate a MANDATORY instruction (something that you **must** do).

Wear protective gloves

Guards must be in position before starting

Ear protection zone

Figure 1.8 Blue mandatory signs tell you what you **must** do

Look for boxes with this symbol throughout the book for MANDATORY instructions you should be aware of:

> **Safe working**
> Following the replacement of any engine components, the car should be road tested to ensure correct function and operation.

Yellow signs

Yellow signs indicate a WARNING (hazard or danger).

Danger of fire

Tripping hazards

Electrical hazard

Figure 1.9 Yellow warning signs tell you about dangers that are nearby

Look for boxes with this symbol throughout the book for WARNINGS you should be aware of:

> **Safe working**
> When checking for an electrical short circuit, only bypass the fuse with an electrical consumer like a bulb. Using other electrical components could cause a sudden discharge of electricity that may burn you.

13

Level 3 Light Vehicle Technology

Look for boxes with this symbol throughout the book for SAFE conditions you should be aware of:

> **+ Emergency**
> If the low battery symbol appears on a digital multimeter screen, replace the battery straight away. Otherwise, you might get inaccurate readings that could lead to electric shock or personal injury.

Green signs

Green signs indicate a SAFE condition (useful information). These are often used for signs indicating where to go in emergency situations.

First aid box

Safe refuge point for wheelchair

Fire exit

Figure 1.10 Green safe condition signs often give information that you will need in an emergency

Fire safety

Careful consideration should be given to fire safety in the workplace. An automotive workshop contains many flammable materials and ignition sources. A separate risk assessment should be carried out for fire hazards, and safety measures must be put in place.

Fire extinguishers

Fire extinguishers should be provided in your workplace and maintained by your company, for safety in case there is a fire. The primary function of a fire extinguisher is to enable you to create an escape route. In any other circumstance, you should only attempt to tackle a fire if it is safe to do so and if you have had adequate training.

A number of different fire extinguishers are available depending on the type of fire to be tackled. Every fire extinguisher has a colour-coded label with a description of its contents and a list of the types of fire it is designed to be used on. Table 1.2 lists the types of fire extinguishers and their uses.

Figure 1.11 Fire exit sign

> **Did you know?**
> More information on health and safety can be found in Chapter 1 of the *Level 2 Principles of Light Vehicle Maintenance and Repair Candidate Handbook*, or on the Health and Safety Executive website – to visit this website, please go to hotlinks and click on this chapter.

Figure 1.12 Fire blanket and four types of fire extinguisher

1 Introduction to light vehicle technology diagnosis & rectification

Table 1.2 Fire extinguishers and their uses

Extinguisher	Class of fire	When to use	Notes
Red label extinguisher – water	Class A – wood, paper, textiles, etc.	Use it by pointing the jet at the base of the flames and keeping it moving across the area of the fire. Make sure that all areas of the fire are out. It works mainly by cooling the burning material.	Do not use on burning fat or oil or on electrical appliances.
Black label extinguisher – carbon dioxide	Class E – electrical equipment Class A – wood, paper, textiles, etc. Class B – flammable liquids: petrol, oil, etc. Class C – flammable gases: LPG, propane, etc.	Mainly used on electrical fires but can also be used on class A, B and C fires. Use it by pointing the jet at the base of the flames. Keep the jet moving across the area of the fire. It works because carbon dioxide gas smothers the flames by replacing oxygen in the air.	Do not use on chip or fat pan fires. Do not use on flammable metal fires. This type of extinguisher does not cool the fire very well and you need to watch that the fire does not start up again. The fumes can be harmful if used in confined spaces – ventilate the area as soon as the fire has been controlled.
Blue label extinguisher – dry powder	Class A – wood, paper, textiles, etc. Class B – flammable liquids: petrol, oil, etc. Class C – flammable gases: LPG, propane, etc. Class D – metal, metal powder, etc. Class E – electrical equipment	Use it by pointing the jet or discharge horn at the base of the flames. With a rapid sweeping motion, drive the fire towards the far edge until all the flames are out. If the extinguisher has a shut-off control, wait until the air clears and, if you can still see the flames, attack the fire again.	Not recommended for use on chip or fat pan fires.
Cream label extinguisher – foam	Class A – wood, paper, textiles, etc. Class B – flammable liquids: petrol, oil, etc.	For fire involving liquids, do not aim the jet straight into the liquid. Where the liquid on fire is in a container, point the jet at the inside edge of the container or on a nearby surface above the burning liquid. Allow the foam to build up and flow across the liquid. It works by forming a fire-extinguishing film on the surface of a burning liquid.	Can be used on class C fires if the gas is in liquid form.

15

Level 3 Light Vehicle Technology

> **Did you know?**
>
> There are currently no set rules for the number of first-aiders required by a company but the Health and Safety Executive makes the following recommendations for high risk occupations such as automotive workshops:
>
> - less than 5 employees – at least one appointed person (not necessarily a trained first-aider)
> - 5 to 50 employees – at least one first-aider trained in emergency first aid at work (EFAW)
> - more than 50 employees – at least one first-aider for every 50 employees, trained in first aid at work (FAW).

Basic first aid

An automotive workshop is a high risk environment and no matter what precautions are taken, there is always the possibility of accidents occurring which may lead to personal injury. The following advice is not a substitute for first aid training, and will only give you an overview of the action you may need to take. You should take care when you attempt to administer first aid that you do not place yourself in danger. Be very careful about what you do, because the wrong action can cause more harm to the casualty.

Good first aid always involves summoning appropriate help; many companies will have a trained first-aider on site and all companies must have a suitably stocked first aid box.

First aid box

The minimum level of first aid equipment that will be found in a suitably stocked first aid box should include:

- a guidance leaflet
- 2 sterile eye pads
- 6 triangular bandages
- 6 safety pins
- 3 extra large, 2 large and 6 medium-sized sterile un-medicated wound dressings
- 20 sterile adhesive dressings (assorted sizes)
- 1 pair of disposable gloves (as required under HSE guidance).

It is your employer's or the designated first-aider's responsibility to ensure that the contents of the first aid box are in date and are sufficient, based on their assessment of your workplace's first aid needs.

The law does not state how often the contents of a first aid box should be replaced, but most items, in particular sterile ones, are marked with expiry dates.

Other equipment such as eye wash stations must also be available if the work being carried out requires it.

Getting help

If you need to call for assistance, the main emergency services can be contacted by calling 999 free of charge from any landline or mobile phone.

Figure 1.13 Every workplace needs to have a well-stocked first aid box

When calling the emergency services, make sure you give the following information:

- your telephone number
- the location of the incident
- the type of incident
- the gender and age of the casualty
- details of any injuries you have observed
- any information you have observed about hazards, for example power cables, fog, ice, gas leaks you could smell.

The recovery position

When dealing with health emergencies, you may need to place someone in the recovery position. In this position a casualty has the best chance of keeping a clear airway, not inhaling vomit and remaining as safe as possible until help arrives. You should not attempt to put someone in the recovery position if you think they might have back or neck injuries, and it may not be possible if any limbs are fractured.

Figure 1.14 The recovery position

Putting a casualty in the recovery position

1. Kneel at one side of the casualty, at about waist level.
2. Tilt the head back – this opens the airway. With the casualty on their back, make sure that their limbs are straight.
3. Bend the casualty's near arm so that it is at right angles to the body. Pull the arm on the far side over the chest and place the back of the hand against the opposite cheek.
4. Use your other hand to roll the casualty towards you by pulling gently on the far leg, just above the knee. This will bring the casualty onto their side.
5. Once the casualty is rolled over, bend the leg at right angles to the body. Make sure the head is tilted well back to keep the airway open.

Action in a health emergency

Table 1.3 overleaf gives guidance on recognising and taking initial action in a number of health emergencies. Remember that you will need to seek professional help in all emergencies.

To find out more about first aid at work, visit the first aid section of the HSE website – to visit this website, please go to hotlinks and click on this chapter.

CHECK YOUR PROGRESS

1. List the five steps involved in a risk assessment.
2. State the four colours that are used to describe the types of fire extinguishers. Name the contents of each type and what type of fire it should be used against.

Table 1.3 Action to take in a health emergency

Emergency type	Symptoms or initial steps	What to do	Other advice
Electrical injuries	If someone receives an electric shock, the electricity passes straight through them because of the large water content within the body. This may cause the skin to burn or look pale or bluish, and the person may not have a pulse. In this instance, the person may not be breathing and the heart may have stopped. DO NOT touch the casualty. Switch off the current source.	Seek help. Stand on dry insulating material (wood, rubber or lino). Isolate the casualty by using material that does not conduct electricity, such as wood or plastic (for example, a wooden broom handle). If you are a qualified first-aider, follow your training for dealing with electric shock. Obtain emergency medical assistance for the casualty.	Electric shock from either an AC source or a DC source is equally dangerous although, because of its nature, a lower AC voltage poses a higher risk of electric shock than DC. Depending on the frequency of the AC voltage an electric shock from an AC source may also produce rapid muscle contractions in the victim, leading to them gripping the electric source without being able to let go.
Bleeding and severe bleeding	If you or a colleague has a cut, the qualified first-aider will be able to issue plasters from the first aid box. Always make sure you wash the cut carefully with soap and running warm water, then dry with a clean paper towel. If possible, wear disposable gloves. If this is not possible, cover any areas of broken skin with a waterproof dressing. Wash your hands thoroughly using soap and water before and after treatment. Take care with any needles or broken glass in the area. If the casualty requires mouth-to-mouth resuscitation and is bleeding from the nose or mouth, make sure you use a mask.	Apply pressure to the wound that is bleeding, using a sterile dressing if possible or any absorbent material. Apply direct pressure over the wound for 10 minutes to allow the blood to clot. Do not try to take any object from the wound, but apply pressure to the sides of the wound. Lay the casualty down and raise the affected part if possible. Make sure the casualty is warm, comfortable and safe. Obtain emergency medical assistance for the casualty.	You should take steps to protect yourself when dealing with casualties who are bleeding. Skin provides an excellent barrier to infections, but you must take care if you have any broken skin such as a cut, graze or sore. Seek medical advice if blood comes into contact with your mouth or nose, or gets into your eyes. Blood-borne viruses (such as HIV and hepatitis) can be passed on only if the blood of someone who is already infected comes into contact with your broken skin.
Shock	Shock occurs because blood is not being pumped around the body efficiently. It can be the result of severe bleeding or burns or a heart attack. The casualty: • will look pale and grey • will be very sweaty, with cold, clammy skin • will have a very fast pulse and may be breathing very fast • may feel sick or vomit.	Keep the casualty warm but not with direct heat – use a blanket or extra clothing. Lay the person down on the floor and raise their feet off the ground. This will help maintain the blood supply to the important organs. Loosen any tight clothing. Watch the person carefully. Check their pulse and breathing regularly. Obtain emergency medical assistance for the casualty.	Do not: • allow the casualty to eat or drink • leave the casualty alone, unless it is essential to do so briefly in order to summon help.

CONTINUED ▶

1 Introduction to light vehicle technology diagnosis & rectification

Emergency type	Symptoms or initial steps	What to do	Other advice
Loss of consciousness	Loss of consciousness can occur through fainting or as the result of injury, a fall or illness. The person will lack awareness and be unresponsive.	Make sure that the person is breathing and has a clear airway. Maintain the airway by lifting the chin and tilting the head backwards. Place the casualty in the recovery position. Obtain emergency medical assistance for the casualty.	DO NOT: • attempt to give anything by mouth • attempt to make the casualty sit or stand • leave the casualty alone, unless it is essential to do so briefly in order to summon help. Find out if the casualty has an existing condition such as diabetes or epilepsy (they may have an ID band with this information); this could help you give relevant information to the emergency services.
Minor burns or chemical injuries	Burns may occur in the workshop from contact with chemicals or hot components or from welding equipment.	Cool down the burn by flooding it with cold water for 10–20 minutes. If it is a chemical burn, flooding with cold water needs to be done for 20 minutes. Make sure that the contaminated water used to cool a chemical burn is disposed of safely. Cover the burn if possible with a clean dressing or even cling film. If there are no clean dressings available, leave it uncovered. Obtain emergency medical assistance for the casualty if it is a significant burn.	Do not use ointment as this could contaminate the wound if the skin is broken.
Objects in the eye	Even though wearing goggles will prevent most eye injuries, there may be incidents where an object becomes lodged in the eye. In the event of an object entering the eye, it is helpful to have a colleague assist with flushing it out. You will need: • good lighting • a bottle of eye wash solution or clean tepid water • some absorbent towel or a handkerchief.	Ask the casualty to lie down if possible, otherwise stand beside or behind them. Wash your hands with soap and water and rinse thoroughly. Locate the object by gently lifting the eye lid and asking the casualty to look up and down and to the side until you see it. Tilt the head to the side that is being flushed and place a towel or cloth against the ear to prevent the eye wash from entering the ear. Using prepared eye wash or, if not available, tepid water, pour the liquid from the inner corner of the eye so that it drains onto the towel. It may be possible to dislodge grit or dust by gently using a moist clean swab or tissue. If after two flushings the object is not dislodged, stop and seek medical help.	Never touch anything that is stuck in the eye but seek medical help.

Level 3 Light Vehicle Technology

Advanced technology topics

The Level 3 Diploma focuses on the diagnosis and repair of light vehicle advanced technology systems. It is expected that by this level you will already have a foundation understanding of light vehicle mechanical and electrical systems.

Chapters 2–5 of this book are intended to give you an overview of electrical and electronic control systems. Many of these systems have been designed to improve performance, comfort and safety for all road users, but their complexity gives rise to complicated and involved diagnostic and repair routines. You will need to develop logical thought processes and a **systematic** approach to your work in order to keep diagnostic times to a minimum while ensuring that safety is maintained and repairs lead to a **first time fix**.

These chapters also provide information about new and emerging technologies. Many of the new technologies aim to reduce the impact of motoring on the environment while maintaining fuel economy and performance.

Some of the subjects included are shown in Table 1.4 opposite.

> **Key terms**
>
> **Systematic** – planning a job so that you do things in order.
>
> **First time fix** – getting the repair right first time.

Structure of each chapter

Chapters 2–5 of this Candidate Handbook have recurring learning outcomes which are stated in the *Before you start* and *Before you finish* sections at the start and end of each chapter. They cover themes such as:

Before you start

- Safe working
- Electronic and electrical safety procedures
- Operation of electrical and electronic systems and components
- Information sources
- Tooling

Before you finish

- Recording and making suitable recommendations

These recurring outcomes contain information and examples that are relevant to each chapter. They will help you to prepare for diagnosing and repairing chassis, engine, electrical auxiliary and transmission systems and to record your work and recommendations.

Other features

Other features found in the chapters include:

CHECK YOUR PROGRESS

These are questions at the end of a subject section to confirm your underpinning knowledge.

1 Introduction to light vehicle technology diagnosis & rectification

Table 1.4 Subjects included in the technical light vehicle chapters of this book

Chapter	Example subjects
Chapter 2 Diagnosis and rectification of light vehicle chassis system faults	Steering technology: power assistance and geometry Braking and active safety: • EPB – electronic parking brake • ABS – anti-lock braking systems • EBD – electronic brake force distribution • TCS – traction control systems • Regenerative braking Active suspension and safety: air suspension, hydro-pneumatic suspension, dynamic stability control
Chapter 3 Diagnosis and rectification of light vehicle engine faults	Pressure charging and variable valve control Engine restoration and repair Electronic ignition and fuel systems Engine management and diagnosis Emission control Alternative propulsion: LPG, compressed natural gas (CNG), bioalcohol/ethanol, biogas, biofuel, electric, solar, liquid nitrogen, hybrid and fuel cell Climate control
Chapter 4 Diagnosis and rectification of light vehicle auxiliary electrical faults	Diagnostic tooling and testing Multiplex and networks Batteries and charging, including hybrid high voltage systems Advanced external lighting Comfort and convenience: central locking, electric windows, screen demisting, windscreen washers and wipers, electric mirrors and security In-car entertainment and satellite navigation Safety systems, including SRS
Chapter 5 Diagnosis and rectification of light vehicle transmission and driveline faults	Electronic clutch control, dual clutch systems, torque converters and dual mass flywheels Manual gearbox control, sequential and seamless shift transmission Automatic transmission, including CVT Limited slip differentials and electronic final drive control

Action
These are activities to extend your learning throughout each chapter.

Case studies

These are examples of good workshop practice that can be adapted to different diagnostic situations.

Level 3 Light Vehicle Technology

NEW TECH
This feature contains examples of new or emerging technologies that manufacturers are now including in their vehicle designs. Many of these technologies are designed to reduce the impact of motoring on the environment or increase safety and passenger comfort.

FINAL CHECK
These are multiple-choice end-of-chapter questions to help confirm your underpinning knowledge.

Skills for work
Working in the motor industry requires that you develop a set of personal skills to ensure that you work in a professional manner. Throughout the chapters you will find a series of scenarios to help you reflect on your own abilities and plan actions to aid your personal development.

Routes to diagnosis

You need to develop a logical approach to your diagnostic routine in order to cut down on time, cost and frustration. The information that follows can be applied to any diagnostic routine, whether mechanical, electrical or electronic.

Remember that correct diagnosis is important to ensure a first time fix.

The flow chart shown below in Figure 1.15 gives an example of a generic routine that should be used when diagnosing any vehicle fault.

Faults and symptoms

Always make sure that you fix the **fault** and not the **symptom**.

With any diagnosis, it is important to listen to the symptoms described by the driver and work out a logical diagnostic routine before starting to work on the vehicle. Don't charge into a diagnosis, as this can lead to errors and misdiagnosis.

Before you begin any diagnosis, you need to gather as much information as possible. Begin by carefully questioning the customer – they have first-hand knowledge of their car and the fault.

Make sure you ask the customer general, open questions. It is a good idea to write down the information that the customer gives you. If you design a pre-diagnostic questionnaire with carefully thought out questions, such as the one shown in Figure 1.17 on page 24, this may save you having to contact the customer after they have left the garage.

- Fill in a pre-diagnosis questionnaire and take it with you to the job.
- Start with the easiest part of the system to access and follow a logical sequence to diagnose the fault.
- Test the system and not the fault.

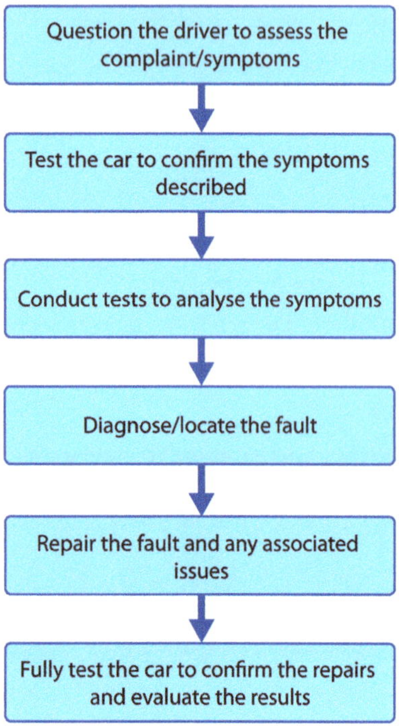

Figure 1.15 Example of a generic routine to use when diagnosing any vehicle fault

Flow chart steps:
1. Question the driver to assess the complaint/symptoms
2. Test the car to confirm the symptoms described
3. Conduct tests to analyse the symptoms
4. Diagnose/locate the fault
5. Repair the fault and any associated issues
6. Fully test the car to confirm the repairs and evaluate the results

Key terms
Fault – something that is responsible for an undesirable situation or event.

Symptom – a sign that indicates a certain situation, especially an undesirable situation.

Case study

Bad practice

This case study shows an example of bad practice where symptoms are cured but not the fault, which may well lead to further issues:

A car is presented to a garage with a partially blocked radiator, caused by poor maintenance and because the coolant has not been replaced at the specified interval. The customer has complained that the engine is overheating as indicated by the temperature gauge on the dashboard — this is the symptom.

The workshop mechanic conducts a number of random tests including: thermostat operation, water pump operation, radiator electric cooling fan operation and a block test to see if the head gasket has failed.

The results of the tests were inconclusive so the mechanic decides to bypass the electric cooling fan so that it runs continuously.

As the cooling fan is running all of the time, the car no longer overheats. This has cured the symptom but the fault of a partially blocked radiator still remains.

With the radiator fan running continuously, the engine will not reach its optimum running temperature, leading to increased wear on engine components, reduced fuel economy and excessive exhaust emissions.

By fixing the symptom, the mechanic has simply increased the likelihood of problems resulting from the original fault.

Skills for work

You have just started a new job as a diagnostic technician. The garage operates a very strict probationary period to ensure that new staff members are reliable and fulfil the job role for which they have been employed.

In order to prove to your employers that you have what it takes to do the job well and to pass your probationary period, you need to exhibit particular personal skills. Some examples of these skills are shown in Table 6.1 at the start of Chapter 6, on pages 304–305.

1. Using the examples given in Table 6.1, choose one skill from each of the following categories that you think you need to demonstrate in your new job.
    - General employment skills
    - Self-reliance skills
    - People skills
    - Customer service skills
    - Specialist skills
2. Now rank these skills in order of importance, starting with the one that it is most important for you to have.
3. Which of the skills chosen do you think you are good at?
4. Which of the skills chosen do you think you need to develop?
5. How can you develop these skills and what help might you need?

Figure 1.16 Discuss issues, symptoms and faults with your customer

Mick's Auto Repair

Inspector's name: _____

Customer's name		Model and model year	
Driver's name		V.I.N	
Date vehicle brought in		Engine model or type	
Registration No.		Odometer reading	km/miles

Symptoms	Engine does not start	Engine does not crank No initial combustion Incomplete combustion	
	Difficult to start	Engine cranks slowly Other	
	Poor idling	No fast idle Idling speed: High Low (rpm) Rough idling Other	
	Poor driveability	Hesitation Backfiring Misfiring Explosion Surging Knocking Other	
	Engine stalls	Soon after starting After accelerator pedal is depressed After accelerator pedal is released Shifting gear During A/C operation Other	
	Others		

Date(s) of problem occurrence	
Frequency of problem occurrence	Constant Sometimes (times per day/month) Once only Other

Conditions	Weather	Clear Cloudy Rainy Snowy Various/other
	Outdoor temperature	Hot Warm Cool Cold (approx. °C / °F)
	Place/road conditions	On motorway Suburbs Inner city Other Uphill Downhill Rough road
	Engine temperature	Cold Warming up Normal Other
	Engine operation	Starting Just after starting Idling Racing Driving Constant speed Acceleration Deceleration Other

Condition of "CHECK ENGINE" lamp		Always on Flickers Does not light up	
Diagnostic code check	1st time (pre-check)	Normal code Malfunction code(s) (	)
	2nd time	Normal mode Normal code Malfunction code(s) (	)
		Test mode	

Figure 1.17 Pre-diagnostic questionnaire

1 Introduction to light vehicle technology diagnosis & rectification

10-minute rule

Many common problems can recur within a certain vehicle range. It may be that after the driver has described the symptoms, you have a good idea of what the fault is before you start work on the car. Knowing about these recurring problems within a vehicle range can lead you to a quick fix, but you should not assume that they are always the correct diagnosis.

It is good practice to set yourself a 10-minute rule. Allow 10 minutes for checking the easy things first, but don't get drawn deeper and deeper into a false diagnostic routine by frustration when the quick fix doesn't work.

Having used up your 10 minutes, step back from the task and reassess your diagnostic routine. If expensive diagnostic equipment needs to be used, contact the customer and inform them of the situation so that they may be prepared for any extra time and cost involved. Some garages have a different labour rate for complex diagnosis as this will help reimburse the investment they have made in equipment and training.

Keep it simple

Break down your diagnosis into small steps, starting with easy to access components. Don't start with the most complex diagnosis technique due to lack of understanding (for example: 'I don't understand how the system operates, so it must be the electronic control unit (ECU) that is at fault.').

Even the most complex systems such as ECUs can be systematically checked:

- If **sensor** information coming into the ECU is incorrect, the fault lies with the sensor.
- If sensor information is correct, but ECU output to the **actuator** is incorrect, the fault lies within the ECU.
- If sensor information is correct and ECU output to the actuator is correct, the fault lies with the actuator.

By using ECU **pin data**, and diagnostic equipment such as an oscilloscope or multimeter, you do not need to test the ECU itself using dedicated equipment. This is because the ECU's correct function and operation can be seen by the previous logical thought process.

Keep a record

Write information down as you go along. Many diagnostic routines are time-consuming and may be interrupted mid flow. Recording the information as you go along will allow you to return to your diagnostic routine at any point. Also, if a similar problem occurs in the future, you can refer to your notes and this might save you diagnostic time and effort.

> **Did you know?**
>
> An open question is one without a straightforward yes or no answer. It allows the customer to give you information about the fault in their own words. Examples of open questions are:
>
> - When does the problem occur?
> - What are the symptoms you are experiencing?
> - Why have you brought the vehicle into the workshop today?

> **Key terms**
>
> **Sensor** – an electronic component that sends information about a particular vehicle system.
>
> **Actuator** – an output component that performs an action.
>
> **Pin data** – technical data about voltages and current found at ECU connection plugs.

Make sure that your diagnostic routine builds up 'the big picture' so you have an overall view of system operation. When part of the big picture is missing, this will normally point you to the root of the fault.

Keep the information you have gathered, and build your own library. The more systems that you test, the more voltage references you record and the more fault codes you diagnose, the greater your reference list. You can use this information to aid your memory and prompt your diagnostic routine. Spending a quarter of an hour checking through your notes and finding previous results can save hours of wasted time and money. You don't have to remember the entire job – just that you had a similar problem to diagnose before.

Diagnostic routine checklist

- ✓ Listen to the customer's vehicle complaint.
- ✓ Question the customer to find out the symptoms.
- ✓ Road test or operate the vehicle to confirm the symptoms.
- ✓ Gather information before you start and take the information with you to the vehicle.
- ✓ Devise a diagnostic strategy.
- ✓ Check the quick things first (10-minute rule).
- ✓ Contact the customer for authorisation if further testing is required.
- ✓ Keep it simple – start with the easiest parts of the system first.
- ✓ Conduct as much diagnosis as possible without stripping down.
- ✓ Locate the root cause of the problem.
- ✓ Fix the fault and not the symptom.
- ✓ Correctly reassemble any dismantled components or systems.
- ✓ Clear any diagnostic trouble codes and **adaptions**.
- ✓ Thoroughly test the systems to ensure correct function and operation.

> **Key term**
>
> **Adaptions** – automatic changes made by an engine management system, in order to maintain an acceptable operation, regardless of engine wear and tear.

CHECK YOUR PROGRESS

1. What is the difference between a fault and a symptom?
2. What is the difference between a sensor and an actuator?

1 Introduction to light vehicle technology diagnosis & rectification

PREPARE FOR ASSESSMENT

The information contained in this chapter, as well as continued practical assignments, will help you to prepare for both the end-of-unit tests and the diploma multiple-choice tests. This chapter will also help you to develop diagnostic routines that are safe for you and the vehicle.

You will need to be familiar with:

- Health and safety legislation
- Risk assessment
- Fire precautions
- Environmental protection
- Basic first aid
- Diagnostic routines

This chapter has given you a brief overview of health and safety legislation that will help you with both theory and practical assessments. It is possible that some of the evidence you produce may contribute to more than one unit. You should ensure that you make best use of all your evidence to maximise the opportunities for cross-referencing between units.

You should choose the type of evidence that will be best suited to the type of assessment that you are undertaking (both theory and practical). These may include:

Assessment type	Evidence example
Workplace observation by a qualified assessor	Safely carrying out a diagnostic routine on an advanced vehicle system
Witness testimony	A signed statement or job card from a suitably qualified/approved witness, stating that you have correctly conducted housekeeping duties in a motor vehicle workshop
Computer-based	A printout showing that you have accessed and researched the health and safety requirements relating to the Health and Safety (Display Screen Equipment) Regulations 1992
Audio recording	A timed and dated audio recording of a statement you have created to record an accident or near miss that has occurred in your workshop
Video recording	Short video clips showing you conducting a workshop risk assessment
Photographic recording	Photographs showing the safety data information supplied with chemicals that are considered hazardous to health (supplied under the COSHH regulations)
Professional discussion	A recorded discussion with your assessor (e.g. during a review) about any health and safety concerns you have in your workplace
Oral questioning	Recorded answers to questions asked by your assessor, in which you explain how you safely isolated a high voltage hybrid system in order to work on the vehicle's CVT transmission
Personal statement	A written statement describing how you conducted a workshop fire risk assessment on behalf of your employer

CONTINUED ▶

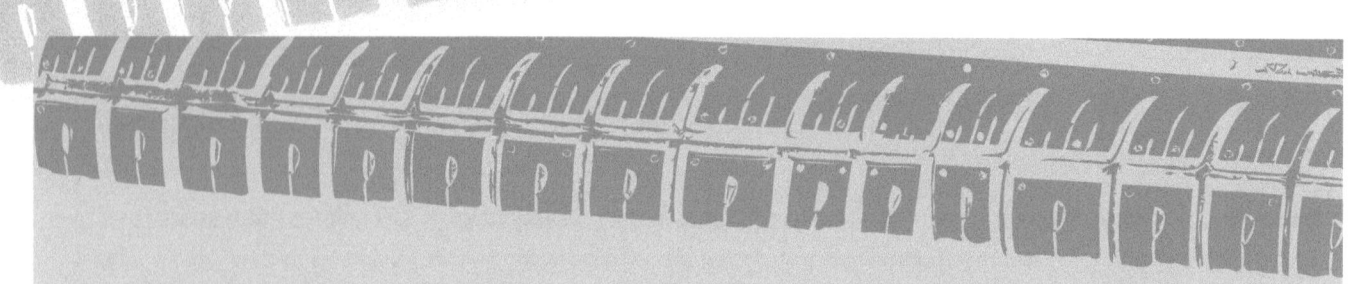

Assessment type	Evidence example
Competence/skills tests	A practical task arranged by your training organisation, asking you to conduct a systematic diagnosis of an advanced technology system
Written tests	A written answer to an end-of-unit test, checking your knowledge and understanding of the safety precautions that should be taken when working near a hybrid vehicle's high voltage system
Multiple-choice tests	A multiple-choice test set by your awarding body to check your knowledge and understanding of advanced light vehicle technology systems
Assignments/projects	A written assignment arranged by your training organisation requiring you to show in-depth knowledge and understanding of a particular electronic system and the logical thought processes involved with diagnosis

Before you attempt a theory end-of-unit or multiple-choice test, make sure you have reviewed and revised any key terms that relate to the topics in that unit. Ensure that you read all the questions carefully. Take time to digest the information so that you are confident about what each question is asking you. With multiple-choice tests, it is very important that you read all of the answers carefully, as it is common for two of the answers to be very similar, which may lead to confusion.

For practical assessments, it is important that you have had enough practice and that you feel that you are capable of passing. It is best to have a plan of action and work method that will help you.

Make sure that you have the correct technical information, in the way of vehicle data, and appropriate tools and equipment. It is also a good idea to check your work at regular intervals. This will help you to be sure that you are working correctly and to avoid any problems developing as you work.

When you are undertaking a practical assessment, always take care to work safely throughout the test. Light vehicle systems are dangerous and precautions should include making sure that you:

- observe all health and safety requirements
- use the recommended personal protective equipment (PPE) and vehicle protective equipment (VPE)
- use tools correctly and safely.

Good luck!

2 Diagnosis & rectification of light vehicle chassis system faults

This chapter will help you to gain an understanding of diagnosis and diagnostic routines that lead to the rectification of braking, steering and suspension system faults. It also explains and reinforces the need to test light vehicle chassis systems and evaluate their performance. It will support you with knowledge that will aid you when undertaking both theory and practical assessments. It will help you develop a systematic approach to complex diagnosis of the light vehicle chassis system.

This chapter covers:

- Operation of electronic ABS and EBD braking systems
- Steering geometry for light vehicle applications
- Power-assisted steering (PAS)
- Suspension systems

BEFORE YOU START

Safe working when carrying out light vehicle diagnostic and rectification activities

There are many hazards associated with the diagnosis and repair of advanced chassis systems. You should always assess the risks involved with any diagnostic or repair routine before you begin and put safety measures in place. You need to give special consideration to the possibility of:

- **Crush or bump injuries:** Many advanced chassis systems are active and could move without warning.
- **Coming into contact with chemicals such as brake fluid:** The complex nature of advanced braking systems means that fluid may be under extreme pressure in the hydraulic circuit. You must take special precautions when maintaining or repairing such systems. Always use the correct tools and follow the manufacturer's instructions.

You should always use appropriate personal protective equipment (PPE) when you work on chassis systems. Make sure that your selection of PPE will protect you from these hazards.

Electronic and electrical safety procedures

Working with any electrical system has its hazards and you must take safety seriously. When you are working with light vehicle electrical and electronic systems, the main hazard is the possible risk of electric shock. (For information on basic first aid for electrical injuries, see Table 1.3 in Chapter 1, page 18.) Although most systems operate with low voltages of around 12V, an accidental electrical discharge caused by incorrect circuit connection can be enough to cause severe burns. Where possible, isolate electrical systems before conducting the repair or replacement of components.

If you are working on hybrid vehicles, take care not to disturb the high voltage system. You can normally identify the high voltage system by its reinforced insulation and shielding, which is often brightly coloured. These systems carry voltages that can cause severe injury or death. If you carry out repairs to hybrid vehicles, always follow the manufacturer's recommendations.

Always use the correct tools and equipment. Damage to components, tools or personal injury could occur if the wrong tool is used or a tool is misused. Check tools and equipment before each use.

If you are using measuring equipment, always check that it is accurate and calibrated before you take any readings.

If you need to replace any electrical or electronic components, always check that the quality meets the original equipment manufacturer (OEM) specifications. (If the vehicle is under warranty, inferior parts or deliberate modification might make the warranty invalid. Also, if parts of an inferior quality are fitted, this might affect vehicle performance and safety.) You should only carry out the replacement of electrical components if the parts comply with the legal requirements for road use.

Information sources

The complex nature of advanced light vehicle chassis systems requires you to have a comprehensive source of technical information and data. In order to conduct diagnostic routines and repair procedures, you need to gather as much information as possible before you start. Sources of information include the following:

Information source	Example
Verbal information from the driver	A description of the symptoms that occur on the car under heavy braking
Vehicle identification numbers	Year of manufacture, taken from VIN plate
Service and repair history	A check of the service history that shows when the brake pads were last replaced
Warranty information	Is the car under warranty and is it valid? (Has the required service and maintenance been conducted?)
Vehicle handbook	To confirm how to correctly operate the vehicle's active suspension system before a road test
Technical data manuals	To find the recommended operating pressures of a hydraulic power-assisted steering for diagnostic purposes
Workshop manuals	To find the recommended procedures used when adjusting steering geometry
Safety recall sheets	To confirm which components need to be replaced for safe operation of a hybrid vehicle's regenerative braking system
Manufacturer-specific information	Information on where to measure suspension heights and how to set ride height sensors
Information bulletins	Information on a common fault with an anti-lock braking system on a particular car
Technical help lines	Advice on the correct routine for bleeding the hydraulic braking system of a hybrid vehicle
Advice from master technicians/colleagues	An explanation of how to set up the company's laser equipment for four-wheel alignment
Internet	An Internet forum page where a number of people who had a similar problem with their electronic power steering explain how it was resolved
Parts suppliers/catalogues	A cross-reference of brake pad part numbers, showing that the incorrect brake pads have been recently fitted
Job cards	A general description of the work to be conducted on a customer's suspension system
Diagnostic trouble codes	A fault code showing that the lateral acceleration sensor circuit needs to be tested on a fully active suspension system
Oscilloscope wave forms	A faulty sine wave pattern being produced from a vehicle's wheel speed sensor

Remember that no matter which information or data source you use, it is important to evaluate how useful and reliable it will be to your diagnostic routine.

Operation of electrical and electronic systems and components

The operation of electrical and electronic systems and components related to light vehicle chassis systems:

Electrical/electronic system component	Purpose
ECU	The electronic control unit (ECU) is designed to monitor the operation of light vehicle chassis systems. It processes the information received and operates actuators that control comfort and safety. It is also known as an ECM (electronic control module) or PCM (powertrain control module).
Sensors	The sensors are mounted on various chassis components and they monitor the components against set parameters. As the vehicle is driven, dynamic operation creates signals in the form of resistance changes or voltage, which are sent to the ECU for processing.
Actuators	The actuators are combined in the manufacture of chassis systems. Motors, solenoids, valves, etc., are operated by the ECU to help to control the action of steering, suspension and braking, leading to improved safety and comfort.
Electrical inputs/voltages	The ECU needs reliable sensor information in order to correctly determine the action of the chassis systems. If battery voltage was used to power sensors, its unstable nature would create issues (battery voltage constantly rises and falls during normal vehicle operation). Because of this, sensors normally operate with a stabilised 5-volt supply.
Digital principles	Many vehicle sensors create analogue signals (a rising or falling voltage). The ECU is a computer and needs to have these signals converted into a digital format (on and off) before they can be processed. This can be done using a component called a pulse shaper or Schmitt trigger.
Duty cycle and pulse width modulation (PWM)	Lots of electrical equipment and electronic actuators can be controlled by duty cycle or pulse width modulation (PWM). These work by switching components on and off very quickly so that they only receive part of the current/voltage available. Depending on the reaction time of the component being switched and how long power is supplied, variable control is achieved. This is more efficient than using resistors to control the current/voltage in a circuit. Resistors waste electrical energy as heat, whereas duty cycle and PWM operate with almost no loss of power.
Fibre optic principles	As vehicle safety systems improve, the need for very fast transmission of information has increased. Fibre optics use light signals transmitted along thin strands of glass to provide digital data transmission. (The light source is switched on and off.) In this way, information is transmitted essentially at the speed of light.

Electrical and electronic control is a key feature of all the systems discussed in this chapter.

The electronic control of most chassis systems is achieved using a method of computer networking also known as 'multiplexing'. You will find a description of multiplexing in Chapter 4 on page 198.

Tooling

No matter what task you are doing to a car, you will need to use some form of tooling.

The following table shows a suggested list of diagnostic tooling that could be used when testing and evaluating light vehicle chassis, braking, steering and suspension systems. Due to the nature of complex system faults, you will experience different requirements during your diagnostic routines and so you will need to adapt the list shown for your particular situation.

Tool	Possible use
Oscilloscope	To test the signal produced by a vehicle wheel speed sensor
Multimeter	To test the voltage signal produced by a vehicle ride height sensor
Test lamp/ logic probe	To test the existence of system voltage at a power steering motor during operation (Always use test lamps with extreme caution on electronic systems, as the current draw created can severely damage components.)
Power probe	To power the compressor of an air-assisted suspension system and check its operation
Pressure gauge	To check the correct function and operation of a hydraulic power-assisted steering system

CONTINUED ▶

Tool	Possible use
Code reader/ scan tool	To retrieve diagnostic trouble codes (DTCs) related to ABS, which indicate the circuit to be tested. To clear trouble codes, reset the malfunction indicator lamp and evaluate the effectiveness of repairs.
Brake dynamometer/ rolling road	To test the efficiency of braking systems and evaluate the effectiveness of repairs
Wheel alignment gauges	To check and set wheel toe settings. Graduated turn plates can be used to confirm correct toe-out on turns (TOOT).
Steering geometry equipment	To check the correct manufacturer's alignment settings for camber, caster, swivel axis inclination and scrub radius
Stethoscope	To help identify suspension noises created during operation. Some stethoscopes are electronic and their listening devices can be attached to various system components, then the car can be road tested and noise/wear can be located.
Laser thermometer	A non-contact thermometer, also known as a pyrometer, can help determine binding brakes by identifying overheated brake system components.

Operation of electronic ABS and EBD braking systems

In order to slow down and eventually stop the movement of a car, a braking system is used. **Friction** between brake pads and discs or brake shoes and drums is used to convert **kinetic** movement energy into heat.

The braking system will normally consist of:

- a brake pedal
- a brake servo or brake booster
- a brake master cylinder
- brake pipes and hoses
- brake calipers, discs and pads
- wheel cylinders, brake shoes and drums.

Figure 2.1 Brake system layout

When the brake pedal is pressed, the master cylinder, assisted by the **brake servo**, forces brake fluid through the brake pipes and hoses to the **wheel assemblies**. At the wheel assemblies, brake pads are clamped against brake discs, or brake shoes are forced against rotating brake drums. This creates friction and changes movement into heat.

The slowing down of the vehicle can only be as good as the grip between the tyre and the road surface. If the tyre skids, vehicle control is lost and steering, braking and acceleration are no longer possible. This means that the efficiency of the braking system is **compromised**.

Anti-lock braking system (ABS)

If the hydraulic pressure in the brake lines can be regulated so that the tyres are prevented from skidding, then vehicle control can be maintained. This is the job of an anti-lock braking system.

Anti-lock braking uses electrics and electronics to regulate the pressure in the hydraulic system, which provides an artificial form of **cadence braking**. This is an active safety system that will help the driver maintain control in emergency situations. A modern anti-lock braking system is so efficient that it is able to control the braking of the car within a 10–30 per cent **slip tolerance**. This means that the tyres are kept on the point of skidding, but the wheel should never fully lock up.

A vehicle fitted with anti-lock brakes uses all of the parts normally associated with a hydraulic disc brake system. Three extra components are needed so that anti-lock braking can be achieved:

1. wheel speed sensors
2. an electronic control unit (ECU)
3. an ABS modulator valve block.

These components work in partnership with the rest of the brakes so that an anti-lock system is achieved.

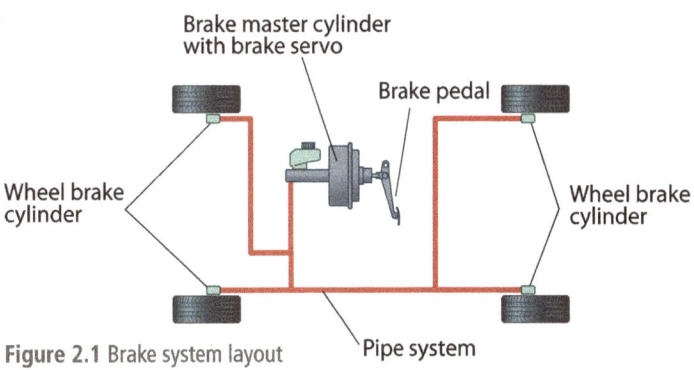

> **Key terms**
>
> **Friction** – the resistance created as one component rubs against another.
>
> **Kinetic** – movement energy.
>
> **Brake servo** – a component that assists the driver with the application of the brakes, meaning that less pedal pressure is required during braking.
>
> **Wheel assemblies** – all of the components found at the wheel, including: wheel, tyre, brake, hub and bearings.
>
> **Compromised** – lower than the desirable standard.
>
> **Cadence braking** – a form of braking in which the pedal is rapidly pressed and released by the driver.
>
> **Slip tolerance** – is a measurement of the tyre traction and grip during braking where 0 per cent means that the tyre is rolling and 100 per cent means that the wheel has fully locked up. On an ABS system maximum brake force is applied between 10 and 30 per cent slip tolerance which is the region where the most efficient braking occurs.

Level 3 Light Vehicle Technology

> **Did you know?**
>
> For many years racing drivers have used a method called 'cadence braking' to help control the vehicle under rapid deceleration. The driver is able to pump the brake pedal on and off so that the hydraulic pressure is regulated and the wheels keep turning. Because the tyres are prevented from skidding on the road surface, the vehicle handling is much improved.

> **+ Emergency**
>
> In case of brake fluid leakage, such as that caused by a burst pipe, a tandem master cylinder is still used. The hydraulic circuitry can be split in a front-to-rear or diagonal formation so that in the event of a sudden loss of hydraulic pressure in one part of the circuit, some braking still remains.

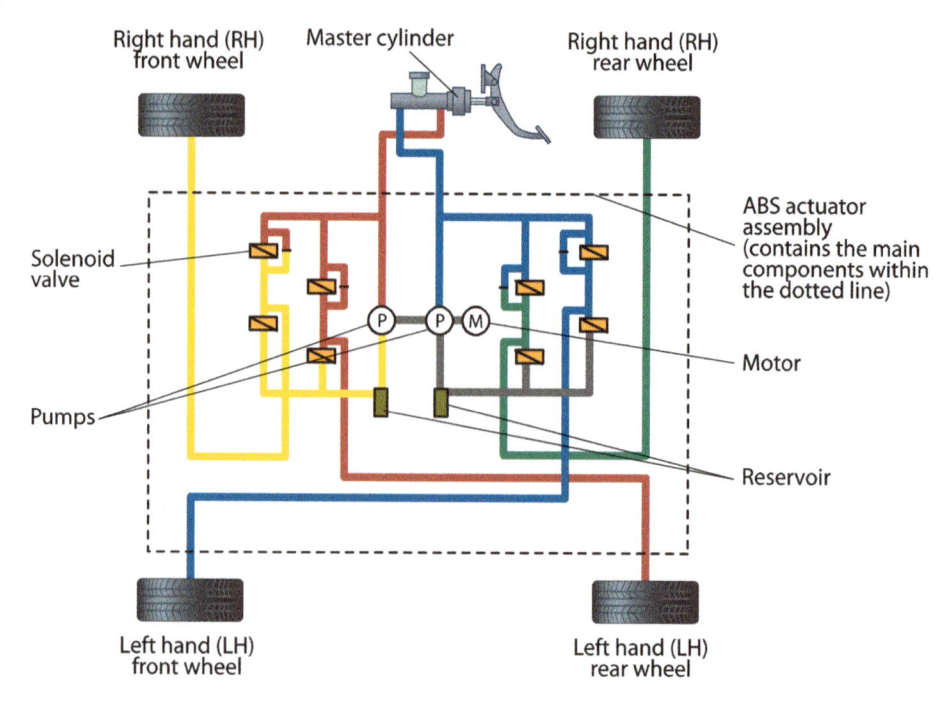

Figure 2.2 ABS schematic

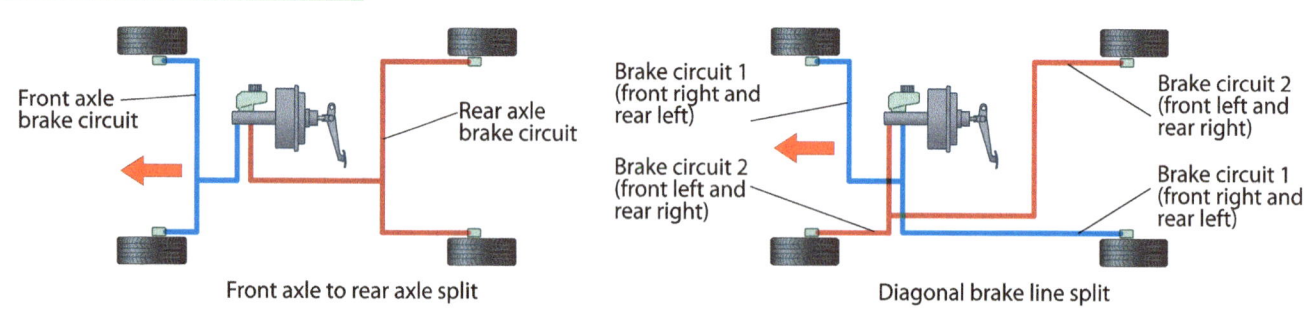

Figure 2.3 Split hydraulic braking system

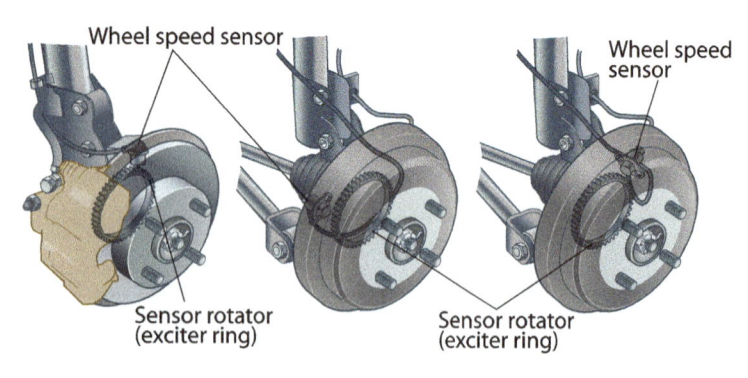

Figure 2.4 Wheel speed sensors

Wheel speed sensors

Although sometimes mounted **inboard** on a driveshaft, most wheel speed sensors are located at the hub assembly of each road wheel. They consist of two main components:

- a toothed wheel called a **reluctor**
- an **inductive magnetic sensor** unit.

The toothed reluctor is mounted on the hub so that it rotates with the wheel.

A permanent magnet inductive sensor is mounted close to the reluctor. It consists of a thin coil of copper wire wrapped around a permanent magnet. While the road wheel rotates, the small magnetic field produced by the inductive sensor is disrupted. As a reluctor tooth comes towards the sensor, the movement of the magnetic field generates a small electric current in the copper winding which is generated in one direction. As a reluctor tooth moves away from the sensor, a small electric current in the copper winding is generated in the opposite direction.

If an oscilloscope is connected to a wheel speed sensor and the wheel is rotated, an AC wave form (**sine wave**) will be seen on the screen. When the wheel speed increases, the amplitude (voltage) or the height of the sine wave will increase and the **frequency** (time signal) will also increase, making the waves appear closer together.

Once the signal from the wheel speed sensor has been created, the **analogue** sine wave is converted into a **digital** signal so that it can be processed by the electronic control unit. Once this is complete, the ECU will detect the frequency and convert this into a reading of speed.

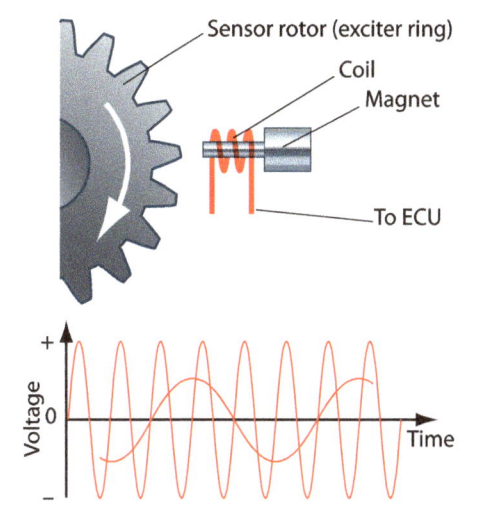

Figure 2.5 Inductive wheel speed sensor

Hall effect wheel speed sensors

Some manufacturers use Hall effect sensors to create the speed signal from the wheel. A rotating drum or magnetic interrupter plate disrupts a magnetic field from a small integrated circuit, which creates a digital or **square wave** form output, as shown in Figure 2.6. The advantages of Hall effect sensors are:

- the signal does not have to be converted before it can be processed by the ECU
- there is less chance of the signal being disrupted by other electronic systems
- as the signal is digital, the information produced can be easily shared with other vehicle systems
- speed of rotation does not affect signal strength, meaning it can operate at very slow speeds
- the sensor gives the system a far greater accuracy because it only has to measure frequency rather than amplitude.

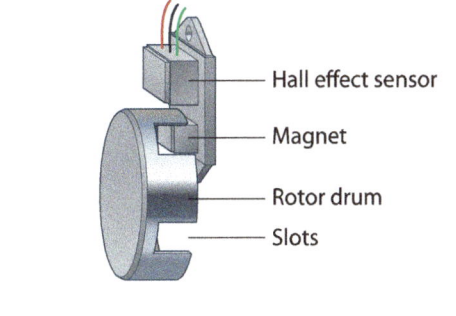

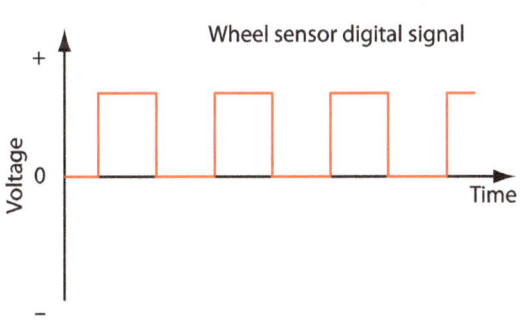

Figure 2.6 Hall effect wheel speed sensor

> **Key terms**
>
> **Inboard** – towards the middle of the car.
>
> **Reluctor** – a toothed wheel used with a sensor to detect speed and position.
>
> **Inductive magnetic sensor** – a sensor that creates electrical voltage when affected by a moving magnetic field.
>
> **Sine wave** – a smooth oscillating (normally up and down) repeating wave form often seen on an oscilloscope.
>
> **Frequency** – how often something happens; the distance/time between the peaks of a wave form.
>
> **Analogue** – a signal with a variable rising and falling voltage.
>
> **Digital** – a signal with a voltage that is switched on and off.
>
> **Square wave** – an angular/square oscillating (normally up and down) repeating wave form often seen on an oscilloscope.

ABS modulator unit

The ABS modulator unit contains a series of electrically controlled valves that regulate the hydraulic pressure in each part of the car's

Level 3 Light Vehicle Technology

> **Key terms**
>
> **Solenoid** – a linear motor (one that moves in a line rather than rotating).
>
> **Armature** – the central shaft of a motor.
>
> **Failsafe** – a system that will still have basic operation if it malfunctions.
>
> **Accumulator** – a storage area or reservoir.

braking system. The valves are of a **solenoid** type – they contain a small coil of copper wire which, when energised, with electricity, creates an invisible magnetic field. This magnetic field moves an **armature**, which operates a valve that restricts the flow of brake fluid to an individual brake.

When the solenoid valve is switched off, a return spring moves the valve back to the open position. This means that the system is **failsafe** in the event of electrical malfunction. If the solenoids are not energised, they will be held in the open position and standard braking is achieved.

Many modulators also contain an electric motor and pump which are used to return fluid to the master cylinder via an **accumulator**.

Electronic control unit (ECU)

The electronic control unit (ECU) is a small computer that is used to monitor the speeds of the individual wheels and to make calculations to determine whether one of the wheels is about to lock up or skid. It then acts as a switch to turn the solenoid valves in the ABS modulator unit on and off. By doing this, it enables the ABS system to provide an electronic version of cadence braking.

ABS operation

During normal driving, the ECU monitors the speed from each wheel speed sensor but remains passive because the brake pedal has not been pressed. The ABS ECU is able to monitor the operation of the brake pedal via the brake light switch. As soon as the driver presses the brake, the ABS system becomes active. Any rapid deceleration from an individual wheel, when compared with the others, may cause that particular wheel to lock up or skid. If this happens, the ABS will take the following three actions:

1. ABS pressure holding phase
2. ABS pressure reduction phase
3. ABS pressure increase phase.

Figure 2.7 Combined electronic control unit and modulator

1. ABS pressure holding phase

In the event of rapid deceleration from an individual wheel, the ABS ECU will operate the solenoid valve that relates directly to the wheel that is rapidly decelerating. Using pulse width modulation (PWM), the ECU regulates the current flowing to the solenoid valve. This regulated current partially operates the valve into a position where it blocks the input from the master cylinder. Pressure after the valve now remains constant, still allowing the wheel to slow down. No matter how hard the driver presses the brake pedal, hydraulic pressure in the caliper part of the system will not increase.

Figure 2.8 Pressure holding

2 Diagnosis & rectification of light vehicle chassis system faults

2. ABS pressure reduction phase

If the signal from the wheel speed sensor continues to indicate that the wheel is still decelerating too rapidly, the ECU will fully energise the solenoid valve. This will still block any extra pressure from the brake master cylinder, but will now open up a passageway, allowing the fluid in the brake caliper circuit to release and flow into an accumulator. As the accumulator fills up, a pump attached to the modulator unit returns the excess fluid to the master cylinder circuit. This will reduce pressure in the caliper circuit and allow the wheel to speed up again.

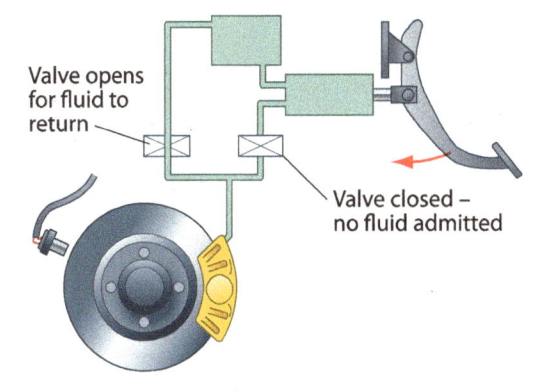

Figure 2.9 Pressure reduction

3. ABS pressure increase phase

As soon as the ECU receives a signal indicating that the wheel is once again speeding up, it switches off the current to the modulator valve. A return spring resets the valve to the open position. This means that pressure from the master cylinder circuit is allowed to increase and normal braking is resumed. If the wheel begins to slow rapidly again, the process is repeated. This will happen over and over many times a second for any individual brake that gives indications that it is about to lock.

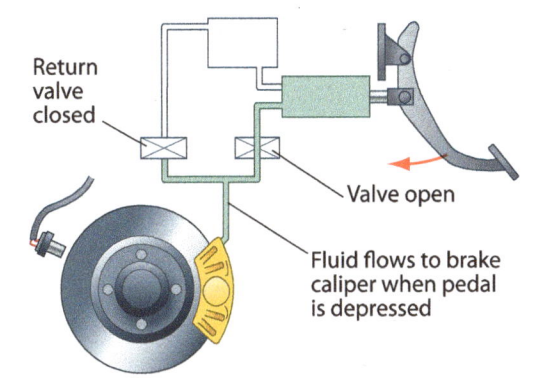

Figure 2.10 Pressure increase

NEW TECH

Brake-by-wire

Some manufacturers are now including brake-by-wire ABS systems in their vehicle design. Instead of the master cylinder applying pressure directly to the brake caliper system, it operates as a pressure measurement sensor. This sensor simulates correct pedal feel so that brake operation feels normal to the driver. The signal from the brake pressure sensor is processed by an ECU, which operates a motor in a secondary master cylinder unit. In conjunction with the wheel speed sensors, the secondary master cylinder applies the maximum brake force required to slow the wheels without allowing them to skid. Because a brake-by-wire system does not use a modulator valve block assembly, the driver does not feel the characteristic pulsing and buzzing through the brake pedal when it is operating.

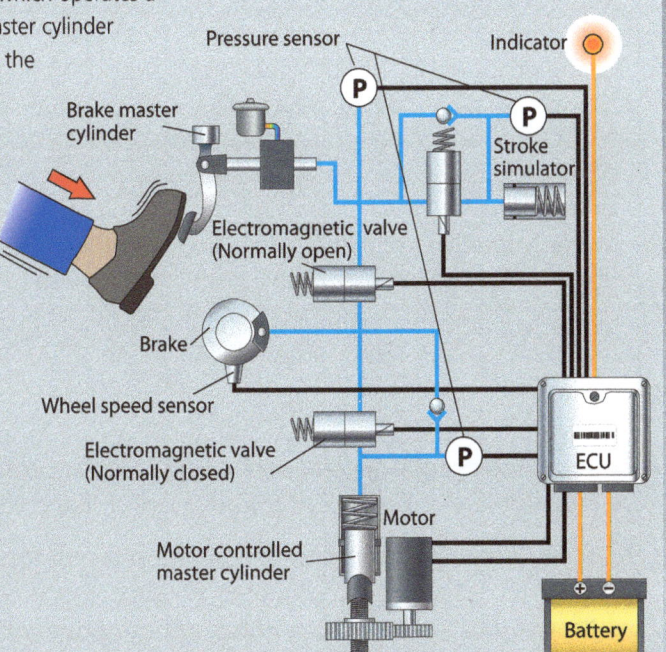

Did you know?

When the ABS system operates, the driver can often feel a buzzing pulsation through the brake pedal. This is caused by the modulator pump returning brake fluid to the master cylinder circuit under pressure – it is a normal function of ABS operation.

Emergency

With brake-by-wire technology, the system must be failsafe. If a fault occurs in the system or power is lost, a safety valve will open. This allows standard braking with the master cylinder acting directly on the caliper circuit.

39

Level 3 Light Vehicle Technology

Fault diagnosis

You should begin any fault diagnosis of an ABS system by inspecting the mechanical units. Because ABS is an electronic system, this stage is often overlooked. Table 2.1 gives an overview of what you need to inspect.

Table 2.1 Inspection of mechanical systems

Mechanical system	Possible fault
Wheels and tyres	Inspect wheels and tyres for: • Tread wear/pattern: Is it suitable for the road surface? • Wheel and tyre size: Different size wheels and tyres on the same vehicle will give inaccurate wheel speed signals. • Tyre pressures: These can cause issues with grip and overall wheel diameter. • Correct wheel bearing adjustment: Loose wheel bearings can cause spacing issues with wheel speed sensors and reluctor rings.
Steering and suspension	Inspect steering and suspension joints for: • Worn ball joints and suspension bushes: These will give unpredictable wheel movements that may reduce grip. • Worn springs and dampers: These may allow the tyre to leave the road surface and lose grip. • Worn steering gear and track rod ends: Play in steering systems will give unpredictable wheel movements that may reduce grip. • Check wheel alignment and steering geometry: Incorrect wheel alignment and steering geometry can allow the tyre to scrub across the road surface and reduce grip.
Brakes	Inspect the brake system by: • Checking the pedal assembly: Check pedal height, **free play**, the operation of the pedal and ensure that the brake light switch operates correctly from the pedal mechanism. Remember that the ABS system relies on the brake light switch to show that the brakes are being operated. • Checking the hydraulic system: Ensure that there are no leaks and that the fluid level is correct. Braking efficiency can be lost if fluid contaminates brake friction surfaces. Ensure that there is no air in the hydraulic system causing the brakes to feel spongy. • Using a specialist tool, conduct a test on the boiling point or moisture content of the brake fluid. If the moisture content of the brake fluid is too high, the boiling point will be reduced and this may lead to vapour lock during prolonged braking. Vapour lock may cause partial or total loss of hydraulic braking operation. • Checking brake assemblies: Conduct a visual inspection of brake assemblies, ensuring that all friction material is in good condition. If the friction material is worn low, it may not be able to dissipate heat correctly and as a result may overheat and cause brake fade.

> **Key terms**
>
> **Free play** – a small amount of clearance or movement between two components.
>
> **Generic** – applicable to a wide range of different situations.

You need to decide on a systematic diagnostic routine and stick to it. Examine the symptoms and take appropriate steps to locate the fault. (See the diagnosis routine in Figure 1.15 on page 22 of Chapter 1).

Table 2.2 gives an indication of some of the symptoms that may be experienced when working on a braking system. This list is not exhaustive and you should conduct a thorough diagnostic routine to ensure that you have correctly located the fault.

2 Diagnosis & rectification of light vehicle chassis system faults

Table 2.2 Symptoms of faults in braking systems

Symptom	Possible fault
Inefficient brakes	• Worn friction material: overheating causing brake fade • Incorrect servo operation: possible vacuum leak • Hydraulic fluid leak: tandem master cylinder only operating on one part of the system • Air in a hydraulic system: causing spongy brakes and loss of leverage at the pedal
Pulsation at brake pedal	• Anti-lock braking system cutting in too early: may be caused by worn tyres which slip on the road surface • Excessive wheel bearing play: too much movement in the wheel bearing may cause the brake discs to move the pads back in the caliper. When the brakes are applied, the pedal has to travel much further before the friction material grips the disc. • Excessive run-out: may be caused by a warped brake disc, or brake disc thickness variation known as 'Martensite deposits' (see below)
Tyre skid/ABS warning light illuminated	• Excessively worn tyres: causing them to slide on the road surface even though the ABS system is in operation • Fault with anti-lock braking system: causing the malfunction indicator lamp to be illuminated when the engine starts or when the vehicle moves. (This will usually put the system into a failsafe mode that allows normal brake operation but with no ABS facility.)

If the ABS warning light is illuminated, you should connect a suitable diagnostic scanner to the diagnostic link connector and retrieve the fault codes. These diagnostic trouble codes are not **generic** and so you will have to refer to the manufacturer's data.

> **Did you know?**
>
> Martensite deposits are a hard crystalline structure formed in steel during manufacture. Martensite is named after the German metallurgist Adolf Martens. When brake discs are manufactured, these Martensite deposits cause high spots to form on the surface of the disc during normal wear and tear. This can lead to a pulsation of the brake pedal, similar to the vibrations caused by a warped disc.

Figure 2.11 A technician using a scan tool to retrieve diagnostic trouble codes

Level 3 Light Vehicle Technology

> **Action**
>
> Although many manufacturers are now using the code letter 'C' to indicate that the ABS fault code is related to the chassis system (similar to those shown in Table 2.8 on page 74), some manufacturers are sill using code descriptions of their own design.
>
> Using the sources of information available to you, look up the ABS fault codes from three different manufacturers.
>
> Do they use 'C' codes or their own format?

Table 2.3 gives some examples of diagnostic trouble codes for the ABS system. Remember that ABS fault codes will be vehicle specific.

Table 2.3 Examples of diagnostic trouble codes

Fault code	Possible cause(s)
16, Solenoid valve, F.L.	Fault in circuit
17, Solenoid valve, F.R.	Fault in circuit
19, ABS relay	Voltage under minimum
25, Wheel speed sensors	Wrong signal, Toothed ring/teeth
28, Solenoid valve, R.L.	Fault in circuit
29, Solenoid valve, R.R.	Fault in circuit
31, Tachometer	Missing signal
35, Hydraulic pump	Relay control circuit electrical, Engine, Wires/connector
37, Stop light switch	Defective/circuit failure
39, Wheel speed sensor F.L.	Wrong signal, Toothed ring/teeth, Air gap
41, Wheel speed sensor F.L.	Circuit failure, Wires/connector
42, Wheel speed sensor F.R.	Wrong signal, Toothed ring/teeth, Air gap
43, Wheel speed sensor F.R.	Circuit failure, Wires/connector
44, Wheel speed sensor R.L.	Wrong signal, Toothed ring/teeth, Air gap
45, Wheel speed sensor R.L.	Circuit failure, Wires/connector
46, Wheel speed sensor R.R.	Wrong signal, Toothed ring/teeth, Air gap
47, Wheel speed sensor R.R.	Circuit failure, Wires/connector
48, Voltage supply	Voltage too low
49, Voltage supply	Voltage too high
52, ABS fault indicator	Circuit failure
55, ECU	Defective
65, Traction control	Fault/ECU error, Programming error
66, ECU	Fault in circuit
67, ECU	Fault in circuit

> **Did you know?**
>
> It is sometimes possible to diagnose if an ABS fault lies within a wheel speed sensor circuit. The malfunction indicator lamp is designed to undertake a self-check process when the ignition key is turned on. It will usually illuminate for around 30 seconds and then switch off. If this happens, no failure from start-up has been detected. When the vehicle starts to move, if the malfunction indicator lamp then illuminates, this can give an indication that a problem exists in the sensing of wheel speed.

It is good practice to read and record any fault codes present, clear the fault codes, road test the car and base your diagnostic routine on any fault codes that have returned. Remember to diagnose the circuit, not guess the fault. Once you have repaired the faults or replaced components, you should fully road test the vehicle and recheck the fault codes. Be sure that you follow up any pending fault codes, which may be an indication of emerging problems.

2 Diagnosis & rectification of light vehicle chassis system faults

Figure 2.12 Dashboard warning system

If the ABS system is not functioning correctly and the malfunction indicator light is not illuminated, check the following:

1. Power supply to the control unit (including the fuse)
2. Earth connection of the control unit
3. Actuator/modulator power supply
4. Actuator/modulator earth connection
5. Actuator/modulator operation
6. Correct alignment and air gap of the wheel speed sensor.

Wheel speed sensor alignment

In order for the wheel speed sensors to operate correctly and accurately, there must be no misalignment between the reluctor and the sensor pickup. Many manufacturers recommend a specific air gap that should be maintained between the sensor unit and the reluctor ring. This air gap is often adjustable, sometimes using elongated mountings or **shims**.

It is often possible to check wheel speed sensors using an ohmmeter.

Did you know?

Because Hall effect sensors contain an integrated circuit, it is not possible to check their resistance using an ohmmeter. A resistance check is only possible on inductive wheel speed sensors.

An oscilloscope test can be conducted on both Hall effect and inductive sensors.

Key term

Shim – a thin, accurately-sized piece of material used to fill a gap.

How to check wheel speed sensors using an ohmmeter

Checklist			
PPE	**VPE**	**Tools and equipment**	**Source information**
• Steel toe-capped boots • Overalls • Latex gloves	• Wing covers • Steering wheel covers • Seat covers • Foot mat covers	• Spanners • Screwdrivers • Jack • Axle stands • Ohmmeter	• Technical data • Job card

43

Level 3 Light Vehicle Technology

1. Switch off any power to the system and disconnect the component from the electrical circuit in order to correctly test for resistance.

2. Set the multimeter to the lowest ohms scale, and connect the test probes together to calibrate the meter.

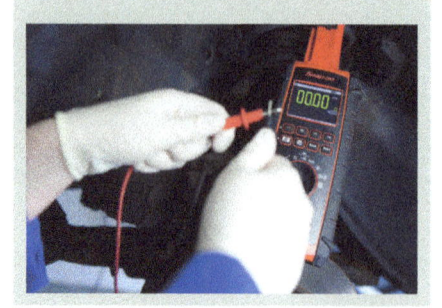

3. Connect the test probes to the circuit wires from the inductive wheel speed sensor, and use the selector dial to achieve the most appropriate reading.

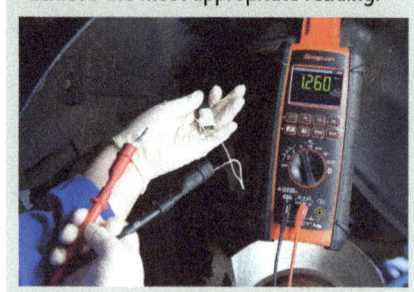

> ⚠️ **Safe working**
>
> You must take precautions when bleeding the hydraulic system of an ABS-equipped car, following the manufacturer's guidelines. It is possible that, unless a maintenance routine is selected using dedicated manufacturer equipment, brake fluid may not correctly pass through a hydraulic actuator unit.

4. Compare this reading with that shown in the manufacturer's technical data.

Although the use of an ohmmeter in diagnosis is helpful, the information provided is limited. It is best to check the correct function and operation of a wheel speed sensor using an oscilloscope. This allows you to check the wheel speed sensor **in situ** throughout its normal operating range.

> **Key term**
>
> **In situ** – kept in place (without removing).

How to check wheel speed sensors using an oscilloscope

Checklist			
PPE	**VPE**	**Tools and equipment**	**Source information**
• Steel toe-capped boots • Overalls • Latex gloves	• Wing covers • Steering wheel covers • Seat covers • Foot mat covers	• Spanners • Screwdrivers • Jack • Axle stands • Oscilloscope	• Technical data • Job card

1. Connect the negative lead of the oscilloscope to a good source of ground or earth.

2. Connect the sensor probe of the oscilloscope to the signal wire of the wheel speed sensor. You can do this at the wheel speed sensor plug or at the ECU.

3. Once you have selected the appropriate scale on the oscilloscope, drive the car or raise the wheel and spin it by hand. Compare the wave form with that found in the manufacturer's technical data.

2 Diagnosis & rectification of light vehicle chassis system faults

4. If you are using a multichannel oscilloscope, you could check all four wheel speed sensors at the same time.

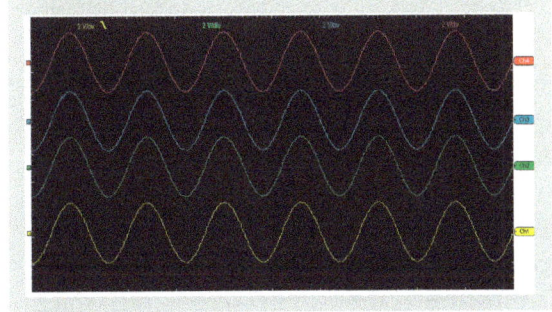

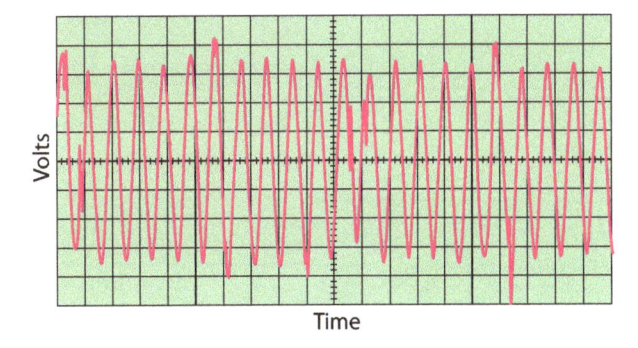

Figure 2.13 Wheel speed sensor wave form

Case study

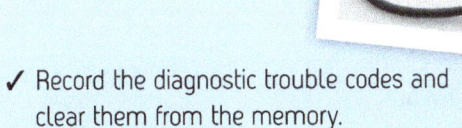

Mrs Rahman brings her car to your garage and complains that the ABS warning light on the dashboard stays on. Your boss asks you to check it out.

When you turn on the ignition and start the car, the light illuminates for around 30 seconds and then goes out. You think you may need to road test the car to investigate the problem fully, so you go back to the reception to check that it's okay to do so.

Here's what you do:

- ✓ Listen to the customer's description of the fault.
- ✓ Question the customer carefully to find out the symptoms.
- ✓ Road test the car to confirm the symptoms described by Mrs Rahman.
- ✓ Gather information from technical manuals (including how to retrieve fault codes) before you start, and take the information to the vehicle.
- ✓ Devise a diagnostic strategy.
- ✓ Check the quick things first (10-minute rule) using visual inspection.
- ✓ Ask the customer for authorisation if further testing is required.
- ✓ Conduct as much diagnosis as possible without stripping down.
- ✓ Using a scan tool, read the diagnostic trouble codes (00287 Wheel speed sensor RR — wrong signal, wires/connector, toothed ring/teeth, air gap).
- ✓ Record the diagnostic trouble codes and clear them from the memory.
- ✓ Road test the car and re-scan to check that the code has returned.
- ✓ Locate the root cause of the problem (wheel speed sensor open circuit caused by a broken wire).
- ✓ Test the component and circuit.
- ✓ Connect an oscilloscope to the right-hand rear wheel speed sensor and check for correct signal.
- ✓ Keep the customer informed of progress and costs.
- ✓ Replace the faulty wheel speed sensor.
- ✓ Correctly reassemble any dismantled components/systems.
- ✓ Clear any diagnostic trouble codes and adaptions.
- ✓ Thoroughly test the system to ensure correct function and operation.

Electronic brake force distribution (EBD) and emergency brake assist

Once the vehicle has been equipped with an anti-lock braking system, it is a very small step to provide other electronic safety devices.

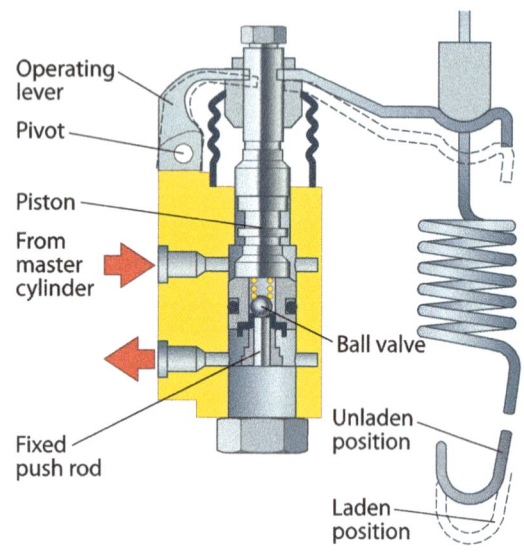

It was once common practice to fit vehicles with a brake pressure proportioning valve (see Figure 2.14). During heavy braking, weight transfer over the front wheels of the vehicle reduced the grip between the rear tyres and the road surface. This often led to the rear wheels entering a skid or creating oversteer. The purpose of the brake proportioning valve was to reduce hydraulic pressure in the rear brakes circuit, which reduced the possibility of a rear wheel skid. If extra weight was placed in the rear, grip to the rear tyres would be increased and the proportioning valve would allow more hydraulic pressure to the rear brakes.

In modern vehicles, electronic brake force distribution (EBD) is used instead of a proportioning valve (see Figure 2.15). EBD makes use of the ABS system to direct hydraulic fluid pressure to the wheels with the most grip, regardless of weight distribution. In this way, stopping distances are reduced and vehicle handling and control are maintained.

Figure 2.14 Brake proportioning valve

Action
Using resources available to you, research manufacturer vehicle makes and models that use electronic brake force distribution and emergency brake assistance. Try to find out when these systems started to appear as safety options. Compare your results with others, to see who can find the earliest examples.

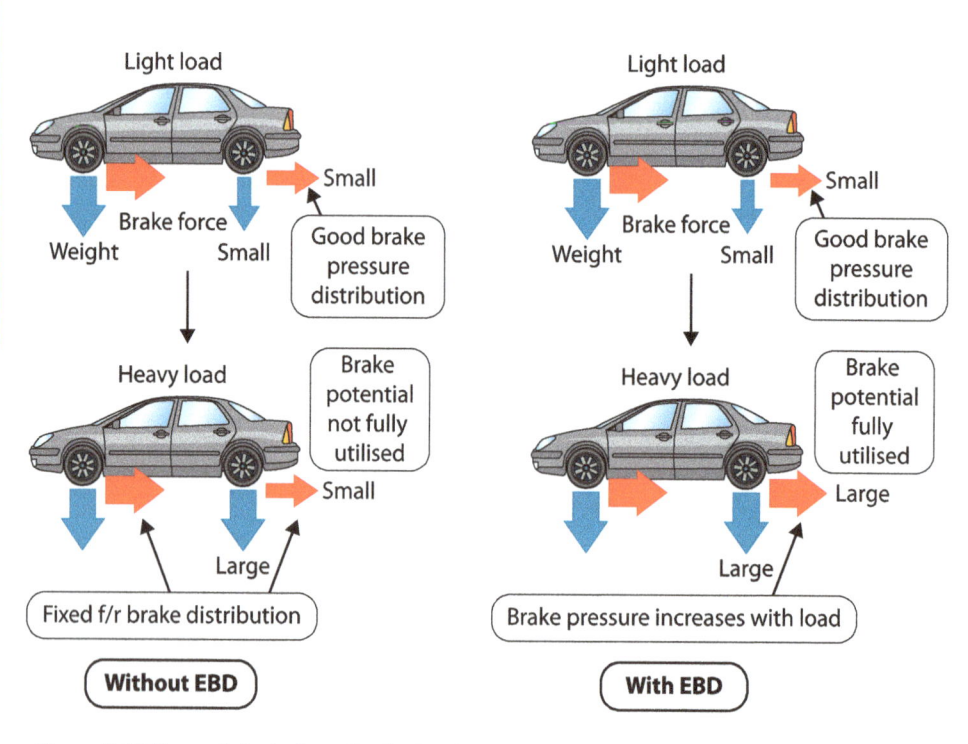

Figure 2.15 Electronic brake force distribution EBD

An emergency brake assistance system can increase pressure in the hydraulics using a solenoid attached to the brake servo unit, or pump actuated accumulator. The emergency brake assistance system senses the speed and rate at which the driver presses the brake pedal in an emergency situation. It responds by directing the extra hydraulic pressure to the wheels with the most grip in conjunction with the ABS system.

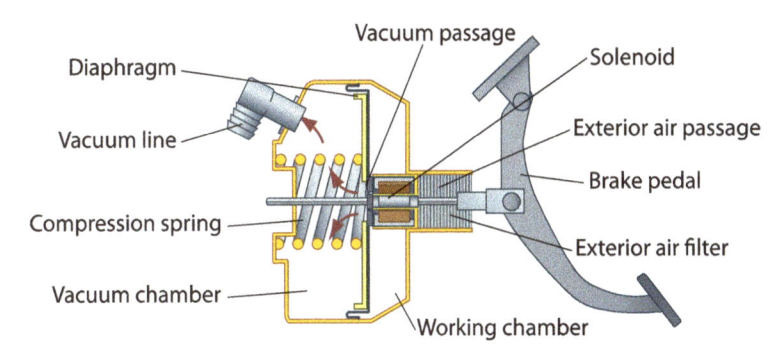

Figure 2.16 Solenoid controlled brake servo

Electronic parking brakes (EPB)

Many manufacturers are now including electronic parking brakes in the design of their cars. Two main types are in use:

- **Cable actuated** – this system uses a motor to draw on a standard cable-operated handbrake mechanism. Sensors detect how much force must be applied to correctly hold the brakes.

Figure 2.17 Cable puller type EPB system

- **Caliper type** – this system has a mechanical actuator integrated into the brake caliper. When this is operated, it applies pressure to the brake pads to hold the vehicle stationary.

Both systems have sensors to measure the amount of force needed to correctly operate the parking brake, and an ECU to manage the brake mechanisms. They may also make use of the ABS wheel speed sensors to detect unwanted movement and reapply pressure to the parking brake mechanism.

To release the parking brake, the driver normally needs to switch on the ignition, apply the foot brake and press the release button. Some systems have an automatic drive away release. As the car is placed in gear and drive is taken up, sensors signal the ECU to release the parking brake and allow the car to drive away. An advantage of electronic parking brakes is that they can be integrated with other chassis management systems, such as ABS and traction control, to improve driver comfort, convenience and safety.

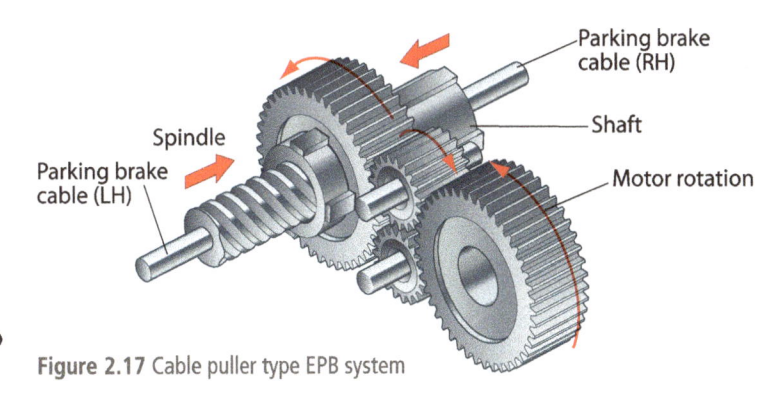

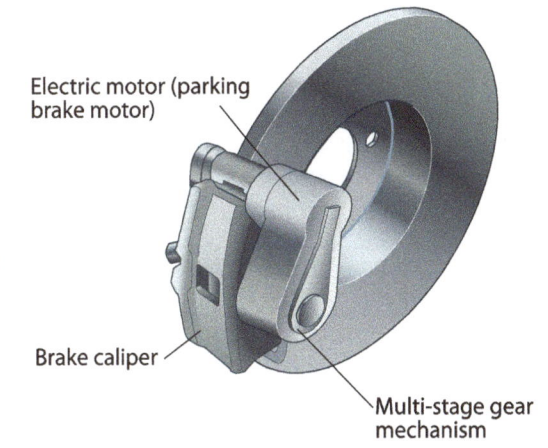

Figure 2.18 Caliper type EPB system

NEW TECH

Emergency braking facility
Some electronic parking brake systems include an emergency braking facility. In the event of a failure in the standard foot brake system, the electronic parking brake can be operated to bring the car to a controlled stop. At speeds greater than 4 mph, if the EPB button is pressed and held, the electronic stability program applies the brakes to bring the car to a controlled stop. When the vehicle is stationary and the button is released, the parking brake is then applied.

Level 3 Light Vehicle Technology

Inspection, testing and repair of electronic parking brake systems may require special procedures and tooling. The efficiency of many designs can be tested in a standard brake rolling road. If the ECU detects that only the two wheels operated by the parking brake are rotating at low speed, it will enter a diagnostic inspection mode. You can then press the EPB button a number of times to gradually apply the brakes and check their operation.

When diagnosing faults with EBP systems, you will need to use a scan tool to retrieve diagnostic trouble codes. Record any fault codes present, clear codes and then operate the system to see if any faults return. Many scan tools can run active tests, where you are able to observe the operation of the parking brake mechanism and make decisions on repair procedures.

Traction control system (TCS)

Another active safety system that can make use of some of the ABS components is traction control system (TCS). This system is designed to reduce the wheel spin during hard acceleration or manoeuvring. If the ABS wheel speed sensors detect a difference in rotational speed, particularly in the driven wheels of a car, it will interpret this information as wheel spin. Traction control knows the difference between wheel spin that is caused by acceleration and slip that is caused by braking because it is also monitoring the operation of the brake light switch.

If wheel spin occurs, a number of actions can be taken by the traction control unit:

- Hydraulic brake pressure can be applied to the spinning wheel.
- Engine torque can be reduced by closing the throttle butterfly actuator.
- Ignition timing can be retarded.
- Fuel injection can be reduced.

Electronic stability control (ESC), electronic stability programs (ESP) and active yaw control (AYC)

Other safety systems can make use of the function and operations of traction control and anti-lock braking systems.

Electronic stability control (ESC) and electronic stability programs (ESP) try to maintain the **attitude** of the vehicle suspension during cornering and manoeuvring procedures. Working in conjunction with active suspension (see page 72), engine torque/drive and ABS, wheel spin and slip are monitored and adjusted where necessary to help improve overall vehicle handling.

Another advantage of ESP is its ability to intervene during understeer or oversteer. When the electronic stability program detects a loss of steering control during a manoeuvre, it automatically applies the brakes to help direct the vehicle where the driver intends to go. Brakes are automatically applied to the wheels individually, such as when the brakes are applied to the outer front wheel to overcome oversteer or the inner rear wheel to overcome understeer. Some manufacturers incorporate a method that

Figure 2.19 Brake roller testing

Safe working

The electronic parking brake mechanism remains active, even when the ignition is switched off. When you are servicing the brake system, you must disable the EPB using a special service tool/procedure or scan tool. Once disabled, you can repair a cable-operated system using standard procedures, but a caliper-operated system will normally need a special service tool to wind back the piston. Following the replacement of brake pads, you may need to use a scan tool to adjust the electronic parking brake mechanism to remove any initial free play.

Safe working

Ensure that that the EPB system is isolated before starting any work, as the brakes may operate unexpectedly and cause injury.

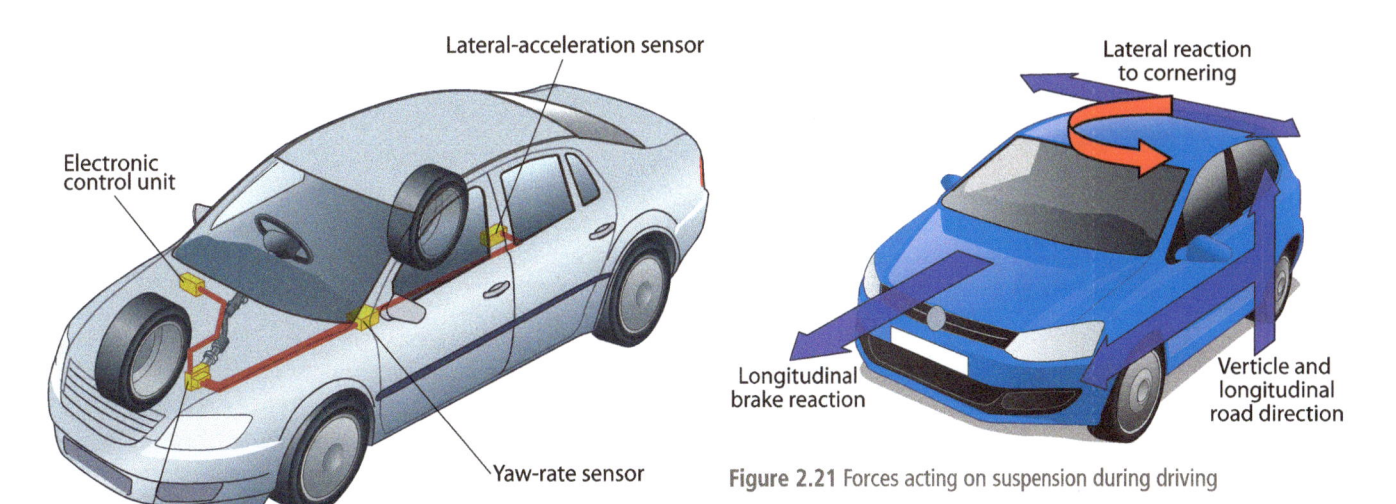

2 Diagnosis & rectification of light vehicle chassis system faults

Figure 2.20 Electronic stability program (ESP)

Figure 2.21 Forces acting on suspension during driving

allows the ESP to be manually switched off, so that it does not intervene in an unwanted way during high performance driving.

The ESP system includes a malfunction indicator lamp on the dashboard, which illuminates to alert the driver if a fault occurs or the system has been switched off.

Active **yaw** control (AYC) is an electronic system that helps reduce the vehicle pivoting around a central point caused by driving, handling or dynamic road conditions. It uses traction control and ABS to achieve this.

Skills for work

You are checking the operation of the brakes on a customer's car and you need assistance from one of your colleagues to press the brake pedal. This task is quite urgent because the customer is waiting in the reception as they need their car to collect children from school. All of your colleagues are very busy with their own jobs and you need to ask one of them to help you now.

This situation requires that you use particular personal skills. Some examples of these skills in Table 6.1 at the start of Chapter 6, on pages 304–305.

1. Using the examples given in Table 6.1, choose one skill from each of the following categories that you think you need to demonstrate in order to persuade one of your colleagues to help you.
   - General employment skills
   - People skills
   - Self-reliance skills
   - Specialist skills
2. Now rank these skills in order of importance, starting with the one that it is most important for you to have in this situation.
3. Which of the skills chosen do you think you are good at?
4. Which of the skills chosen do you think you need to develop?
5. How can you develop these skills and what help might you need?

Did you know?

ABS, EBD, TCS, ESP and AYC are active safety systems. Active safety systems are designed to reduce the possibility of an accident occurring. These systems have also reduced the need for limited slip differentials (see pages 289–292) in many vehicle designs.

Key terms

Attitude – suspension height and angle which affect the position of the vehicle body.

Yaw – a twist or oscillation around a vertical axis.

Safe working

Due to the advanced nature of modern braking systems, traditional methods of bleeding the hydraulics by pumping the brake pedal are no longer acceptable. Specialist tools and routines are often required in order to carry out maintenance and repair. To conduct brake bleeding, you may require:

- pressure bleeders
- vacuum bleeders
- diagnostic scan tools.

Always follow the manufacturer's recommended guidelines and maintenance routines.

49

Level 3 Light Vehicle Technology

Key term

Regenerative – a method of braking in which energy is extracted from the parts braked, to be stored and reused.

Action

To see a simple example of the processes involved in turning kinetic energy into electricity and slowing systems down, try the following:

- Run a vehicle engine in your workshop.
- Allow the engine to tick over and the idle to become stable.
- Now switch on a number of electric circuits, such as headlights, heated rear window and wiper systems.

The extra electrical loading will cause the alternator to generate more electricity and slow the engine slightly. The slowing of the engine tickover is a form of regenerative braking.

(The fall in engine tickover speed should very quickly return to a stable idle, due to the control of the engine management system.)

Did you know?

Some regenerative braking systems are known as KERS, which stands for kinetic energy recovery system.

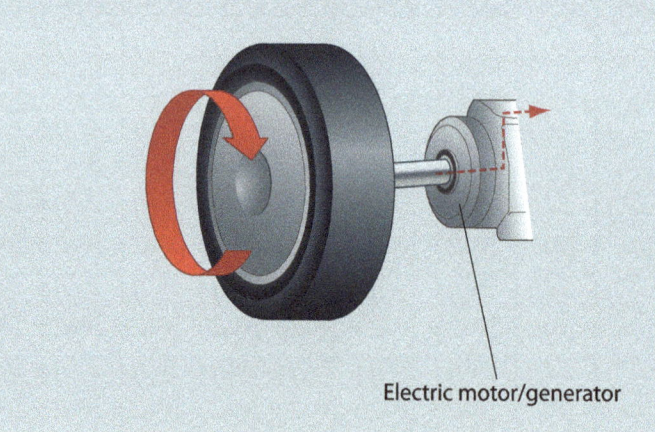

NEW TECH

Hybrid vehicle regenerative braking

A standard braking system uses friction to convert the kinetic (movement) energy of a vehicle into heat. Hybrid vehicles often use a different method for braking and slowing down, which is called **regenerative** braking. This is a highly efficient process that turns some of the vehicle's kinetic energy into electricity that can be used to help charge the high voltage battery system.

A hybrid vehicle uses a combination of an internal combustion engine and electric motor to provide drive. If the electric motor is driven mechanically, for example during braking, it can be converted into an electrical generator. The conversion of kinetic energy into electricity actually slows the vehicle down. Any extra deceleration required by the driver that is not covered by the regenerative braking is handled by a brake-by-wire system that operates brake calipers and pads against discs. A sophisticated electronic control system is used to calculate the amount of braking required and splits the operation between the generator and the brakes.

The limitations of regenerative braking include:

- The regenerative braking effect drops off at lower speeds.
- Most road vehicles with regenerative braking only have power on some wheels, for example a two-wheel drive car. The regenerative braking power only applies to the drive wheels because they are the only wheels linked to the drive motor. In order to provide controlled braking under difficult conditions, such as wet roads, friction-based braking is necessary on the other wheels.
- If the batteries or capacitors are fully charged, no regenerative braking takes place.

Electric motor/generator

Figure 2.22 Hybrid vehicle regenerative braking

CHECK YOUR PROGRESS

1. Name the three main components of an ABS system.
2. Describe an advantage of using an electronic parking brake system.
3. State two tyre problems that could cause the ABS malfunction warning light to illuminate.

Steering geometry for light vehicle applications

In order for a vehicle to **manoeuvre** and handle correctly, all four road wheels must work together to provide true rolling motion. If this doesn't happen, safety may be compromised and rapid tyre wear could occur. The two main issues associated with incorrect steering **geometry** are:

1. **Oversteer:** The front tyres have more grip than the rear tyres. During a cornering procedure, the back end of the car can break away.
2. **Understeer:** The front tyres have less grip than the rear tyres. As the car is turned into a bend, it tries to continue moving in a straight line.

The Ackerman steering principle

The Ackerman steering principle allows one wheel to turn at a greater angle than the other when cornering. This is achieved by creating angles between the steered pivot points (suspension ball joints, for example) and the steering rack or track rod.

- If the rack is placed behind the steered wheels, then it will be shorter in length than the distance between the two pivot points of the steered wheels.
- If the steering rack is placed in front of the steered wheels, then it will be longer in length than the distance between the two pivot points of the steered wheels.

The steering rack is connected to the front hub by two short, angled steering arms. The angle of the steering arms is determined by the length of the **wheel base** of the vehicle. If an imaginary line is projected from the front suspension ball joints to the centre of the rear axle, this angle can be determined, as shown in Figure 2.25.

Figure 2.23 Oversteer

Figure 2.24 Understeer

> **Key terms**
>
> **Manoeuvre** – to steer or change direction of travel.
>
> **Geometry** – the measurement of points, lines, surfaces and angles.
>
> **Wheel base** – the distance between the front and rear axles.

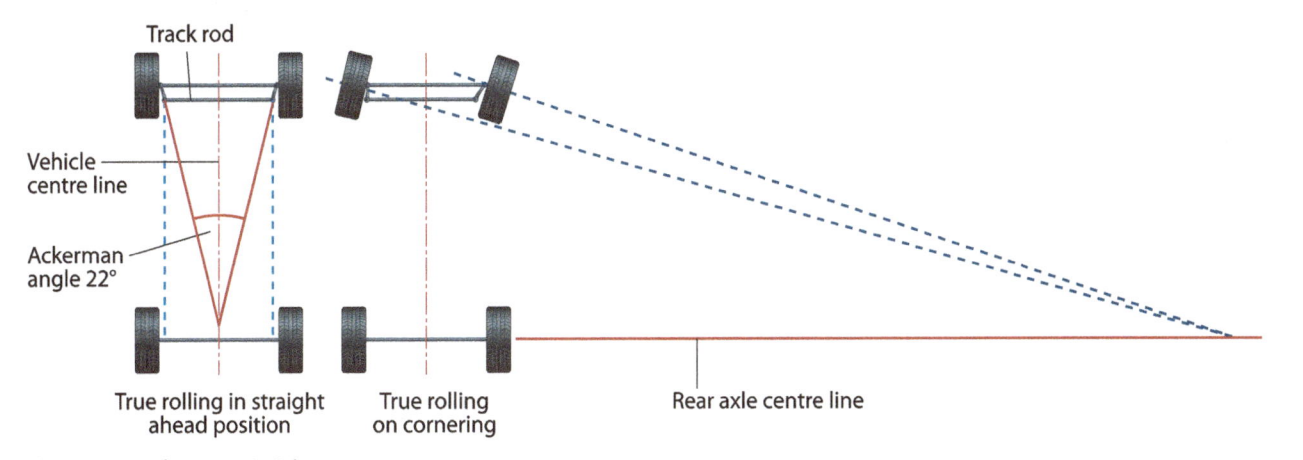

Figure 2.25 Ackerman principle

Level 3 Light Vehicle Technology

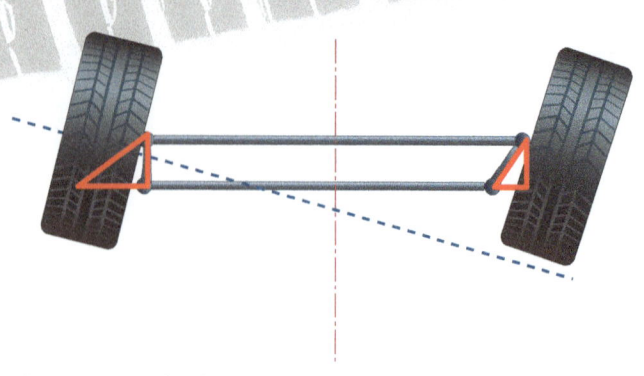

Figure 2.26 Angles that create toe-out on turns (TOOT)

The angle of the steering arms creates two imaginary triangles at each wheel. When the steering is turned to one side, one of these triangles is squashed while the other is stretched, ensuring that the wheel on the inner part of the bend is turned at a greater angle than the outer wheel. If two imaginary lines are projected at right angles to the front steered wheels, they would meet up at a point on an imaginary line projected at right angles from the rear axle. This creates a steering geometry setting known as toe-out on turns (TOOT) and provides true rolling motion, so that neither tyre scrubs across the road surface as it is driven forwards.

Swivel axis inclination (SAI) and king pin inclination (KPI)

The swivel axis created by the front suspension ball joints (or king pin on a beam axle) is tilted inwards at the top. This creates a steering geometry setting known as swivel axis inclination (SAI) or king pin inclination (KPI) as shown in Figure 2.27.

As the steering is turned from the straight ahead direction, the stub axle upon which the wheel is mounted is forced downwards, due to the swivel axis inclination as shown in Figure 2.28. This has the effect of raising the front of the vehicle upwards. When the turn has been completed, the weight of the vehicle pushing downwards on the front stub axles helps return the wheels to the straight ahead position.

Swivel axis inclination will also create a geometry setting known as offset as shown in Figure 2.29.

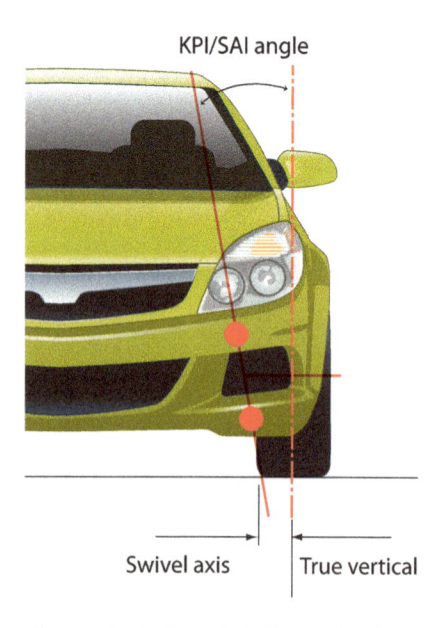

Action

Raise a car on a vehicle hoist or ramp in your workshop.

With the help of an assistant and using a piece of string:

- Hold one end of the string against the front suspension/steering ball joint.
- Stretch out the string to a point in the middle of the rear axle.
- Compare the angle created by the string with the angle of the steering arms on the vehicle. This will show you the Ackerman steering angles.

Figure 2.27 Swivel axis inclination (SAI)

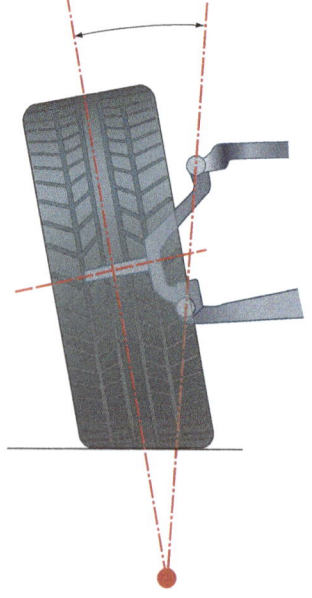

Figure 2.28 Stub axle angle affected by SAI

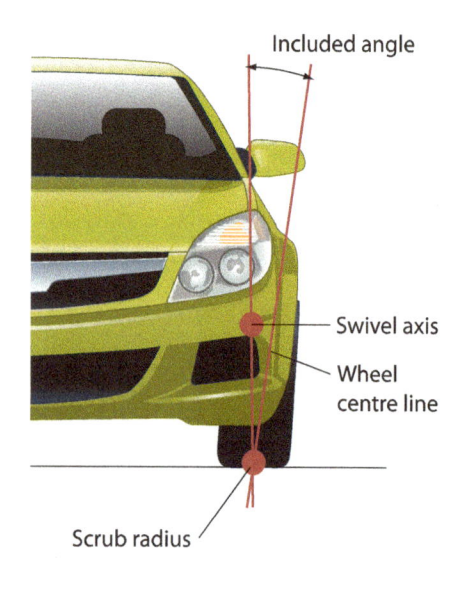

Figure 2.29 Offset

2 Diagnosis & rectification of light vehicle chassis system faults

A small amount of swivel axis inclination in the suspension ball joints will create a large offset (known as positive offset). A large offset has the effect of creating drag on the tyre as it rolls along the road. This drag wants to make the steering toe out.

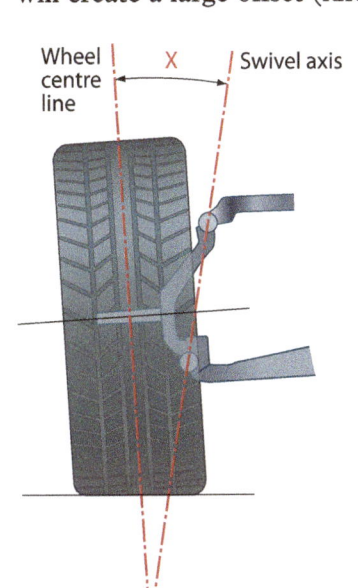

Figure 2.30 Positive offset

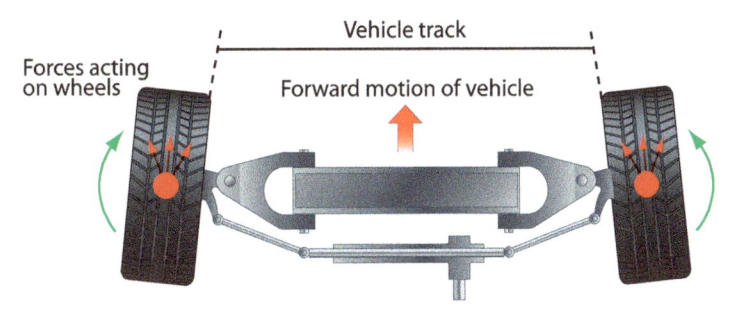

Figure 2.31 Rolling forces creating toe-in

A large amount of swivel axis inclination in suspension ball joints will create a small amount of offset (known as negative offset). Negative offset has the effect of creating drag on the tyre as it rolls along the road. This drag wants to make the steering toe in.

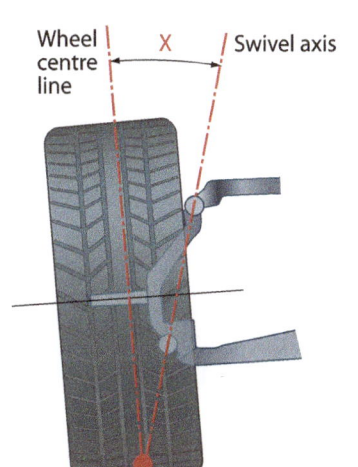

Figure 2.32 Negative offset

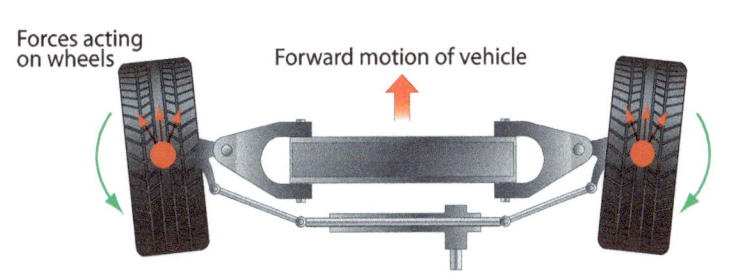

Figure 2.33 Rolling forces creating toe-out

Offset can have a desirable effect on steering action. The drag acting on the tyre helps the driver as they turn the steering wheel from the straight ahead direction. This is known as a 'turning moment'. Also, if a tyre suffers a blowout or part of the braking system fails, offset provides more control to the driver in an emergency situation.

The toe-in and toe-out created by steering offset must be compensated for when wheel alignment is set up. This is usually accomplished by adding or subtracting an initial toe setting from a parallel tracking position.

Level 3 Light Vehicle Technology

Camber angle

Camber angle is the leaning inwards or leaning outwards of the road wheel and tyre when viewed from the front or rear.

- If the road wheel leans outwards at the top, this is positive camber.
- If the road wheel leans inwards at the top, this is negative camber.

Negative camber can produce good road holding during cornering, but can create a very heavy steering feel.

Positive camber can give light and nimble steering, but sometimes handling can be compromised.

Excessive positive or negative camber can create rapid wear on the tyre tread.

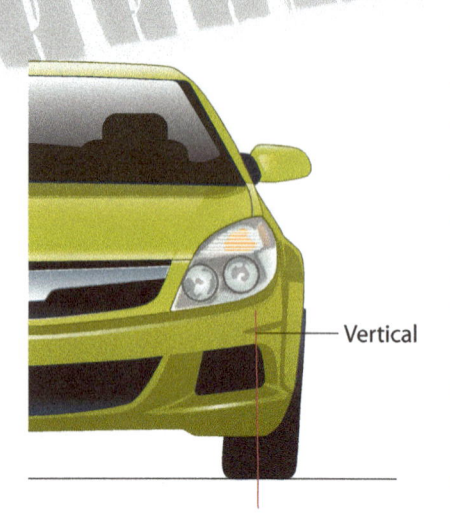

Figure 2.34 Camber angle viewed from the front

Did you know?

On early vehicles with large wheels and narrow tyres, it used to be common practice to have a large positive camber. This helped put the pivot point of the tyre (where it touches the road surface) closer to the steering axis point, creating a lighter, more positive steering feel. It also helped place most of the vehicle load under the large inner wheel bearings.

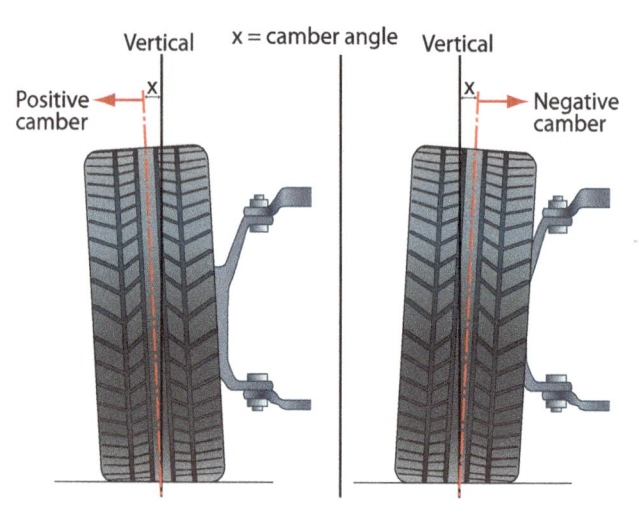

Figure 2.35 Positive and negative camber

Negative camber produces wear on the inner shoulder of the tyre.

Positive camber produces wear on the outer shoulder of the tyre.

Most manufacturers now try to achieve zero rolling camber. This means that when the vehicle is in motion, the road wheel and tyre are vertical. To allow suspension movement during normal operation, an initial camber setting is often included from the manufacturer.

Figure 2.36 Wear on the inner shoulder

Figure 2.37 Wear on the outer shoulder

Caster angle

Caster angle is a steering geometry setting that tilts the angle of the front suspension from the vertical when viewed from

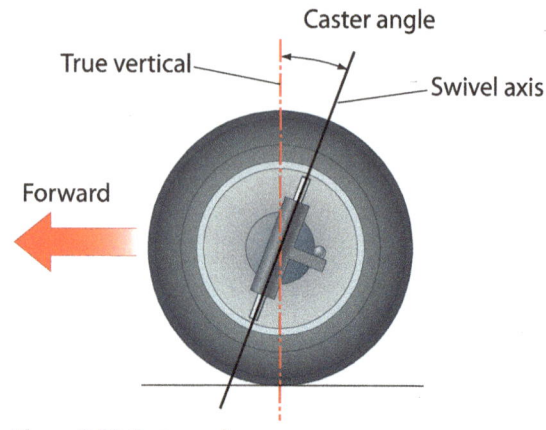

Figure 2.38 Caster angle

the side. This is very similar to the angle created by the front suspension forks on a motorcycle. The steering angle can project the weight of the vehicle to a point in front of the centre of the contact patch of the tyre where it meets the road surface. As the vehicle is in motion, this will create drag from the road surface which will try to pull the wheel and tyre into a straight ahead direction. This will help create a self-centring action when combined with swivel axis inclination.

Toe angle

Toe angles are a measurement of the inwards or outwards attitude of a pair of wheels on the same axle. This is often referred to as 'tracking'.

- When the front of the wheels in the direction of travel are pointing inwards, this is known as 'toe-in'.
- When the front of the wheels in the direction of travel are pointing outwards, this is known as 'toe-out'.

Toe setting is required so that when the wheels are in motion, they run as parallel as possible in order to reduce tyre scrub and wear. Because the dynamic motion has a tendency to splay the steered wheels outwards on a rear-wheel drive car, and a tendency to pull the steered wheels inwards on a front-wheel drive car, many manufacturers include an initial toe setting in their technical data to allow for this.

The tracking measurement is taken at different positions on the road wheels depending on the country of origin.

- In Europe, the tracking measurement is taken at the front edge of the road wheel rim.
- In the United States, the tracking measurement is taken at the front edge of the tyres.
- In Japan, the tracking measurement is taken from the centre of the tyres.

Toe setting measurement can either be described in millimetres or in degrees, and will usually include a plus or minus **tolerance**.

Table 2.4 overleaf gives a summary of the steering geometry settings and angles and their purposes.

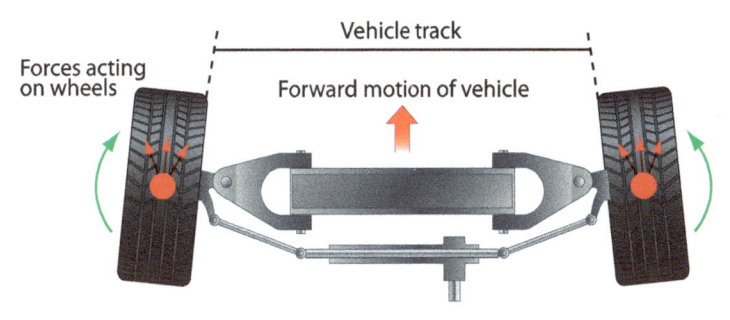

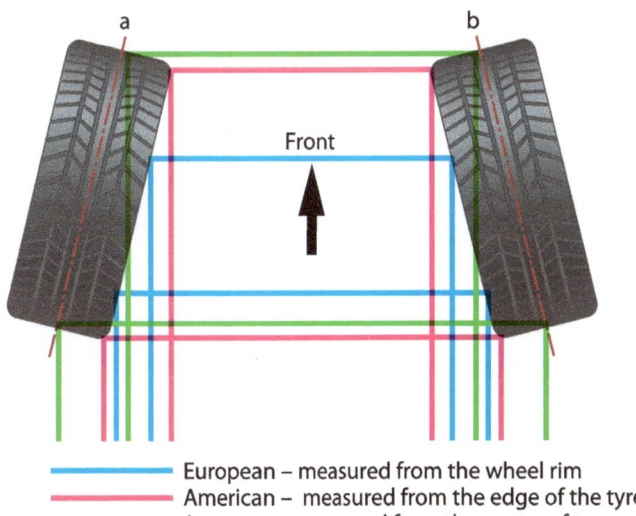

Figure 2.39 Toe-in

Figure 2.40 Toe-out

Key term

Tolerance – an allowable difference from the optimum measurement.

European – measured from the wheel rim
American – measured from the edge of the tyre
Japanese – measured from the centre of tyre

Figure 2.41 Tracking measurements

55

Level 3 Light Vehicle Technology

> **Action**
>
> Choose two vehicles: one front-wheel drive and one rear-wheel drive. Look up the manufacturer's initial toe settings in the technical data for each vehicle.
>
> - What is the initial toe setting for the front-wheel drive vehicle?
> - What is the initial toe setting for the rear-wheel drive vehicle?
> - What is the total tolerance given for the front-wheel drive vehicle?
> - What is the total tolerance given for the rear-wheel drive vehicle?

Table 2.4 Summary of steering geometry

Steering geometry setting or angle	Purpose
Camber angle (viewed from the front or rear of the car)	Positive camber angle can help reduce steering effort. Negative camber angle can help improve cornering performance.
Caster angle (viewed from the side of the car)	Caster angle helps to provide the steered wheels with a self-centring capability. This gives good straight line stability, helping to reduce the possibility that the car will wander from side to side.
Swivel axis inclination (SAI)	Swivel axis inclination, when combined with caster angle, will help provide a self-centring capability of the steered wheels, giving directional stability. In an emergency situation such as a tyre blowout, this reduces the amount that the vehicle will pull to one side.
Toe-out on turns (TOOT)	Toe-out on turns helps provide true rolling motion as the steered wheels travel around a bend and reduces tyre wear during cornering.
Front toe setting	Front toe setting helps determine the direction of travel of the front of the vehicle and reduces tyre wear.
Rear toe setting	Rear toe setting helps to determine the direction of travel of the rear of the vehicle and reduces tyre wear.

> **Skills for work**
>
> You have never used the company's four-wheel alignment equipment to set the steering geometry before. You need to ask advice from your mentor, but he is busy dealing with a customer. Your query won't take long and if you can quickly interrupt with your question you can get on and finish the job.
>
> This situation requires that you use particular personal skills. Some examples of these skills are shown in Table 6.1 at the start of Chapter 6, on pages 304–305.
>
> 1. Using the examples given in Table 6.1, choose one skill from each of the following categories that you think you need to demonstrate in order to get the advice that you need.
>   - General employment skills
>   - Self-reliance skills
>   - People skills
>   - Specialist skills
> 2. Now rank these skills in order of importance, starting with the one that it is most important for you to have in this situation.
> 3. Which of the skills chosen do you think you are good at?
> 4. Which of the skills chosen do you think you need to develop?

> **CHECK YOUR PROGRESS**
>
> 1. What type of tyre wear could be caused by excessive positive camber angle?
> 2. On what point of the wheel and tyre is the tracking measurement taken in Europe?
> 3. What does the term true rolling mean?

Power-assisted steering (PAS)

As the design of vehicles has improved, the introduction of wider, low profile tyres has placed strain on the vehicle steering system and on the driver. To help improve driver comfort and control, many modern vehicles incorporate a power-assisted steering (PAS) mechanism. These mechanisms are designed to reduce driver effort during slow-moving manoeuvres and to help maintain control of the vehicle at speed.

Early systems used **hydraulics**, fed from an engine-powered pump to drive pistons in the steering mechanism, which helped the driver to undertake manoeuvres. More modern systems use either electro/hydraulic or fully electric power assistance. When this is linked with active safety systems, such as electronic stability programs (ESP), it improves handling and vehicle control.

No matter which type of power-assisted steering system is used, it must be failsafe. This means that the vehicle can still be safely steered if the power assistance becomes faulty.

> **Key term**
>
> **Hydraulic** – using fluid pressure to provide a source of mechanical force or control.

> **Did you know?**
>
> Many steering columns incorporate a safety feature in case the vehicle is involved in an accident. Most columns are manufactured in two pieces and can collapse in on themselves like a telescope if an impact occurs. This reduces the possibility of the driver being impaled on the column.

Hydraulic power assistance

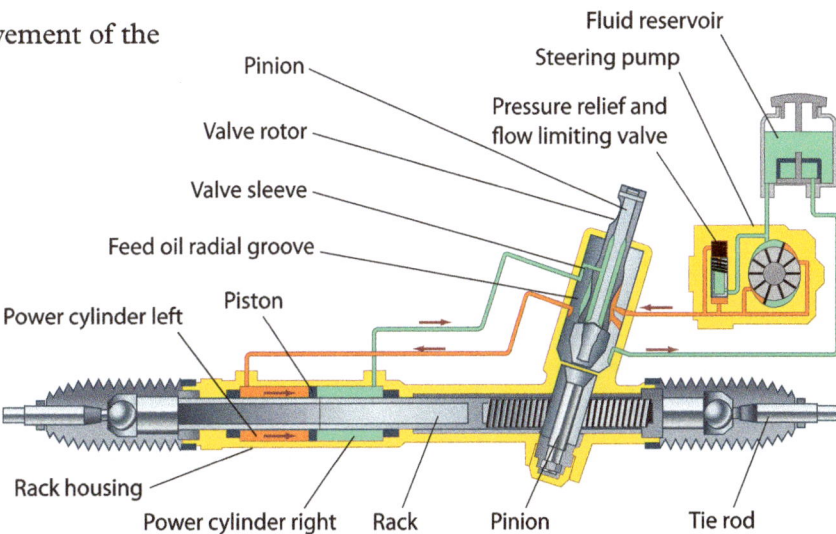

Hydraulic power-assisted steering needs three main components to operate:

- a hydraulic pump, to provide a pressure source
- a control valve, to direct hydraulic fluid pressure to the steering mechanism
- a hydraulic piston, to assist the movement of the steering mechanism.

Hydraulic pump

The hydraulic pump is usually a vane type oil pump, operated by an auxiliary drive belt connected to the engine's crankshaft. When the engine is running, the vane pump is turned and it provides hydraulic pressure to operate the power-assisted steering mechanism. The pump is usually attached to a fluid reservoir which can maintain the correct quantity and level of hydraulic fluid within the system.

Figure 2.42 Hydraulic power-assisted steering

As engine speed increases, the hydraulic pump will rotate faster and faster. This means that hydraulic pressure increases as the engine speeds up. As the greatest amount of power assistance is required during slow speed operations, a method of controlling hydraulic pressure is needed. The outlet of the hydraulic pump will normally contain a flow control valve. The purpose of the flow control valve is to reduce hydraulic pressure as engine speed increases. This means that the greatest assistance will occur at slow speed and almost no assistance will be given during high-speed manoeuvres; because of this, 'steering feel' is maintained.

Control valve

At the lower end of the steering column, a rotary control valve is used to direct the hydraulic pressure to the correct side of the hydraulic piston in the steering mechanism. As the driver turns the steering wheel, the control valve covers and uncovers passageways that direct the fluid through pipes to the piston mechanism.

This control valve may be linked to the steering column using a **torsion bar spring**. The amount of flex created in this spring reflects the amount of effort that the driver is putting in to turn the steering. As a result, the control valve is able to vary the fluid pressure in accordance with steering effort – it provides the most power assistance when the greatest effort is required. A safety mechanism will be incorporated between the steering column and rotary control valve in case the torsion bar should fail. Because of this, if the torsion bar breaks, the steering column will still be able to turn the steering mechanism and the system is failsafe.

Hydraulic piston

The hydraulic piston is formed inside the steering unit and is attached to the output gear. The piston creates two sealed chambers into which hydraulic fluid can be directed, via the rotary control valve. With the engine running and the steering in the straight ahead position, the control valve directs equal pressure to both chambers. When there is equal pressure either side of the hydraulic piston, no power assistance is given. As the steering is turned, passageways formed in the control valve direct the hydraulic fluid pressure to one side of the piston and open the other side up to the return system and reservoir. The higher hydraulic pressure found on one side of the piston provides the power assistance to help the driver steer. The amount of assistance provided is controlled so that it is **proportional** to the amount of effort the driver is applying to the steering mechanism.

Speed sensitive steering

Some hydraulic systems incorporate a solenoid valve in the delivery line to the rotary control valve. It is connected to the output from an ECU, which is able to vary the pressure and flow rate of the power steering fluid depending on vehicle speed. A vehicle speed sensor (VSS) sends a signal to the ECU. This information is processed and used to actuate the solenoid in the delivery line. Duty cycle is then used to control the amount of power assistance provided. In this way, the greatest amount of assistance can be given when the car is performing low speed manoeuvres, such as parking, and assistance is reduced when the vehicle is moving at speed, for example on a motorway.

Did you know?

When the hydraulic power-assisted steering is turned on full lock, the load on the engine-driven pump can cause the engine idle speed to fall to a point where steering assistance is reduced or the engine may stall. A pressure switch, mounted in the hydraulic pipes, allows the engine management system to monitor pressure and loading so that it can intervene and raise the engine speed as required.

Key terms

Torsion bar spring – a sprung metal rod which, when twisted, tries to return to its original shape.

Proportional – when two amounts change at the same rate. In this case, the assistance provided is proportional to the effort the driver applies to the steering mechanism.

2 Diagnosis & rectification of light vehicle chassis system faults

Hydraulic power-assisted steering fault diagnosis

Before you assume that there is a fault with the hydraulic power-assisted steering system, you should first consider other factors, as shown in Table 2.5.

Table 2.5 Factors affecting power-assisted steering and associated faults

Factors affecting power-assisted steering	Possible fault
Road surface and conditions	The type of road surface that a vehicle is travelling over has an effect on the amount of grip produced at the tyre. Weather conditions such as ice or rain can also affect tyre grip. Such conditions can lead to symptoms of very light steering, giving the impression that too much power assistance is being provided.
Tyre pressures	Tyre pressures that are too low produce heavy steering; tyre pressures that are too high produce light steering.
Tyre condition	As tyre tread wears, depending on the conditions and road surface, grip may be lost, which leads to symptoms of light steering. This can give the impression that too much power assistance is being provided.
Wheel alignment and steering geometry	Incorrect wheel alignment or steering geometry can make the steering feel heavy or light. Always check wheel alignment and steering geometry before any diagnosis of the power-assisted steering system.
Steering and suspension linkages	Excessive play or partial seizure of steering and suspension linkages and components can lead to heavy or light steering. This may be confused with power-assisted steering faults.

Table 2.6 describes some hydraulic power-assisted steering symptoms and faults.

Table 2.6 Hydraulic PAS symptoms and faults

Power assistance too low	Power assistance too high
Power steering fluid level too low	Flow control valve stuck closed
Broken/slack drive belt	Rotary control valve torsion bar spring broken
Worn hydraulic pump	Speed sensitive solenoid valve stuck open
Flow control valve stuck open/faulty	Vehicle speed sensor (VSS) signal too low
Hydraulic piston leaking	Speed sensor ECU fault
Speed sensor control system failed	

> ⚠️ **Safe working**
>
> If you have to take the steering wheel off during the repair of a steering system, it is important that you follow any precautions required for the safe removal of any airbags (see Chapter 4 page 234.) You must also use the correct puller. If the steering column is struck with a hammer during the removal of the steering wheel, it is possible for the safety system to collapse, making the column unusable.

Level 3 Light Vehicle Technology

> **Key term**
>
> **In series** – incorporated as part of the circuit.

To diagnose hydraulic system pressure faults, you must fit a power-assisted steering pressure gauge **in series** with the pump, then run the engine at different speeds and under different operating conditions. The gauge is similar in operation to an engine oil pressure gauge, but it has a manually operated valve that is able to control the flow of fluid in the system. In this way, you can check flow and pressure and compare them with the manufacturer's specifications.

- By connecting the gauge with the valve in the open position, you can check supply pressure when the steering is at full lock.
- By connecting the gauge with the valve in the open position, you can check supply pressure when the engine is operating in its normal rev range (1000 to 3000 rpm) with the steering in the straight ahead position.
- By connecting the gauge with the valve in the closed position, you can check the maximum supply pressure available from the pump.

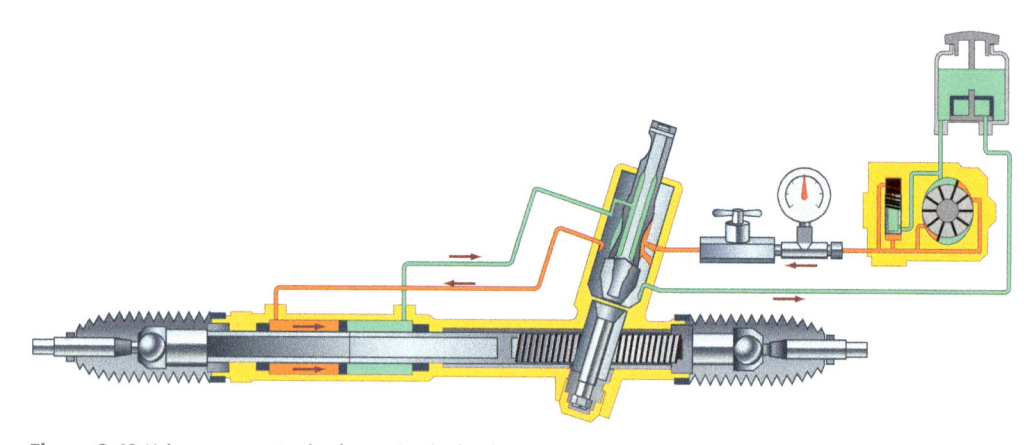

Figure 2.43 Using a gauge to check steering hydraulic pressures

> **Safe working**
>
> Do not run the hydraulic pump for more than five seconds with the valve on the pressure gauge in the closed position, as this may cause the pump to overheat and can cause damage.
>
> Power-assisted hydraulic fluid pressures can reach around 80 bar (1160psi) so you must take care when working on these systems.

> **Action**
>
> Examine a vehicle in your workshop and identify the main component parts of a hydraulic power-assisted steering system.

> **NEW TECH**
>
> **Steer by wire**
>
> Some manufacturers are experimenting with a system of steer by wire. In this set up, the steering wheel is replaced with a joystick that can control not only steering but also acceleration and braking. The main drawbacks of this system are the difficulty in making it failsafe and the reluctance of drivers to give up control of the steering wheel.

Table 2.7 Typical operating pressures that may be found during hydraulic testing

Pressure gauge valve open			Pressure gauge valve closed	
Engine speed 1000 rpm	Engine speed 3000 rpm	Engine speed 1000 rpm, steering turned to full lock	Engine run (maximum 5 seconds)	Symptom and possible fault
40 bar	35 bar	80 bar	80 bar	Correct operation
25 bar	25 bar	80 bar	80 bar	Power assistance too low Flow control valve stuck open/faulty
40 bar	35 bar	60 bar	80 bar	Power assistance too low Faulty control valve or hydraulic piston
40 bar	35 bar	60 bar	60 bar	Power assistance too low Worn pump
65 bar	65 bar	80 bar	80 bar	Power assistance too high Flow control valve stuck closed/faulty

Noise and bleeding

Noise from a hydraulic power-assisted steering system may be caused by:

- loose drive belt
- air in the hydraulic system, known as **cavitation**
- pump bearing wear.

If air has entered the system, it can be bled by topping up the fluid reservoir with the correct grade of power steering fluid and turning the steering from lock to lock with the engine running. This will help fluid to fully circulate through the steering system and air to escape once it has reached the reservoir.

Speed sensitive solenoid diagnosis

To diagnose the correct operation of the speed sensitive system, you need to connect an oscilloscope to the speed sensor input at the power steering ECU. (You can find out how to connect and use an oscilloscope in Chapter 4, pages 194–195.).

With the vehicle in motion, you should see a wave form with a frequency proportional to vehicle speed on the screen of the oscilloscope. This signal may be analogue (see Figure 2.44) or digital (see Figure 2.45). If you don't see this wave form, check the sensor and wiring.

If the input to the ECU is correct, you should then connect the oscilloscope to the output for the power steering solenoid. At slow speed, you should see a duty cycle or pulse width modulation (PWM) on the screen of the oscilloscope that reduces as vehicle speed increases.

> **Key term**
>
> **Cavitation** – the creation of bubbles in a hydraulic fluid.

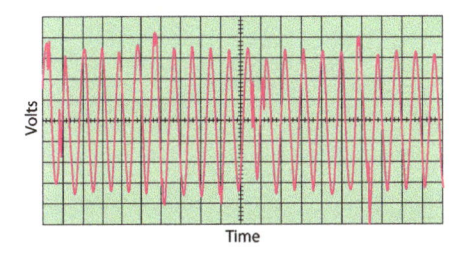

Figure 2.44 Analogue speed signal

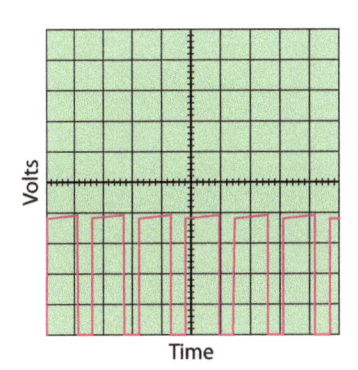

Figure 2.45 Digital speed signal

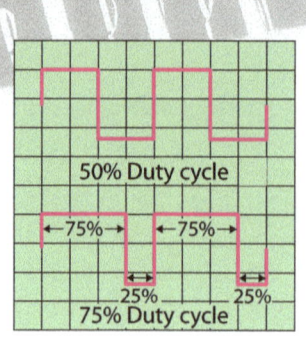

Figure 2.46 Duty cycle wave form

If the output signal is correct, then you should check the wiring and solenoid for correct function and operation.

If the input to the electronic control unit is correct but the output is missing or incorrect, you should suspect a fault with the ECU.

Electro-hydraulic power-assisted steering

An alternative to the engine-driven hydraulic pump used in power-assisted steering is for the pump to be driven using an electric motor. The electric motor, pump and fluid reservoir can then be mounted in a position away from the engine and activated when required. Sensors mounted on the steering column are able to measure the amount of turning effort applied to the steering system by the driver and send this information along with a signal from the vehicle's speed sensor to the steering electronic control unit (ECU). The ECU is then able to operate the electric motor to turn the pump and supply a pressure to the piston in the steering rack that is proportional to vehicle speed and the steering effort applied.

The advantages of this type of system are:

- Fewer loads are placed on the engine during operation, so fuel economy and engine emissions are improved.
- The pump and reservoir can now be mounted anywhere away from the engine, which increases the scope for vehicle design.
- If the engine stalls/cuts out, power assistance can be maintained.
- The pump can be controlled by a switch to provide even greater assistance when parking.

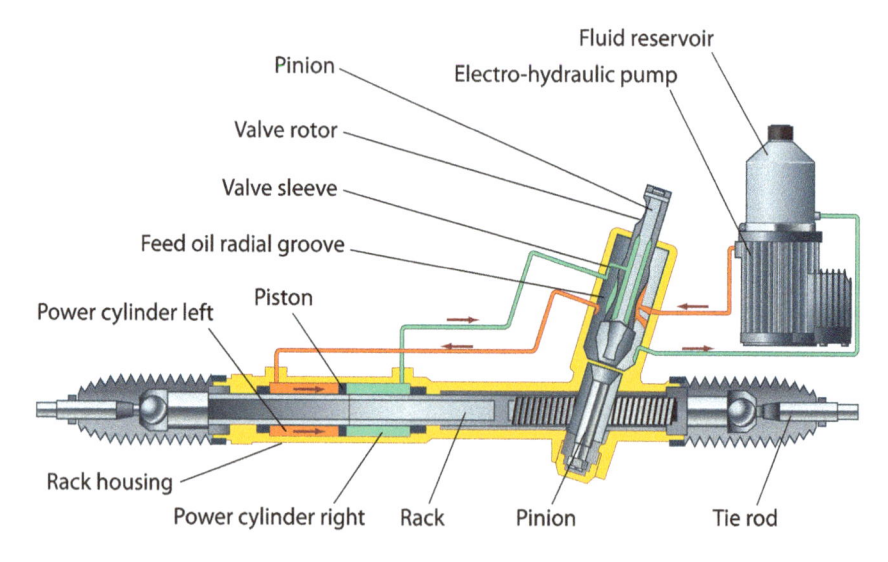

Figure 2.47 Electro-hydraulic power steering

Electronic power-assisted steering (EPS)

Many modern vehicles use electric motors to drive steering mechanisms and provide power assistance instead of hydraulics. A reversible direct current electric motor can be connected to the steering rack or column depending on manufacturer design. As the driver applies effort at the steering wheel, movement and turning effort are registered by a torque sensor mounted on the steering column. The information provided by this sensor, as well as other inputs such as vehicle speed, is then sent to the steering electronic control unit (ECU). The ECU is then able to operate the electric motor in the desired direction with a force controlled by duty cycle, which is proportional to vehicle speed and the steering effort applied.

The advantages of this system are:

- The motor is only operated when the steering is turned – this reduces loads, improves fuel economy and reduces engine emissions.
- The motor and control system is very compact and can be used unobtrusively, even on small cars.
- Assistance can be easily varied to provide greater help when parking.
- Less maintenance is needed as there is no fluid system or leakage.
- If combined with a vehicle radar system, it can be used to provide a self-parking function.

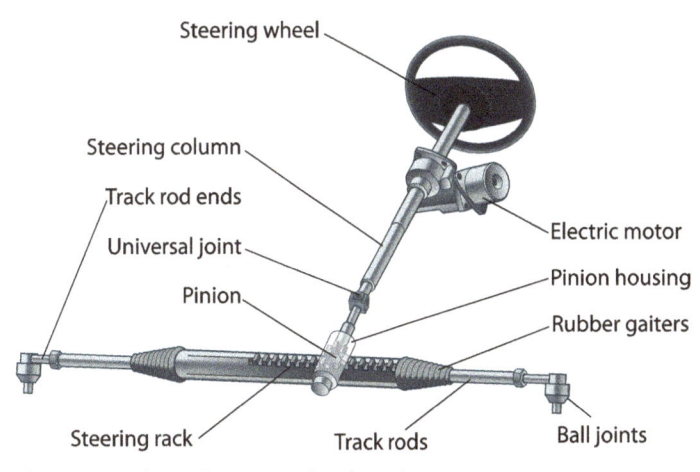

Figure 2.48 Electronic power-assisted steering

To diagnose faults with electronic power-assisted steering systems, you can often use a scan tool to retrieve diagnostic trouble codes (DTCs), which you can then use to direct your fault finding routines.

NEW TECH

Self-parking cars

Parallel parking is a slow speed manoeuvre that many drivers find difficult. As a response to consumer demand, car manufacturers are starting to design and sell self-parking cars. Advantages of self-parking cars include:

- Choosing a parking space is not restricted by the driver's skill at parallel parking.
- A self-parking car can often fit into smaller spaces than most drivers can manage on their own, which allows the same number of cars to take up fewer spaces.
- Parking takes less time, which helps to keep traffic moving.
- Minor damage created by parking is reduced.

Many systems operate with the driver controlling vehicle speed and direction with the normal driving controls. They have sensors distributed around the front and rear bumpers of the car, which act as both transmitters and receivers. These sensors send out signals, which bounce off objects around the car and reflect back to them. The car's ECU then uses the amount of time that it takes those signals to return to calculate the location of the objects. The electronic power-assisted steering then manoeuvres the car into the parking space.

Manufacturers are now designing vehicles that are completely autonomous and will control the drive as well as the steering. This means that the driver simply has to select an appropriate parking spot and position the vehicle close to the space. Having pressed a button, the car can then park completely by itself.

Action

Investigate vehicles in your workshop and state if they use:

- hydraulic power-assisted steering
- electro-hydraulic power-assisted steering
- electronic power-assisted steering.

Level 3 Light Vehicle Technology

> **Key terms**
>
> **Compliance** – the willingness to follow instructions or commands (such as steering movements).
>
> **Voids** – holes manufactured in rubber bushes to allow them to flex in a certain direction.

> **Action**
>
> Examine some vehicles in your workshop and try to identify the void bushes that would provide compliance steering of the rear axle.

> **Safe working**
>
> Take care when manoeuvring a four-wheel steer car in the workshop, as the rear axle can make the back end of the car move sideways, which could catch people or equipment and cause injury or damage.

Wheels turn in the same direction

Wheels turn in opposite directions

Figure 2.50 Four-wheel steer direction phasing

You need to check the input and output signals from the ECU using an oscilloscope in a similar way to the routine used for checking the speed sensitive system (see pages 61–62).

Four-wheel steering

To improve the manoeuvrability and handling of cars, a system that allows all four wheels to steer can be used.

Many cars use a passive system called **compliance** to provide small amounts of four-wheel steering. Rubber suspension bushes that support the non-steered axle are manufactured with holes known as **voids**. These voids (see Figure 2.49) allow the axle to flex slightly during cornering, giving a very basic form of four-wheel steering.

Figure 2.49 Compliance bush

Some manufacturers produce an active four-wheel steering system where a steering rack is provided for both the front and the rear axles.

The front and rear steering racks are normally connected mechanically or driven electronically by motors. The rear steering rack does not normally pivot the wheels to the same degree as the front rack, but the Ackerman principle and toe-out on turns are still used. When the driver rotates the steering, the front rack is controlled in the normal manner. Depending on vehicle speed and steering angle, the rear rack is operated to provide a degree of rear-wheel steering.

The steering racks are phased, so that when the front wheels are turned, the rear wheels initially pivot in the same direction by a small amount. In this way, when changing direction at speed, the car will have a tendency to crab sideways across the road, giving good directional stability. As the front wheels are turned at a greater angle, the rear wheels straighten up and then begin to turn in the opposite

64

direction to the front wheels. This means that very tight turning circles can be produced for slow speed manoeuvres.

If a fault occurs in a four-wheel steering mechanism, the system is failsafe. The rear axle will return to a neutral position and the steering will act only on the front wheels, as with a standard car.

When diagnosing four-wheel steering system faults, you should ensure that you follow the manufacturer's guidelines and specifications for the set up and adjustment of the front and rear steering geometry settings. Incorrect wheel alignment may cause handling issues and rapid tyre wear.

Wheel alignment

When checking the steering geometry and wheel alignment, it is important that certain conditions are met:

- the vehicle should be placed on a flat/level surface
- the vehicle should not contain any excessive or unevenly distributed load
- steering and suspension should be assessed to ensure that there is no damage or wear that could lead to inaccurate steering geometry measurements being taken
- wheels and tyres should be in good condition and tyres inflated to the correct pressures.

Before any wheel alignment measurements are taken, the steering should be assessed for the correct setting of toe-out on turns (TOOT). This can be done by placing the steered wheels on graduated turn plates and rotating the steering from lock to lock. As the steering is turned to the full lock position, the angle of the steered wheels should be recorded with the inner wheel turning at a greater angle than the outer wheel. The steering should then be turned on the opposite lock and the steered angles should swap over to repeat the measurements in the other direction. If toe-out on turns is found to be incorrect, this should be put right before continuing with any other steering angle adjustments. Toe-out on turns can often be corrected by adjusting the length of the arms at the track rod ends so that they are equal on the right- and left-hand sides of the steering rack.

Steering geometry equipment can now be mounted to the vehicle wheels following the manufacturer's guidelines. Measurements should be taken, recorded and compared with manufacturer specifications. If incorrect, alterations can now be made, ensuring that if the toe angle is changed, equal adjustments are made to both sides of the steering so as not to disturb the toe-out on turns.

If at any point during your repairs or adjustments the vehicle is jacked up or moved, the suspension and steering should be settled by bouncing or rolling backwards and forwards before measurements are rechecked.

CHECK YOUR PROGRESS

1. Name the three main components of a hydraulic power-assisted steering system.
2. State two advantages of electronic power-assisted steering when compared to hydraulic power-assisted steering.
3. Describe the direction that the rear wheels are angled as the steering is turned on a four-wheel steering system.

Level 3 Light Vehicle Technology

> **Key terms**
>
> **Inertia** – the tendency for a body to resist acceleration.
>
> **Mass** – the quantity of inertia possessed by an object, often referred to as its weight.
>
> **Bump** – the upwards movement of suspension.
>
> **Rebound** – the downwards movement of suspension.
>
> **Friction** – a measurement of grip or how hard it is to slide one surface against another.

Suspension systems

The purpose of a light vehicle suspension system is to ensure that the wheels and tyres of the car remain in contact with the road as it travels over an uneven surface, while also providing a comfortable ride for the occupants. Suspension systems need to react against **inertia** forces created due to **mass** and acceleration.

A set of springs is used between the wheel and hub assemblies and the vehicle chassis. Springs are designed to absorb the impact of bumps and potholes, reacting with an upwards movement on **bump** and a downwards movement on **rebound**.

If left unchecked, the bouncing created by the springs of the suspension system would make the car very unstable and would compromise handling. As a result, suspension dampers, sometimes called shock absorbers, are connected between the suspension and the chassis to help take the bounce (oscillation) out of the spring. Many of these dampers use hydraulics to convert the bounce or oscillation to heat, using the principles of **friction**.

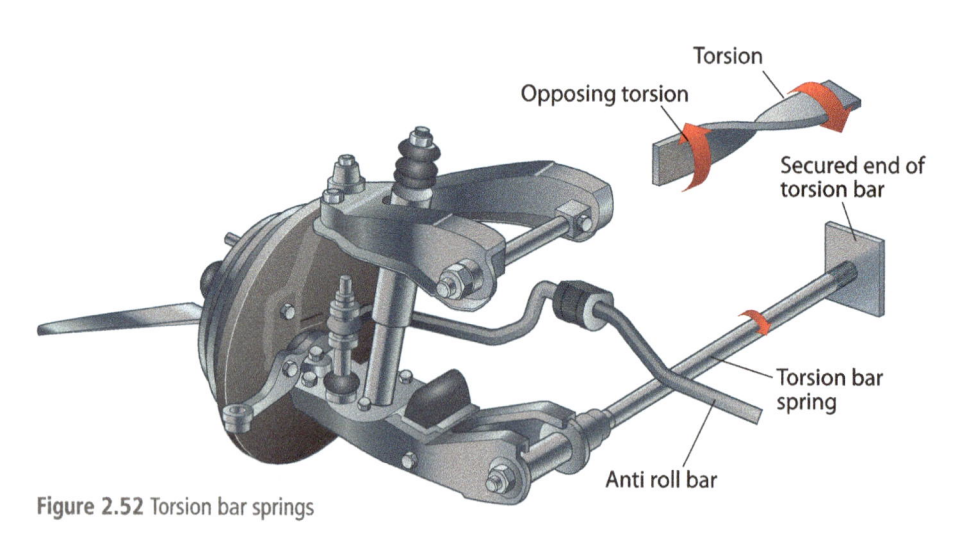

Constant rate coil spring   Variable rate coil spring

Taper coil spring

Figure 2.51 Coil springs

Figure 2.52 Torsion bar springs

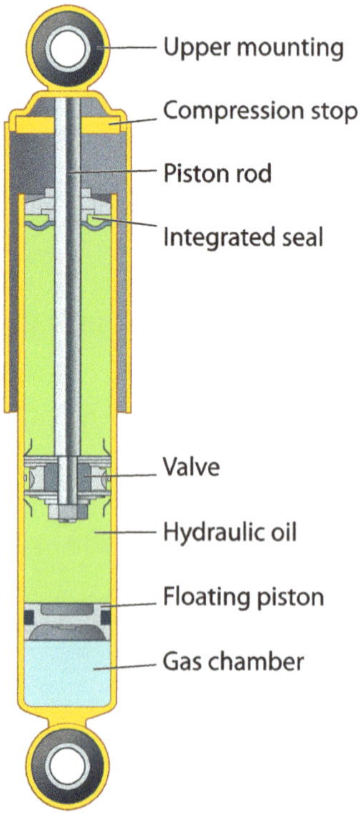

Figure 2.53 Suspension damper

More information on suspension springs and dampers can be found in *Level 1 Diploma in Light Vehicle Operations Candidate Handbook* (Chapter 6) and *Level 2 Diploma in Light Vehicle Maintenance & Repair Candidate Handbook* (Chapter 2).

2 Diagnosis & rectification of light vehicle chassis system faults

All cars have suspension, but standard systems can lead to undesirable vehicle body movement such as:

- **Pitching:** the alternate rising and dipping of the front and rear of the vehicle body on uneven road surfaces.

Figure 2.54 Pitching

- **Dive:** downward movement of the front and upward movement of the rear as the brakes are applied.

Figure 2.55 Dive

- **Drive/squat:** upward movement of the front and downward movement of the rear during acceleration, or downward movement of the chassis and upward movement of the wheels after travelling over large bumps.

Figure 2.56 Drive/squat

- **Roll:** leaning movement of the vehicle body when turning a corner.
- **Yaw:** the tendency for the vehicle body to try to pivot or rotate around its centre axis.

Figure 2.57 Roll

To help overcome undesirable vehicle body movements created during operation, some manufacturers are fitting active suspension systems. These try to reduce the effect of these movements while maintaining comfort and handling. There are two main types of active suspension system:

- self-levelling suspension
- ride-controlled suspension.

Figure 2.58 Yaw

67

Table 2.8 Examples of basic suspension system symptoms and faults with descriptions of how they can be diagnosed

Symptom	Possible fault	How to diagnose
Suspension too hard	Worn springs allowing the suspension to contact the bump stops	Measure the suspension ride height and visually inspect the condition of the road springs
Excessive spring oscillation	Leaking dampers	Conduct a bounce test by pushing down and releasing the suspension and assessing the damping action. Visually inspect the dampers for leaks.
Noise	Excessively worn suspension bushes	Raise the vehicle and using a pry-bar, check for excessive play/movement in the suspension bushes
Ride height too low	Broken road spring	Measure the ride height and visually inspect the springs for damage

Self-levelling suspension

As forces are placed on suspension springs, usually due to the loading of weight inside the car, they are compressed and become shorter. This means that the effectiveness of the spring has been reduced, as its operating length has been compromised and it cannot move through its intended design distance as it travels over a bump. If the weight placed inside the car is distributed unevenly, this will also have an effect on vehicle attitude (see page 48). The car may lean to one side, or be lower at the front or rear. As well as affecting the handling, this may have an effect on other systems, such as headlamp aim.

A self-levelling suspension system uses conventional springs and damper units, but also provides additional support to the springs when the vehicle is loaded. The additional support is designed to react against the loading of weight inside the car and return it to its original design ride height. A number of methods can be used to automatically maintain suspension spring ride height, including:

- self-energising suspension
- air-assisted suspension.

Self-energising suspension

A self-energising system is a type of self-levelling system which uses a component that looks similar to a large telescopic suspension damper. It is mounted at the back of the vehicle, between the rear axle and the car body. If weight is placed in the rear of the car, the load will compress the rear suspension. This will lower the rear of the car and raise the front. When the vehicle is driven, and the car moves over bumps in the road, the self-energising unit acts like a pump, forcing gas between internal chambers and reacting against hydraulic fluid. This has the effect of raising the rear suspension to its original ride height without affecting normal suspension movement.

A self-energising suspension unit is a sealed component and can only be visually assessed for damage or leaks. When diagnosing faults in this type of system, pay particular attention to the mounting points of the self-energising unit. Excessive movement in bushes or connecting ball joints may cause noise during operation.

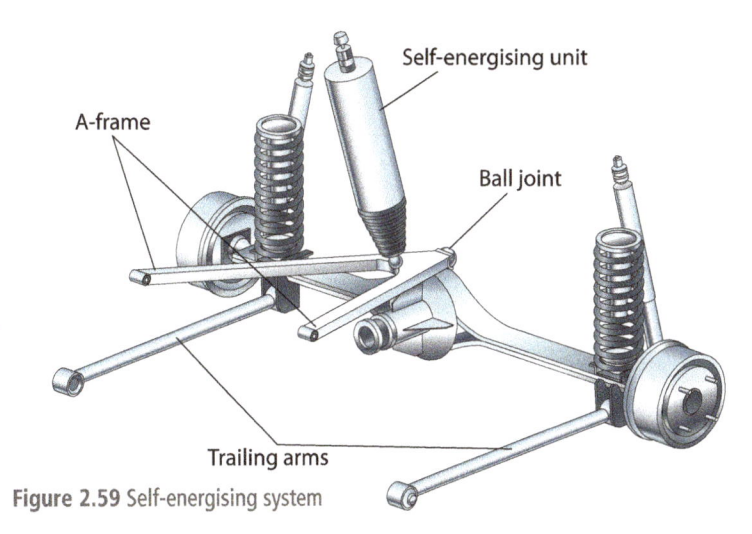

Figure 2.59 Self-energising system

Case study

Mr Vincent brings his car to your garage and complains that when he drives around at night, oncoming cars keep flashing their lights at him. Your boss says he knows what the problem is, and tells Mr Vincent that his headlamp aims are too high; oncoming drivers think that he is driving with his main beam headlights on and that's why they are flashing him.

Your boss tells you to take the car into the workshop and realign the headlamps.

Once the car is in the workshop and you have set up the headlamp aligning equipment, you can't see anything wrong with the aim. After thinking about what else might cause this problem, you go back to the reception area to ask for permission to test the self-energising rear suspension.

Here's what you do:

- ✓ Listen to the customer's description of the fault.
- ✓ Question the customer carefully to find out the symptoms.
- ✓ Gather information from technical manuals (including ride height) before you start and take it to the vehicle.
- ✓ Devise a diagnostic strategy.
- ✓ Check the quick things first and conduct a visual inspection for damage, tyre pressures, etc.
- ✓ Measure the ride height of the rear suspension with no load placed inside.
- ✓ Place a heavy load in the rear and recheck the ride height.
- ✓ Drive the car a short distance and allow the self-energising unit to operate over a bumpy surface.
- ✓ Check that the ride height has returned to its original position.
- ✓ Remove the load from the rear of the car and the rear suspension should rise.
- ✓ Drive the car a short distance and allow the self-energising unit to operate over a bumpy surface.
- ✓ Check that the ride height has returned to its original position.
- ✓ Locate the root cause of the problem (the self-energising unit).
- ✓ Keep the customer informed of progress and costs.
- ✓ Replace the self-energising suspension unit.
- ✓ Correctly reassemble any dismantled components/systems.
- ✓ Thoroughly test the system to ensure correct function and operation.

Level 3 Light Vehicle Technology

Air-assisted suspension

An air-assisted self-levelling system uses airbags or chambers attached to the spring and damping units of a standard suspension. When weight or load is placed in the car, the compression of the suspension is detected by ride height sensors connected between the vehicle body and the suspension units. The ride height sensors are normally variable resistors or **potentiometers**, connected by a lever arm to the suspension. These sensors send a voltage signal to an ECU that corresponds to how far the suspension has been compressed.

> **Key term**
>
> **Potentiometer** – a variable resistor.

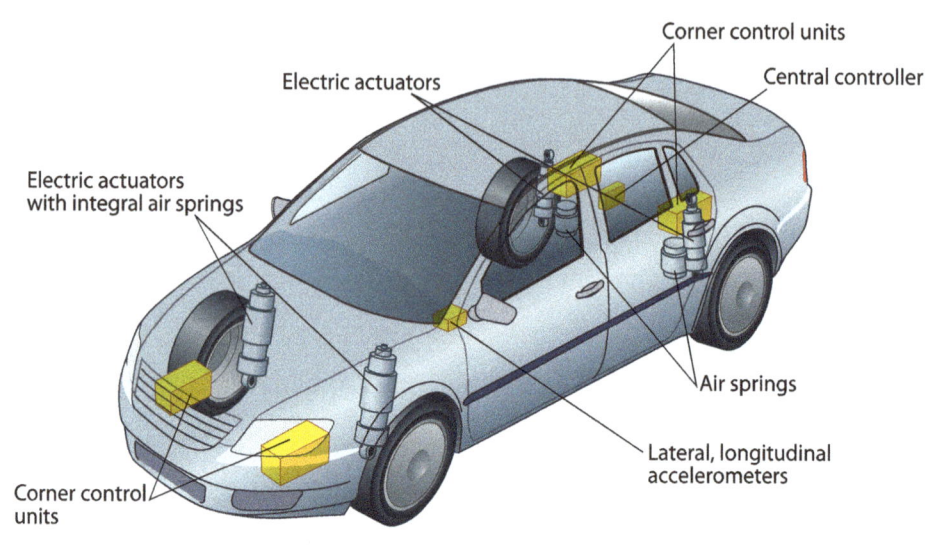

Figure 2.60 Air-assisted self-levelling suspension

The ECU monitors the suspension height and, if needed, operates a small on-board compressor to pump air and direct it to the appropriate self-levelling unit. This assists the springs and returns the car to its original ride height.

If the load is removed and the suspension rises, the ride height sensors signal the ECU and the air is expelled through solenoid valves, allowing the suspension to return to its original height.

To ensure that the air-assisted self-levelling system only reacts to load placed in the vehicle and not to normal suspension movement as the car is driven over bumps, the ECU is programmed with a short time delay before it operates the compressor or solenoid valves. This helps prevent unnecessary activation of the system during normal operation.

Ride height sensors are normally adjustable so that the self-levelling system can be calibrated if any suspension components are replaced.

Testing an air-assisted system for correct operation

1. Measure the ride height of the suspension with no load placed inside.
2. Place a heavy load in the car and recheck the ride height.
3. Start the engine and, after approximately two seconds' delay, the compressor should begin to operate.

> **Did you know?**
>
> The compressor of an air-assisted suspension system will normally contain a filter and dryer unit. The purpose of this is to clean the air entering the system and remove water moisture. If water entered the system, it could corrode the valve system – as water is not compressible, it would stop the system operating correctly.

> **Safe working**
>
> Remember to calibrate any ride height sensors if repairs have been carried out on an air-assisted system. This will ensure the safe and correct operation of the suspension system.

2 Diagnosis & rectification of light vehicle chassis system faults

4. Check that the ride height has returned to its original position (this may take between 20 and 40 seconds to complete).
5. Remove the load from the car and the suspension should rise.
6. After approximately two seconds' delay, the solenoid valves should operate and ride height should begin to fall.
7. Check that the ride height has returned to its original position (this may take between 20 and 40 seconds to complete).
8. Some systems incorporate a self-diagnosis facility which should be used in the event that the system fails to operate correctly. A scan tool should be attached to the diagnostic socket and any fault codes retrieved.
9. Power should be checked at the control system fuse, the compressor and ECU.
10. Ride height sensors can often be checked for correct function by connecting them to an ohmmeter and ensuring that the resistance changes as the suspension height moves up and down.

Ride control suspension

Unlike self-levelling, a ride control suspension system is one that monitors dynamic car body attitude while the vehicle is in motion. It then makes corrections in suspension stiffness to counteract undesirable vehicle body movement.

These systems can be manually switched by the driver, for different driving styles, or monitored by an ECU and automatically adjusted when required.

Variable rate damping

Dampers are fitted to a suspension system to reduce the bounce or **oscillation** of springs. They operate by forcing oil though valves inside the shock absorber. The resistance to flow of the oil through the valves contained in the shock absorber helps slow and control the movement of the suspension by turning the energy into heat.

In a ride control system, variable rate damping can change the flow of the oil through the shock absorber valves to give different damping characteristics under different driving conditions.

The damper valves are connected to the central piston rod of the shock absorber. This piston rod can be rotated by a small **stepper motor**, which changes the position of the valves and controls the oil flow through the damper.

- If oil flow is reduced, the suspension becomes stiffer and handling is improved.
- If oil flow is increased, the suspension becomes softer, giving a more comfortable ride.

Various suspension ride height, steering angle and acceleration sensors monitor the motion of the car and signal an ECU to adjust the stepper motors controlling the dampers.

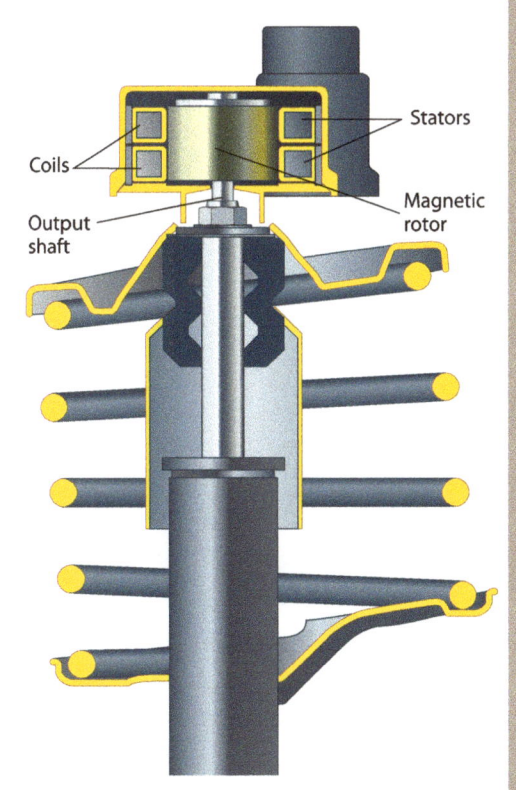

Figure 2.61 Variable rate damper

> **Did you know?**
>
> Some manufacturers have created a variable rate damping system which uses metallic particles, held in suspension in the damper oil. When an electro magnet is energised on the damper, the metal particles line up changing the viscosity of the oil. By varying the strength of the magnetic field, during dynamic suspension operation, the damper can be stiffened or softened depending on driving and handling requirements.

> **Key terms**
>
> **Oscillation** – a continuous backwards and forwards movement.
>
> **Stepper motor** – an electric motor that moves in small stages or 'steps'.

Active air suspension

Some manufacturers produce vehicles with active air suspension. The operation of this system is very similar to the air-assisted self-levelling system (see pages 70–71), but in this design the air is used as the main springing medium.

An active air suspension system consists of the following main components:

- rubber airbag springs
- an air compressor unit
- an air solenoid valve manifold
- an accumulator air tank
- ride height and load sensors
- steering angle and acceleration sensors
- yaw rate and lateral acceleration sensors
- an electronic control unit.

The strong rubber airbags are connected between the car's suspension components and the vehicle chassis. When inflated, the airbags act as the main suspension spring.

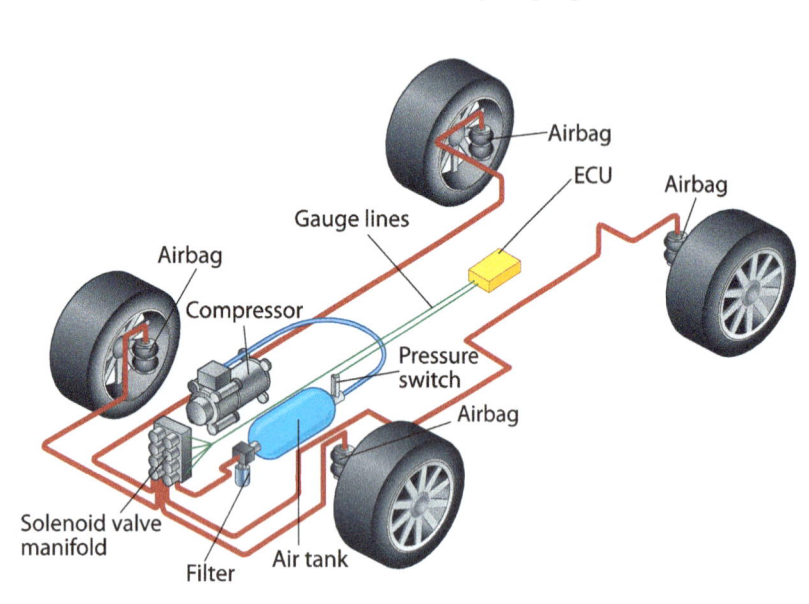

Figure 2.62 Active air suspension

The air compressor unit is able to supply pressurised air to the system via an air filter and drier, so that the suspension airbags or accumulator tank can be kept topped up.

The air solenoid valve manifold is able to direct the compressed air to the appropriate airbag to inflate it or allow air to be exhausted from the system if pressure needs to be reduced.

The accumulator tank acts as a reservoir of pressurised air that can be added to the system at a moment's notice to react to dynamic suspension movements.

Suspension ride height, steering angle and vehicle acceleration forces are monitored by sensors as the vehicle is driven. This information is relayed to the electronic control unit, which can then operate the solenoid valve manifold to add or remove air from the airbag springs. The air is supplied from the air compressor or the accumulator air tank. This way the correct vehicle attitude can be maintained under all driving conditions.

Active hydro-pneumatic systems

A hydro-**pneumatic** system uses a hydraulic liquid to transfer suspension movement to a sealed compressed gas chamber.

An active hydro-pneumatic system consists of the following main components:

- hydro-pneumatic suspension units
- a hydraulic pump
- a solenoid valve block

> **Key term**
>
> **Pneumatic** – using gas to provide a source of mechanical force or control.

> **Did you know?**
>
> The gas used in the suspension of a hydro-pneumatic system is nitrogen. Nitrogen is an inert gas, which means that it is stable and non-reactive. It will not support combustion so is safe in the event of a fire.

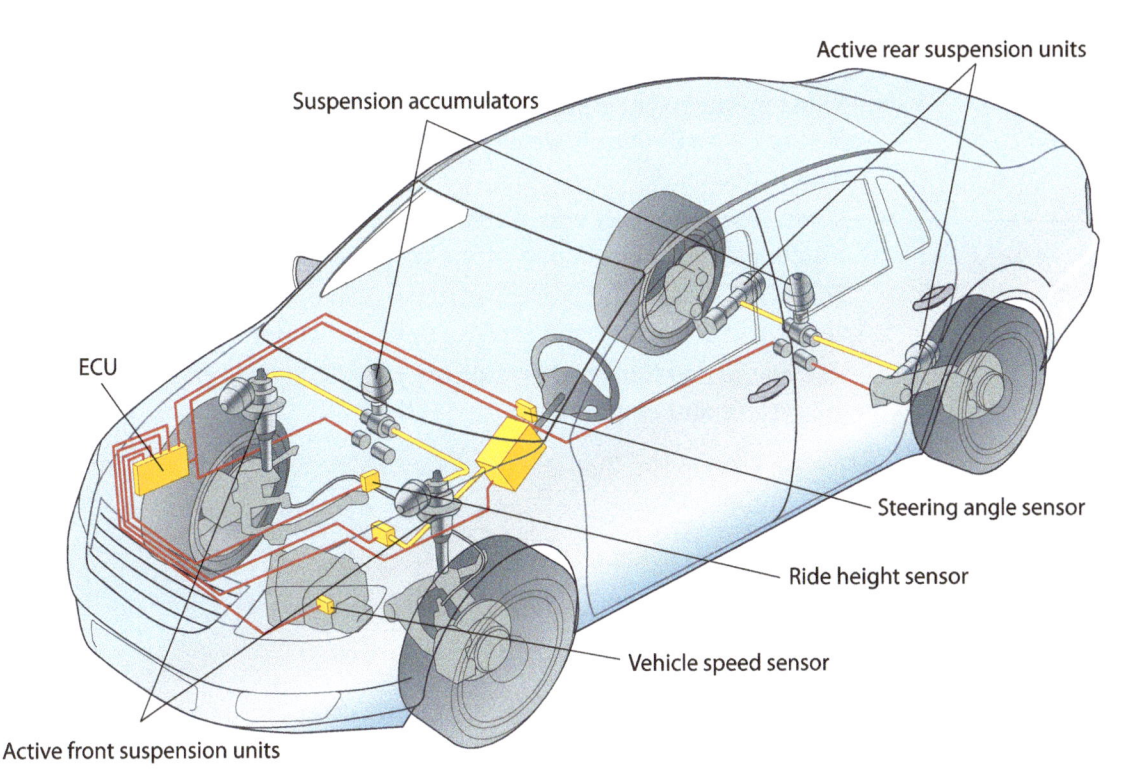

2 Diagnosis & rectification of light vehicle chassis system faults

Figure 2.63 Active hydro-pneumatic suspension

- pressure accumulators
- ride height and load sensors
- steering angle and acceleration sensors
- yaw rate and lateral acceleration sensors
- an electronic control unit.

As the car travels over a bump, the suspension arms operate a piston. This forces fluid through a series of pipes to a sealed compressed gas chamber hydro-pneumatic suspension unit. The fluid pushes against a rubber diaphragm, which compresses the gas still further so that it reacts like a spring. The diaphragm and fluid are forced back in the opposite direction, causing the piston at the suspension unit to react against the bump and provide suspension action.

The engine-driven pump draws fluid from a reservoir and uses it to charge up a pressurised accumulator tank. The pressure accumulators are then able to add extra fluid to the system via the solenoid valve block.

The solenoid valve block is able to control the flow of fluid into or out of the system, therefore adjusting the ride height. If fluid is added, ride height is increased, if fluid is removed, ride height is reduced.

Suspension ride height, steering angle and vehicle acceleration forces are monitored by sensors as the vehicle is driven. This information is monitored by the suspension electronic control unit. When more fluid is required in one part of the suspension system to maintain vehicle attitude, a valve can be opened and fluid is added under pressure. If less fluid is required in one part of the suspension system to maintain vehicle attitude, another valve can be opened and excess fluid is returned to the reservoir.

Self-levelling and ride control system fault diagnosis

- Many modern active suspension systems incorporate a self-diagnostic facility that will illuminate a malfunction indicator lamp (MIL) if a fault is detected.
- If you suspect a fault with a self-levelling or ride control system, conduct a visual inspection of the components for damage or wear.
- Following the manufacturer's instructions, connect a scan tool to the diagnostic connector and retrieve any fault codes.
- Use any stored fault codes to direct your diagnostic routine then test the system and component.

Table 2.9 gives some examples of diagnostic trouble codes for the suspension system.

Table 2.9 Examples of suspension system fault codes

Fault code	Possible cause(s)
C1710	Left front damper actuator short circuit to battery
C1711	Left front damper actuator short circuit to ground
C1712	Left front damper actuator open circuit
C1715	Right front damper actuator short circuit to battery
C1716	Right front damper actuator short circuit to ground
C1717	Right front damper actuator open circuit
C1720	Left rear damper actuator short circuit to battery
C1721	Left rear damper actuator short circuit to ground
C1722	Left rear damper actuator open circuit
C1725	Right rear damper actuator short circuit to battery
C1726	Right rear damper actuator short circuit to ground
C1727	Right rear damper actuator open circuit
C1735	Compressor relay short to battery
C1736	Compressor relay short to GND/Open circuit

It is good practice to read and record any fault codes present, clear the fault codes, road test the car and base your diagnostic routine on any fault codes that have returned. Remember to diagnose the circuit, not guess the fault. Once you have repaired any faults, or replaced components, you should fully road test the vehicle and recheck the fault codes. Be sure that you follow up any pending fault codes that may be an indication of emerging problems.

> **Safe working**
>
> Some active suspension systems may try to correct the ride height if the car is jacked up. This could be dangerous or require a system reset after repairs have been carried out. Always follow the manufacturer's guidelines for disabling automatic ride height adjustment before starting work.

> **CHECK YOUR PROGRESS**
>
> 1. What type of gas is used in a hydro-pneumatic suspension system?
> 2. How can you check the correct operation of an air-assisted self-levelling system?
> 3. How is handling improved in a variable rate damping system?

BEFORE YOU FINISH

Recording information and making suitable recommendations

At all stages of a diagnostic routine, maintenance or repair, you should record information and make suitable recommendations. The table below gives examples of how to do this.

Stage	Information	Recommendations
Before you start	Record customer/vehicle details on the job card. Make a note of the customer's repair request and any issues/symptoms. Locate any service or repair history.	Advise the customer how long you will require the car. Describe any legal, environmental or warranty requirements.
During diagnosis and repair	Carry out diagnostic checks and record the results on the job card or as a printout from specialist equipment. List the parts required to conduct a repair. Note down any other non-critical faults found during your diagnosis.	Inform your supervisor of the required repair procedures so that they can contact the customer and gain authorisation for the work to be conducted.
When the task is complete	Write a brief description of the work undertaken. Record your time spent and the parts used during the diagnosis and repair on the job card. (This information should be as comprehensive as possible because it will be used to produce the customer's invoice.) Complete any service history as required.	Inform the customer if the vehicle will need to be returned for any further work. Advise the customer of any other issues you noticed during the repair.

Remember, any work you conduct on a customer's car should be assessed to ensure that it is conducted in the most cost-efficient manner. You should consider reconditioning, repair and replacement of components within units.

FINAL CHECK

1. The average slip tolerance from an ABS system is approximately:
   a 0% to 5%
   b 5% to 10%
   c 10% to 30%
   d 30% to 50%

2. The wave form created by an inductive wheel speed sensor will be a:
   a sine wave
   b square wave
   c cosine wave
   d digital wave

3. For failsafe operation the ABS solenoid valves are:
   a held closed
   b energised to open
   c energised to close
   d always open

4. The driver is aware that the ABS is not working when:
   a the tyres skid
   b the MIL is illuminated
   c the pedal pulses during braking
   d all of the above

5. The acronym EBD stands for:
   a extra brake distribution
   b even brake distribution
   c every brake distributed
   d electronic brake force distribution

6. Which of the following are **not** types of electronic parking brake (EPB)?
   a cable
   b caliper
   c hydraulic
   d pneumatic

7. Understeer is where:
   a the front slip angle is larger than the rear slip angle
   b the front slip angle is smaller than the rear slip angle
   c the front slip angle equals the rear slip angle
   d there is no slip angle created

8. Camber angle is viewed from:
   a the front of the vehicle
   b the side of the vehicle
   c above the vehicle
   d below the vehicle

9. Caster angle is viewed from:
   a the front of the vehicle
   b the side of the vehicle
   c above the vehicle
   d below the vehicle

10. A hydraulic pressure reading of 65 bar at 1000 rpm on a PAS system is:
    a too high
    b correct
    c too low
    d not possible

2 Diagnosis & rectification of light vehicle chassis system faults

PREPARE FOR ASSESSMENT

The information contained in this chapter, as well as continued practical assignments, will help you to prepare for both the end-of-unit tests and the diploma multiple-choice tests. This chapter will also help you to develop diagnostic routines that enable you to work with light vehicle chassis system faults. These advanced system faults will be complex and non-routine and may require that you work with and manage others, in order to successfully complete repairs.

You will need to be familiar with:

- Diagnostic tooling
- Electrical and electronic principles
- Diagnostic planning and preparation
- Anti-lock braking systems
- Electronic brake distribution
- Traction control
- Steering geometry
- Power-assisted steering
- Active suspension and ride height control
- Electronic stability programs

This chapter has given you an overview of advanced vehicle chassis systems and has provided you with the principles that will help you with both theory and practical assessments. It is possible that some of the evidence you generate may contribute to more than one unit. You should ensure that you make best use of all your evidence to maximise the opportunities for cross-referencing between units.

You should choose the type of evidence that will be best suited to the type of assessment that you are undertaking (both theory and practical). These may include:

Assessment type	Evidence example
Workplace observation by a qualified assessor	Carrying out a two-stage diagnosis on an electronic power-assisted steering system
Witness testimony	A signed statement or job card from a suitably qualified/approved witness, stating that you have correctly tested and reset steering alignment
Computer-based	A printout from a diagnostic scan tool showing the results from a system test to check the function of an electronic stability program
Audio recording	A timed and dated audio recording of you describing the process involved when checking the ride height of a self-levelling suspension system
Video recording	Short video clips showing you carrying out the various stages involved in a brake test on a rolling road
Photographic recording	Photographs showing you carrying out the stages of an ABS repair if the assessor is unable to be present for the entire observation (e.g. parts unavailable at the time). The photos should be used as supporting evidence alongside a job card.
Professional discussion	A recorded discussion with your assessor about how you diagnosed and repaired a suspension ride height sensor

CONTINUED ▶

Assessment type	Evidence example
Oral questioning	Recorded answers to questions asked by your assessor, in which you explain how you diagnosed excessive tyre wear caused by incorrect steering geometry
Personal statement	A written statement describing how you carried out the repair of a traction control system
Competence/skills tests	A practical task arranged by your training organisation, asking you to correctly use an oscilloscope to test a wheel speed sensor
Written tests	A written answer to an end-of-unit test to check your knowledge and understanding of light vehicle chassis systems
Multiple-choice tests	A multiple-choice test set by your awarding body to check your knowledge and understanding of light vehicle chassis systems
Assignments/ projects	A written assignment arranged by your training organisation requiring you to show in-depth knowledge and understanding of a particular chassis system (e.g. ABS)

Before you attempt a theory end-of-unit or multiple-choice test, make sure you have reviewed and revised any key terms that relate to the topics in that unit. Ensure that you read all the questions carefully. Take time to digest the information so that you are confident about what each question is asking you. With multiple-choice tests, it is very important that you read all of the answers carefully, as it is common for two of the answers to be very similar, which may lead to confusion.

For practical assessments, it is important that you have had enough practice and that you feel that you are capable of passing. It is best to have a plan of action and work method that will help you.

Make sure that you have the correct technical information, in the way of vehicle data, and appropriate tools and equipment. It is also wise to check your work at regular intervals. This will help you to be sure that you are working correctly and to avoid any problems developing as you work.

When you are undertaking a practical assessment, always take care to work safely throughout the test. Light vehicle chassis systems are dangerous and precautions should include making sure that you:

- isolate active chassis systems before you start work, to prevent them operating during your repairs
- observe all health and safety requirements
- use the recommended personal protective equipment (PPE) and vehicle protective equipment (VPE)
- use tools correctly and safely.

Good luck!

3 Diagnosis & rectification of light vehicle engine faults

This chapter will help you to gain an understanding of diagnosis and diagnostic routines that lead to the rectification of engine mechanical and electronic control system faults. It also explains and reinforces the need to test light vehicle engine systems and evaluate their performance. It will support you with knowledge that will aid you when undertaking both theory and practical assessments. It will help you develop a systematic approach to complex diagnosis of the light vehicle engine system.

This chapter covers:

- Light vehicle engine systems
- Pressure-charged induction systems
- Valve mechanisms
- Engine mechanical faults
- Ignition systems
- Combustion
- SI fuel systems (petrol)
- CI fuel systems (diesel)
- Engine management
- Alternative fuel vehicles
- Heating, cooling and ventilation

BEFORE YOU START

Safe working when carrying out light vehicle diagnostic and rectification activities

There are many hazards associated with the diagnosis and repair of advanced engine systems. You should always assess the risks involved with any diagnostic or repair routine before you begin, and put safety measures in place. You need to give special consideration to the possibility of:

- **Burns and scalds:** Many engine diagnostic routines and processes require the engine to be at normal operating temperature. Hot components may cause burns and are a source of ignition for flammable materials.
- **Coming into contact with or inhaling exhaust fumes:** Running engines to confirm correct operation is vital to the diagnostic process. If this is conducted in a confined space, such as a workshop, you must always use exhaust extraction.

You should always use appropriate personal protective equipment (PPE) when you work on engine systems. Make sure that your selection of PPE will protect you from these hazards.

Electronic and electrical safety procedures

Working with any electrical system has its hazards, and you must take safety seriously. When you are working with light vehicle electrical and electronic systems, the main hazard is the possible risk of electric shock. (For information on basic first aid for electrical injuries, see Table 1.3 in Chapter 1, page 18.) Although most systems operate with low voltages of around 12V, an accidental electrical discharge caused by incorrect circuit connection can be enough to cause severe burns. Where possible, isolate electrical systems before conducting the repair or replacement of components.

If you are working on hybrid vehicles, take care not to disturb the high voltage system. You can normally identify the high voltage system by its reinforced insulation and shielding, which is often brightly coloured. These systems carry voltages that can cause severe injury or death. If you carry out repairs to hybrid vehicles, always follow the manufacturer's recommendations.

Always use the correct tools and equipment. Damage to components, tools or personal injury could occur if the wrong tool is used or a tool is misused. Check tools and equipment before each use.

If you are using measuring equipment, always check that it is accurate and calibrated before you take any readings.

If you need to replace any electrical or electronic components, always check that the quality meets the original equipment manufacturer (OEM) specifications. (If the vehicle is under warranty, inferior parts or deliberate modification might make the warranty invalid. Also, if parts of an inferior quality are fitted, this might affect vehicle performance and safety.) You should only carry out the replacement of electrical components if the parts comply with the legal requirements for road use.

Information sources

The complex nature of advanced light vehicle engine systems requires you to have a good source of technical information and data. In order to conduct diagnostic routines and repair procedures, you will need to gather as much information as possible before you start. Sources of information include:

Information source	Example
Verbal information from the driver	A description of the symptoms that occur on the car when driving
Vehicle identification numbers	Engine type taken from VIN plate
Service and repair history	A check of the service history that shows when the engine oil and filter were last changed
Warranty information	Is the car under warranty and is it valid? (Has the required service and maintenance been conducted?)
Vehicle handbook	To confirm how to correctly drive the vehicle for economy conditions
Technical data manuals	To find the recommended operating pressures of a common rail diesel fuel system for diagnostic purposes
Workshop manuals	To find the recommended procedures used when setting up camshaft timing
Safety recall sheets	To confirm which components need to be replaced for safe operation of a hybrid vehicle's electric motor system
Manufacturer-specific information	Vehicle specific diagnostic trouble codes relating to the engine's variable valve control system
Information bulletins	Information on a common fault found on turbochargers
Technical help lines	Advice on the correct routine for testing fuel pressure, flow and volume
Advice from master technicians/colleagues	An explanation of how to set up the company's diagnostic scan tool and retrieve fault codes
Internet	An Internet forum page where a number of people who had a similar problem with the failure of an ignition coil explain how it was resolved
Parts suppliers/catalogues	A cross-reference of spark plug part numbers, so that you can make sure that you use the correct heat range to prevent damage
Job cards	A general description of the work to be conducted on a customer's engine system
Diagnostic trouble codes	A fault code showing that the mass airflow (MAF) sensor circuit needs to be tested to ensure correct engine management operation
Oscilloscope wave forms	A faulty current ramp wave form being produced from an electric fuel pump commutator

Remember that no matter which information or data source you use, it is important to evaluate how useful and reliable it will be to your diagnostic routine.

Operation of electrical and electronic systems and components

The operation of electrical and electronic systems and components related to light vehicle engine systems:

Electrical/electronic system component	Purpose
ECU	The electronic control unit (ECU) is designed to monitor and control the operation of light vehicle engine systems. It processes the information received and operates actuators that control engine running and performance.
Sensors	The sensors monitor various engine operating components against set parameters. As the vehicle is driven, dynamic operation creates signals in the form of resistance changes or voltage, which are relayed to the ECU for processing.
Actuators	The actuators are used to control engine functions and performance. Motors, solenoids, valves, transformers, etc., are operated by the ECU to help control the action of the engine, leading to improved performance, emissions and fuel economy.
Electrical inputs/voltages	The ECU needs reliable sensor information in order to correctly determine the action of the engine systems. If battery voltage was used to power sensors, its unstable nature would create issues (battery voltage constantly rises and falls during normal vehicle operation). Because of this, sensors normally operate with a stabilised 5-volt supply.
Digital principles	Many vehicle sensors create analogue signals (a rising or falling voltage). The ECU is a computer and needs to have these signals converted into digital (on and off) before they can be processed. This can be done by using a component called a pulse shaper or Schmitt trigger.
Duty cycle and pulse width modulation (PWM)	Lots of electrical equipment and electronic actuators can be controlled by duty cycle or pulse width modulation (PWM). This works by switching components on and off very quickly so that they only receive part of the current/voltage available. Depending on the reaction time of the component being switched and how long power is supplied, variable control is achieved. This is more efficient than using resistors to control the current/voltage in a circuit. Resistors waste electrical energy as heat, whereas duty cycle and PWM operate with almost no loss of power.
Fibre optic principles	As engine running and emission systems improve, the need for very fast transmission of information has increased. Fibre optics use light signals transmitted along thin strands of glass to provide digital data transmission. (The light source is switched on and off.) In this way information is transmitted essentially at the speed of light.

Electrical and electronic control is a key feature of all the systems discussed in this chapter.

Tooling

No matter what task you are doing to a car, you will need to use some form of tooling. Always use the correct tools and equipment.

The following table shows a suggested list of diagnostic tooling that could be used when testing and evaluating light vehicle engine mechanical and electronic systems. Due to the nature of complex system faults, you will experience different requirements during your diagnostic routines and so you will need to adapt the list shown for your particular situation.

Tool	Possible use
Oscilloscope	To test the signal produced by a crankshaft position sensor
Multimeter	To test the voltage signal produced by an engine coolant temperature sensor (ECT)
Test lamp/ logic probe	To test the existence of system voltage at an engine system power control relay (Always use test lamps with extreme caution on electronic systems, as the current draw created can severely damage components.)
Power probe	To power the compressor clutch of an air-conditioning system and check its operation
Pressure gauge	To check for engine cylinder leakage and mechanical system wear

CONTINUED ▶

83

Tool	Possible use
Code reader/ scan tool	 To retrieve diagnostic trouble codes (DTCs) related to the engine, which indicate the circuit to be tested. To clear trouble codes, reset the malfunction indicator lamp, and evaluate the effectiveness of repairs
Performance dynamometer/ rolling road	 To test the efficiency of the engine system, fine tune performance and evaluate the effectiveness of repairs
Exhaust gas analysers	 To check for correct engine operation and exhaust emissions output. The chemical make-up of engine exhaust gases can be used to: • assist with the diagnosis and repair of engine operating systems • evaluate the effectiveness of repairs – you need to make sure that vehicles operate within the current legal emission standards
Air-conditioning recovery stations	 To check that the vehicle's air-conditioning system is functioning within the specified operating limits. To safely recover and replace fluorinated refrigerant gases in accordance with environmental and legal regulations
Stethoscope	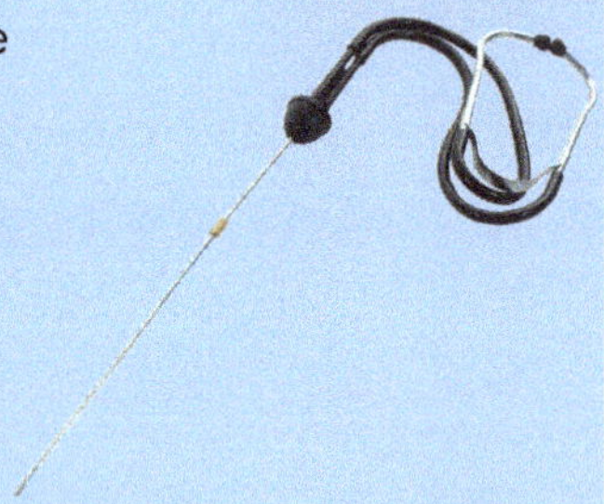 To help identify engine noises created during operation. Some stethoscopes are electronic and their listening devices can be attached to various engine components, then the car can be road tested and noise/wear can be located
Laser thermometer	 A non-contact thermometer, also known as a pyrometer, can help determine the correct operation of an engine's cooling system, with cooler areas indicating poor circulation and possible blockage

3 Diagnosis & rectification of light vehicle engine faults

Light vehicle engine systems

Modern engines are designed to meet very strict manufacturing and performance guidelines. In many cases, engine capacity has reduced in order to improve fuel economy and reduce exhaust emissions. Designers and manufacturers are continually improving their systems in order to keep up with technology.

Pressure-charged induction systems

The performance of an engine is affected by many factors, but one of the most important is **volumetric efficiency**.

Volumetric efficiency is how well the engine cylinder can be filled with a fresh charge of incoming air and fuel in the correct quantities. A **naturally aspirated** engine relies on atmospheric pressure to force air into the cylinder above the descending piston. As the piston descends, a low pressure called a **depression** is created above the piston crown, and air is forced into the cylinder by atmospheric pressure through the open inlet valve.

Due to the speed of operation, it is very difficult to fill the cylinder completely and, as a result, only about 80 per cent of the available space will be filled. As engine speed increases, this figure often falls and volumetric efficiency is reduced, which affects overall engine performance.

Engine manufacturers and designers have come up with a number of methods to help improve volumetric efficiency. These include:

- variable valve timing and lift
- forced air induction.

Turbocharging

To improve the performance of internal combustion engines, air is sometimes forced into the cylinder above the piston. (The more oxygen that is contained in the cylinder, the more fuel that can be burned and therefore more energy can be released.) One method that can be used to raise the pressure of the incoming air is to use a turbocharger. A **turbine** driven by exhaust gas rotates a set of compressor vanes, forcing air through the intake system and into the combustion chamber.

> **Key terms**
>
> **Volumetric efficiency** – how well an engine cylinder can be filled with air and fuel.
>
> **Naturally aspirated** – not turbocharged or supercharged.
>
> **Depression** – a low or negative pressure.
>
> **Turbine** – a set of mechanical rotating blades or vanes.

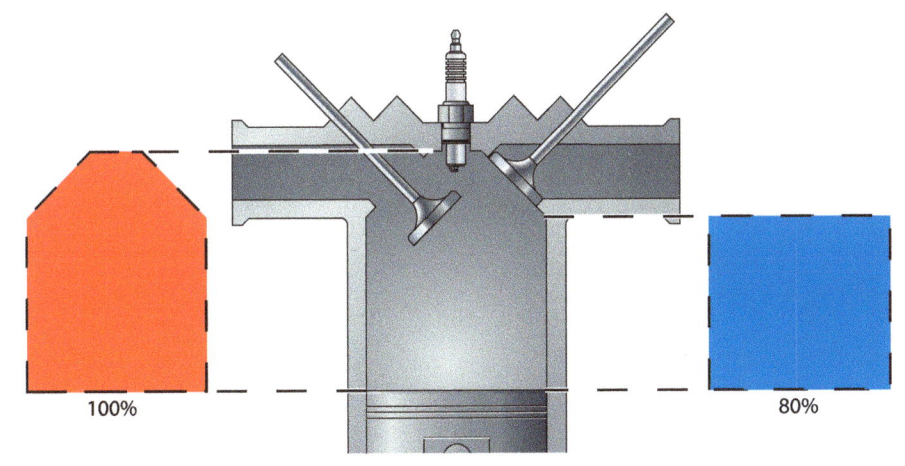

Figure 3.1 Volumetric efficiency of a naturally aspirated engine

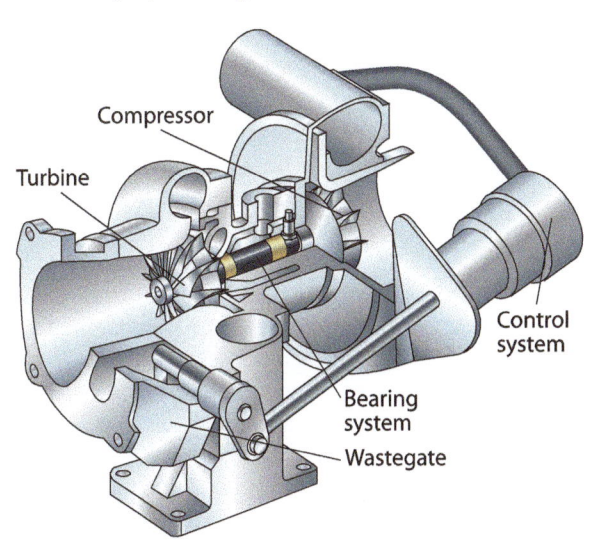

Figure 3.2 Turbocharger

85

Level 3 Light Vehicle Technology

The advantages of using a turbocharger in vehicle design include:
- improved volumetric efficiency of above 100 per cent
- an increase in engine performance
- turbochargers harness and recycle the energy produced by engines, transforming more of the fuel consumed into energy (power), giving greater fuel economy
- smaller capacity engines can be used in vehicle design but output performance is similar to that of larger engines
- no loss of performance at altitude where air pressure is lower
- more responsive drive produced giving safer overtaking
- the turbo increases the amount of air entering the engine, which can help reduce exhaust emissions produced
- the increased performance improves the enjoyment of driving.

A turbocharger has four main disadvantages, as shown in Table 3.1.

Table 3.1 The main disadvantages of turbochargers

Disadvantage	Description	Methods used to overcome turbocharger inefficiency
Lag	The turbine has to be spinning at considerable speed before it produces any usable boost in performance. This leads to a condition called turbo lag, where during the initial acceleration, no noticeable performance increase is felt. As the turbine speed increases, a sudden surge of power is introduced, which is called boost.	Multiple turbochargers can be fitted to an engine. They are manufactured in different sizes, which help to give a smooth delivery of boost pressure. A small turbine will **spool up** to speed quickly but only produce a small amount of boost pressure. If this is combined with a larger turbocharger, as the small turbocharger runs out of boost, the larger turbine is up to speed and can take over. A turbocharger can be designed with variable vane geometry. A small set of moveable blades or vanes are fitted around the outer edge of the compressor turbine. You can vary the angle of these blades or vanes to allow for different amounts of boost at different engine speeds. The variable vane geometry can be controlled by intake system pressure or by ECU actuation.
Overboost	As engine speed rises, the exiting exhaust gas turns the turbine faster and faster. This means that boost pressures may continue to rise above safe limits. Excessive boost pressure can affect engine performance and may lead to engine damage. In a petrol engine, the overcharging of a cylinder with compressed air can raise the effective **compression ratio** to a point where the petrol begins to auto-ignite. With auto-ignition, the fuel no longer requires a spark plug to initiate combustion and therefore acts like a compression ignition diesel engine. If the boost pressure is not limited, the forces acting on mechanical engine components can be so great that premature failure may occur.	To help prevent the turbocharger overboosting and causing detonation of the fuel and engine damage, an exhaust wastegate is often used. The wastegate is a mechanically controlled valve which opens to allow some exhaust gases to bypass the turbine. The wastegate is connected to a **servo** valve, which is acted on by pressure or vacuum created in the inlet manifold. As the engine speeds up and boost increases, the servo is able to operate at a preset value, wasting some of the turbine energy from the exhaust. The actuator arm from the servo is often adjustable to allow the technician to set the wastegate operating pressure. Modern engines often incorporate an electrically operated solenoid valve into the controller unit of the wastegate servo. When actuated by the ECU through a duty cycle, this solenoid valve is able to accurately control the operation of the wastegate servo, and therefore the amount of boost pressure that is available to the engine at any time.

CONTINUED ▶

3 Diagnosis & rectification of light vehicle engine faults

Disadvantage	Description	Methods used to overcome turbocharger inefficiency
Temperature	The temperature of the induction air is raised due to the heat of the turbocharger and the action of compressing the incoming air. This rise in temperature makes the incoming charge less dense, reducing the effect that the oxygen contained has on the combustion process.	To help increase the **density** of oxygen in the compressed air charge from the turbocharger, a method of removing heat is needed. This is normally achieved by using an intercooler. An intercooler is a small radiator placed after the turbocharger, through which the compressed air charge travels. As it passes through the intercooler, some of the heat energy created in the turbocharger is removed. The cooling of the air charge helps increase the density of the oxygen contained in the incoming air. To improve the operation of the intercooler still further, some manufacturers spray water onto the outer surface of the intercooler. As the water evaporates from the surface of the intercooler radiator, **latent heat** created by evaporation reduces the temperature still further.
Backpressure	Backpressures created in the inlet tract have a tendency to try to slow the turbine down and reduce its performance. This is particularly noticeable during gear change. As the driver closes the throttle in order to change gear, a butterfly valve restricts airflow in the intake manifold, creating pressures which slow down the turbine. When the driver resumes acceleration, the turbine has to speed back up before boost is once again usable. This creates lag during gear change and reduces performance.	To ensure that the turbocharger does not lose performance during gear change, some vehicle manufacturers and **aftermarket suppliers** incorporate a component called a **dump valve**. During operation, as the driver lifts their foot off of the accelerator to change gear, a vacuum pipe connected to the inlet manifold operates a servo controller in the boost pressure side of the turbocharger. A valve opens and dumps the boost pressure from the inlet manifold side of the turbocharger, often accompanied by a loud whoosh as the air is allowed to escape. Because boost pressure has now been dumped, there is no backpressure in the manifold to slow the turbine down. Once the gear change has taken place, the driver puts their foot back on the accelerator and a spring closes the servo in the dump valve and boost is returned. As the turbine is still spinning at speed, the boost is almost immediate and turbo lag is reduced.

> **Key terms**
>
> **Spool up** – wind up to speed.
>
> **Compression ratio** – the difference in volume above the piston between when it is at the bottom of its stroke and when it is at the top of its stroke.
>
> **Servo** – a control system that converts mechanical motion into one requiring more power.
>
> **Density** – how closely the molecules of a substance are packed together.
>
> **Latent heat** – the temperature at which a substance changes from a liquid to a vapour or gas.
>
> **Aftermarket suppliers** – companies that manufacture components that can be fitted as alternatives or upgrades to the original equipment manufacturer (OEM) components.
>
> **Dump valve** – a component used to get rid of turbo boost pressure during gear change.

Figure 3.3 Turbocharger dump valve

Level 3 Light Vehicle Technology

> ⚠️ **Safe working**
>
> On vehicles fitted with a turbocharger, it is advisable to allow the engine to idle for a few minutes before switching it off. This allows the turbocharger to slow down. If this isn't done, the turbine of the turbocharger may still be rotating at very high speed (hundreds of thousands of revs per minute). The drive spindle and the bearings of the turbocharger are lubricated with oil from the engine lubrication system. As soon as the engine is switched off, oil pressure to the turbine bearings is lost. Because the turbine continues to spin at very high speed, this can lead to wear in the turbocharger bearings and premature failure.

> **NEW TECH**
>
> **Shut down period**
> Some manufacturers incorporate a shut down period on high performance engines with turbochargers. The driver is often able to switch off and remove the ignition key, and the engine management system will keep the engine running for up to two minutes while the turbocharger speed is allowed to spool down.

Superchargers

A supercharger is a mechanically driven air compressor connected directly or indirectly to the engine's crankshaft. This can be achieved using drive gears, drive chains or drive belts.

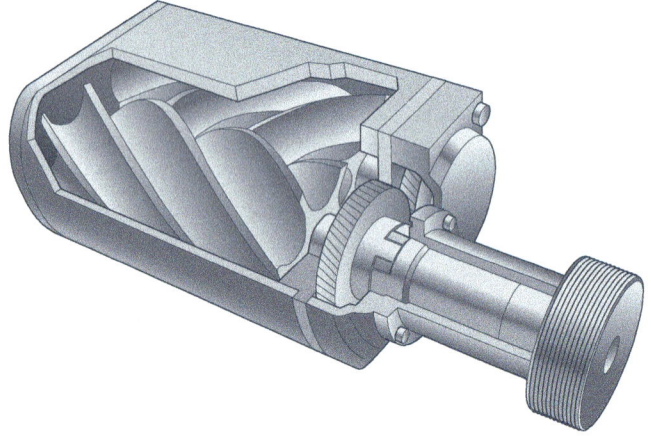

Unlike a turbocharger, which is driven from exhaust gases, the boost provided by a supercharger is instant, with no lag. Unfortunately, a supercharger is considered a **parasite** of engine power. This means that some of the turning effort created at the crankshaft is used to drive the supercharger; as a result, not all the power created by the supercharger is converted into performance. Some manufacturers have created a variable drive supercharger with a clutch that disengages at lower engine revs. This allows performance, economy and emissions to be kept within reasonable limits.

Figure 3.4 Supercharger

A number of different types of supercharger compressor can be used, including:

- **Reciprocating piston:** This uses a piston, connecting rod and crankshaft in a similar manner to a standard **reciprocating** engine.

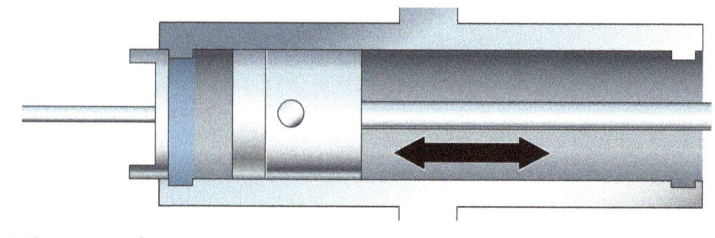

> **Key terms**
>
> **Parasite** – in terms of light vehicles, something that draws energy away from the engine without gain or advantage.
>
> **Reciprocating** – moving backwards and forwards or up and down.
>
> **Tolerance** – an allowable difference from the optimum measurement.

Figure 3.5 Piston supercharger

3 Diagnosis & rectification of light vehicle engine faults

- **Rotary vane:** This system uses two or three rotor gears, machined and connected with very close **tolerances**, to compress the air for the intake charge.
- **Centrifugal:** A centrifugal vane uses a compressor turbine in a similar way to a turbocharger. As it rotates, air is thrown outwards by centrifugal force into a chamber where it is compressed before moving on to the intake manifold.

Figure 3.6 Vane supercharger

Figure 3.7 Centrifugal supercharger

- **Axial:** This uses a set of rotor blades in a similar way to a jet engine. These rotor blades force air through smaller and smaller channels until it has been compressed to a level that will create extra boost for the engine.

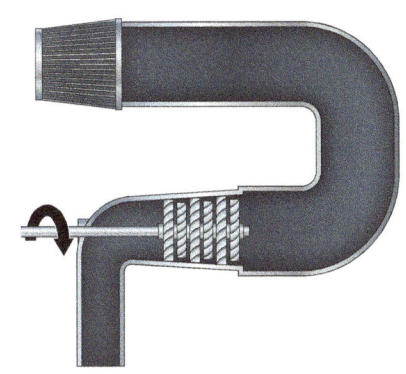

Figure 3.8 Axial supercharger

Common faults associated with turbochargers and superchargers

Table 3.2 lists some of the symptoms and faults associated with pressure charging.

Table 3.2 Faults associated with pressure charging

Symptom	Possible fault
Noise	Noise can be created in both superchargers and turbochargers caused by bearing wear. Because of high rotational speeds, good lubrication is essential. If service and maintenance procedures have been neglected, dirty oil can affect the lubricating properties or block oil galleries leading to the turbo or supercharger. If the engine is switched off immediately, oil pressure is also lost and rapid bearing wear can occur. If the turbine bearings have become worn, play is sometimes apparent in the drive spindle of the turbine. You can usually assess this by moving the turbine spindles with your fingers.
Burning oil	A common fault with turbochargers is the leaking of lubrication oil past the seals around the drive spindle, caused by excessive wear and play in the bearings. Symptoms produced are normally high volumes of oil smoke coming from the exhaust, especially during acceleration. Because of pressure differences between the inlet side and the exhaust side of a turbo, most of the oil is forced into the hot exhaust gases where it is combusted. Methods to confirm this fault include: • Removing spark plugs to check for fouling. If the spark plugs do not show signs of oil contamination, this is usually an indication that oil has entered the exhaust system after the engine. • Checking for contamination or carbon build-up on the turbines, which indicates a leakage of oil past the spindle seals.

CONTINUED ▶

Level 3 Light Vehicle Technology

Symptom	Possible fault
Loss of boost	Loss of boost may be caused by a restriction in the air intake system, leakage on the pressure side of the engine intake system or a wastegate seized in the open position. If a loss of boost occurs on a vehicle, it is important to check these areas carefully before you assume there is a fault with the turbine or supercharger itself.
Overboost	Overboost is normally created by a wastegate that is seized in the closed position. You should carefully examine the wastegate and operate it with the aid of a pressure or vacuum pump to confirm correct operation.

Diagnosis

Many of the symptoms produced by faulty turbo or superchargers will present as a lack of performance. If the boost provided by a forced air induction system is lower than expected, the intake pressure will need to be tested. Specialist gauges and connection pipes are available, which when joined to the intake, are able to measure the amount of pressure provided by the turbo or supercharger and this can be compared to the manufacturer's specification. If performance is low you should check:

- for restriction in the air intake
- if the turbo/supercharger rotates freely
- that the wastegate or drive system is operating correctly
- that there are no induction air leaks.

Valve mechanisms

Intake and exhaust valves are one of the largest restrictions to the movement of air, fuel and exhaust inside an internal-combustion engine. This restriction leads to a reduction in performance.

Many manufacturers now use multiple valve and complex camshaft drive arrangements to help improve performance. Multivalve arrangements have two or more valves of each type (inlet and exhaust) included in the design to make best use of the space in the combustion chamber. This improves the amount of air/fuel that can be drawn into the cylinder and the speed with which exhaust gas can be expelled.

Variable valve control

Another method of reducing the restriction caused by the inlet and exhaust valves is variable valve control. This is able to produce a number of different operating conditions, including:

- **Camshaft phasing:** The operation of the camshaft is **advanced** or **retarded** in order to open the valves earlier or close them later. By altering the operation of valve timing during different engine running conditions, enhanced performance is achieved.
- **Variable valve lift:** The amount that the valve is opened is changed during different engine running conditions. If valves are only opened a small amount during slow speed operation, this achieves smooth

> **Action**
> 1. Research and name two manufacturers that use turbochargers in their vehicle designs.
> 2. Research and name two manufacturers that use superchargers in their vehicle designs.

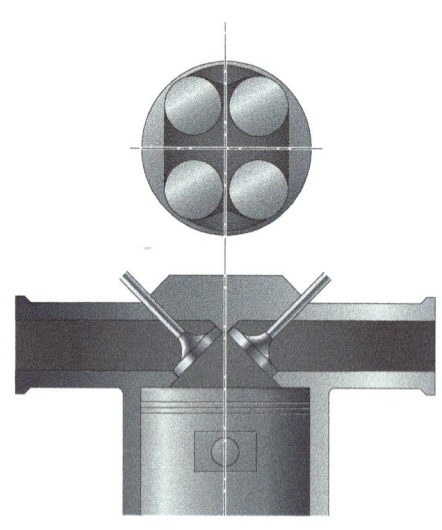

Figure 3.9 Multivalve arrangement

> **Key terms**
> **Advanced** – ahead of time (early).
> **Retarded** – lagging behind (late).

running, fuel economy and low emissions. If the valves are opened a large amount during high speed operation, this achieves enhanced performance.

- **Valve operation speed:** The speed of the camshaft is varied during different engine running conditions. If the camshaft is sped up during engine slow speed operation, the valve is only open for a short period of time, which gives smooth running, fuel economy and lower emissions. If the camshaft is slowed down during engine high speed operation, the valves are held open for longer. This allows more air and fuel to enter the engine, producing enhanced performance.

Variable valve control mechanisms

Different manufacturers use distinct types of valve drive mechanisms on their engine designs. Some examples of variable valve control mechanisms are listed below.

VVT-I

The VVT-I system is a method used to control the phasing of the inlet camshaft(s). A controller mechanism is mounted on one end of the camshaft. The controller connects the camshaft to the timing chain. The VVT-I controller allows a small amount of rotational movement between the camshaft and the timing chain, which means that the timing and operation of the valves can be varied according to engine speed and load.

The VVT-I mechanism is held in a standard timing position by spring pressure. Engine oil is then directed to chambers inside the VVT-I unit. This provides a hydraulic pressure, which rotates the camshaft slightly in relation to the timing chain. Depending on which pressure chamber engine oil is directed to, camshaft timing can be advanced or retarded. If oil pressure is lost, or when the engine is first started, the spring-loaded mechanism inside the VVT-I unit returns the camshaft to a standard timing position.

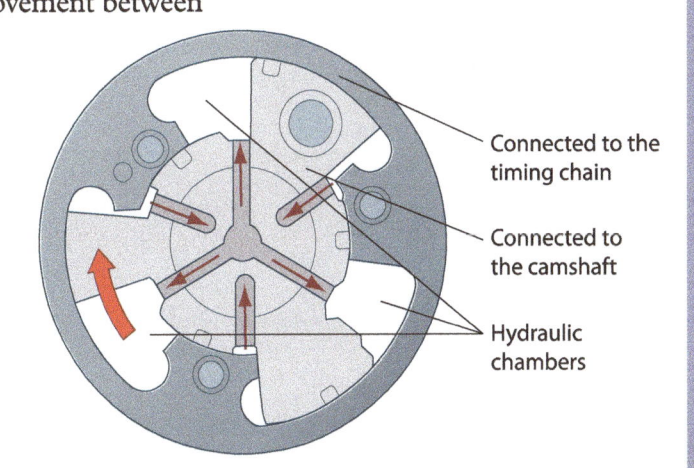

Figure 3.10 VVT-I controller

VANOS

The VANOS system is a method used to control the phasing of the inlet camshaft(s). A VANOS control mechanism is mounted at one end of the camshaft. The controller connects the camshaft to the timing chain.

The VANOS controller (see Figure 3.11) uses a small **intermediate** gear with a spiral **helix** cut into one surface and **splines** cut into the other, to join the camshaft with the timing chain drive mechanism. The intermediate gear is able to slide along the camshaft on the splines. As it does this, the spiral helix acts against the timing chain drive mechanism, causing the camshaft to rotate slightly. The small amount of rotational movement between the camshaft and the timing chain means that the valves can be advanced or retarded according to engine speed and load.

> **Key terms**
>
> **Intermediate** – in-between.
>
> **Helix** – shaped like the spiral of a coil spring.
>
> **Splines** – grooves machined along the length of a shaft.
>
> **Profile** – the outline of a shape.

Hydraulic oil pressure from the engine is directed to one side or the other of the VANOS controller unit, which moves the intermediate gear backwards or forwards along the camshaft to alter the valve timing when required.

VTEC

The VTEC system is a method used to control how far the inlet valves open. Different cam **profiles** are machined on the same camshaft, and rocker arms are used to transfer the movement from the different cam profiles to the inlet valves when required.

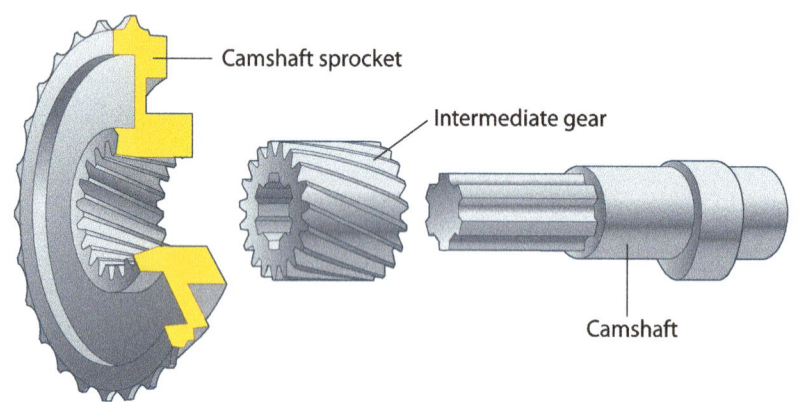

A shallow cam profile will only open the inlet valve a short distance, whereas an aggressive cam profile will open the valve fully. At low engine speed and load, a shallow cam profile is used to provide smooth running, fuel economy and lower emissions. At high engine speed and load, an aggressive cam profile can be used to provide performance.

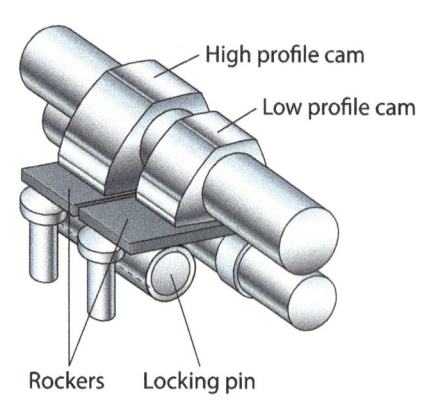

Figure 3.11 VANOS controller

To switch between the different cam profiles, hydraulic oil pressure is used to move a locking pin between two rocker arms. At slow engine speed, the shallow cam profile operates the low lift rocker arm to open the valve; the high lift rocker arm moves freely against a return spring (idles) with no effect on engine operation.

As engine speed increases, hydraulic oil pressure locks the high lift rocker arm to the low lift rocker arm. The aggressive cam profile now takes over, opening the inlet valve fully. As engine speed falls, hydraulic oil pressure is directed away from the locking pin and a return spring is used to unlock the two rocker arms and the low lift cam takes over. Due to the nature and operation of the VTEC system, the change in performance can often be felt by the driver as they accelerate.

Figure 3.12 VTEC control

VVC

The VVC system is a method used to control the speed of operation of the inlet camshaft. If the speed of the camshaft can be varied according to engine performance requirements, the inlet valve can be held open for longer or shorter periods.

At slow engine speed and load, the inlet valve can be held open for a short time, giving smooth running, fuel economy and lower emissions. At high engine speed and load, the inlet valve can be held open for a longer period, allowing more air and fuel to enter the engine and improve performance.

In order to keep the valve timing correct, the camshaft must still rotate at half crankshaft speed. This means the speeding up and slowing down of the inlet camshaft must be completed in the same time that it would normally take to make one rotation. Because of this, the variation of speed is controlled so that during half a rotation of the camshaft it is

3 Diagnosis & rectification of light vehicle engine faults

moving fast, and during the other half rotation of the camshaft it is moving slowly (the camshaft still completes one revolution in the same amount of time as it would in a normal system).

The drive from the timing belt to the camshaft is controlled by an **eccentric** drive pin. The drive pin is able to move in a slot to vary the amount of eccentricity created between the timing belt and the camshaft. This variation of eccentric movement is able to speed up and slow down the operation of the inlet camshaft depending on engine performance requirements.

Because the operation of the eccentric drive would affect all cylinders at the same time, a method is needed of controlling each set of inlet valves separately. In a four-cylinder engine, four separate inlet camshafts are used. Two timing belts (one at the front of the engine and one at the back of the engine) each drive a pair of separate camshafts. Each pair of camshafts has one solid camshaft and one hollow camshaft, with the solid camshaft being driven through the middle of the hollow one. An ECU determines engine speed and load and directs hydraulic oil pressure to the eccentric drive pin mechanism.

In this way, as engine speed and performance requirements increase, the amount of eccentricity on the drive can be controlled so that, as the cam lobe rotates and opens the valve, it is slowed down (so the valve is held open for longer). Then, as the cam lobe rotates away from the valve, it is speeded up so that it still completes one revolution in the same amount of time it normally would.

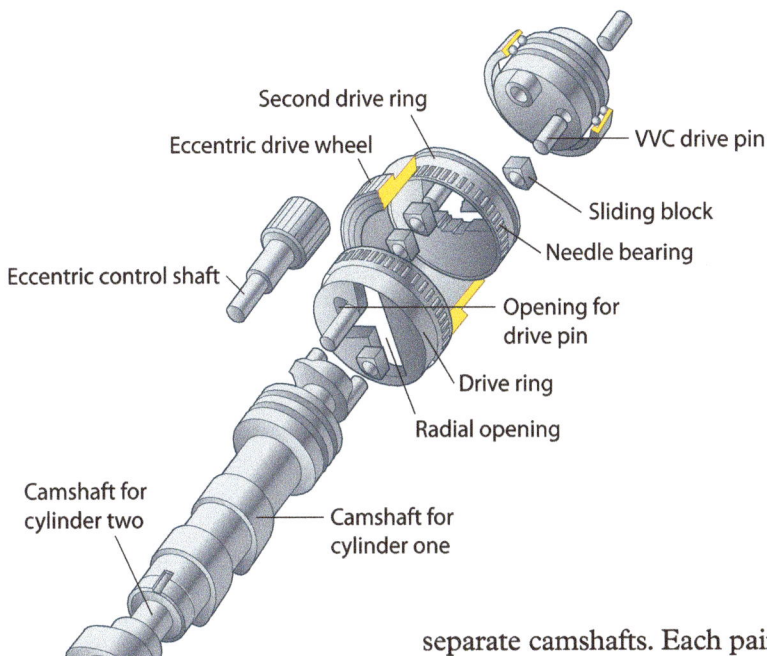

Figure 3.13 VVC control

Valvetronic

An advancement in the use of variable valve control is valvetronic. In this system, the vehicle manufacturer has removed the need for a throttle butterfly valve used to control the amount of air entering the engine. Instead, an electronic system which controls the amount that the inlet valves open is used to regulate the amount of air entering the engine.

An electric motor mounted near the inlet camshaft drives an eccentric gear mechanism. The eccentric gear mechanism is connected to an intermediate shaft that can be moved towards or away from the inlet camshaft. The intermediate shaft is fitted with a roller which follows the opening face of the camshaft profile. The lower edge of the intermediate shaft operates the rocker mechanism connected to the inlet valves.

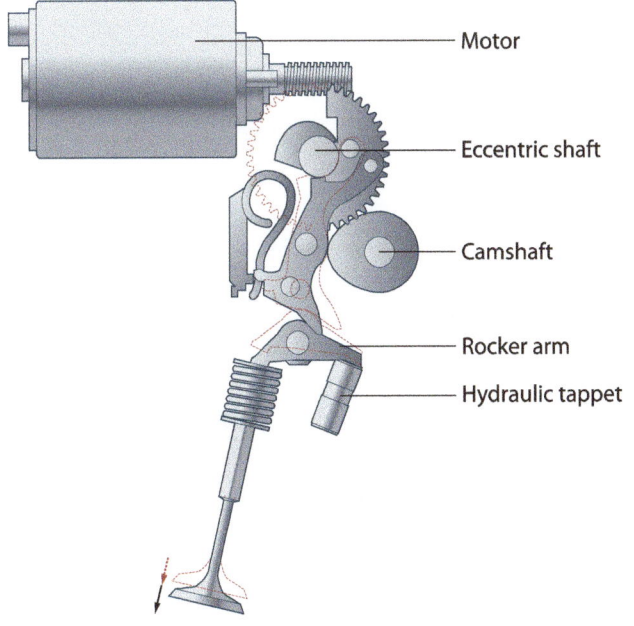

Figure 3.14 Valvetronic

Level 3 Light Vehicle Technology

When the engine is running at slow speeds, the motor and eccentric gear move the intermediate shaft away from the camshaft so that only a small amount of movement is transferred to the rocker and inlet valve mechanism. To increase the engine speed, the driver presses the accelerator pedal and a voltage from the throttle position sensor sends a signal to the ECU, which controls the electric motor. The motor turns the eccentric gear, which moves the intermediate shaft towards the camshaft. In this way, a larger amount of movement is transferred to the rocker, opening the valve further and allowing more air to enter the engine, which increases its speed.

> **Key terms**
>
> **Longitudinally** – running lengthways rather than across.
>
> **Eccentric** – off centre, not in the middle.

Other methods of valve control

Three-dimensional camshaft profiles: The cam lobe is tapered so that if the camshaft is moved **longitudinally** a different profile is able to operate on the valve mechanism controlling lift.

Composite camshafts: Separate cam lobes are mounted on a driveshaft that allows slight rotational or axial movement.

When actuated:

- rotational cam lobes are able to advance or retard the valve timing
- axial cam lobes are able to swap between different profiles and vary the amount of valve lift.

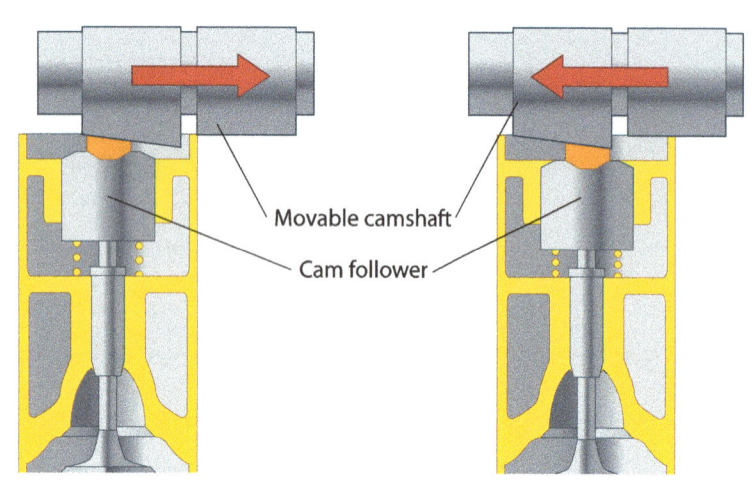

Figure 3.15 Three-dimensional cam profile

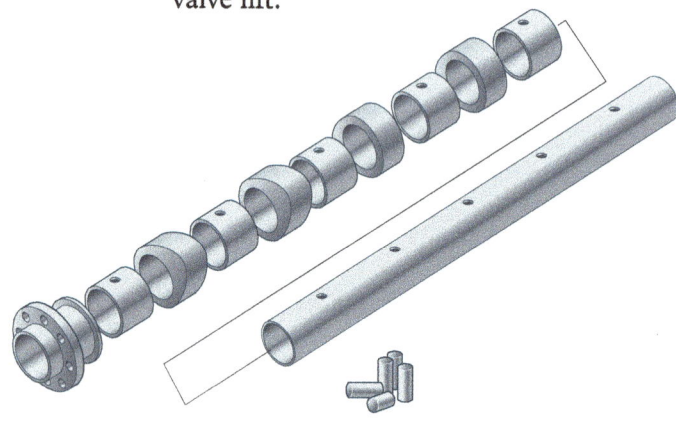

Figure 3.16 Composite camshaft

Solenoid operated valves: Electronically controlled solenoid motors are used to open the inlet and exhaust valves instead of a camshaft. Using solenoid motors gives engine designers complete control over both timing and lift for any engine operating conditions – adaptions can be made to compensate for most driving situations. As this system has no camshaft or valve train to drive, loads placed on the engine are reduced, which leads to improved performance, greater fuel economy and lower emissions.

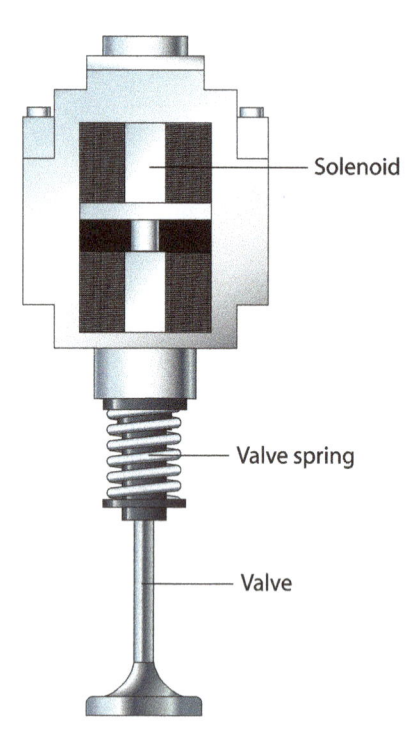

Figure 3.17 Solenoid operated valves

3 Diagnosis & rectification of light vehicle engine faults

Faults associated with variable valve control

Some faults associated with variable valve control are shown in Table 3.3.

Table 3.3 Variable valve control faults

Symptom	Possible fault
Loss of cylinder compression	If inlet or exhaust valves leak or are not fully sealed during engine operation, compression can be lost. This reduces performance and can lead to the valves overheating and burning out. To diagnose issues with leaking valves, you can use a compression test or cylinder leakdown test (see pages 100–101).
No performance change during operation	Many variable valve control systems rely on oil pressure to operate the hydraulic system. If oil pressure is lost to the valve control mechanisms, variable valve timing and lift will no longer be available. To diagnose hydraulic system drive issues you should check oil level, pressure and condition. This should be done before attempting any mechanical strip down and overhaul.
Noise	A number of variable valve timing mechanisms use hydraulic tappets. These hydraulic tappets are designed to take up excess valve clearance during normal operation. If oil pressure, quantity or condition is poor, there may be excessive noise from the valve train mechanism. To diagnose these issues you should use a technician's stethoscope to help isolate the location of any noise before you strip down the engine. Once the general location is established, remove engine covers and check for excessive play or wear.
Oil smoke	As the engine wears, it is common for oil to leak past the valve stem oil seals and get burned within the combustion chamber. This will usually lead to blue smoke emitting from the exhaust system. Oil that is leaking past the valve stem oil seals will be most apparent on start-up or if the vehicle is left at idle for a period. To help diagnose oil smoke issues, the vehicle should be carefully road tested under various driving conditions and the amount of oil smoke assessed.

Figure 3.18 Blue oil smoke

CHECK YOUR PROGRESS

1. List three types of supercharger.
2. What does the term turbo lag mean?
3. Why is a throttle butterfly not required in a valvetronic system?

Level 3 Light Vehicle Technology

> **Key terms**
>
> **Perforations** – a series of small holes in a component.
>
> **Hydro-locked** – when a liquid has entered the combustion chamber; as the piston moves up rapidly on its compression stroke, the engine locks solid due to the fact that liquids are virtually incompressible.

Engine mechanical faults

Because of the high quality of engineering procedures that go into the manufacture of a new engine, straightforward failure of mechanical components is rare. Issues will normally arise from poor maintenance procedures or excessive long-term wear and tear. A complex mechanical engine failure normally affects more than one system, and so your diagnostic procedures should include methods to find the original cause of the fault. Many system failures can produce varied symptoms and it is important that you fix the fault and not the symptom.

Table 3.4 shows examples of engine mechanical damage or failure and suggests possible causes.

Table 3.4 Engine mechanical faults

Symptom	Possible fault
Cylinder liner damage 	Engine cylinder liners may be a dry interference fit (where the inner component is forced into the outer component) and can suffer with hairline cracks, which lead to premature failure or loss of compression. An alternative is a wet liner system, where a series of engine cylinders are fitted loosely in the engine block. When the cylinder head is fitted, the liners are clamped into place to prevent movement. Coolant is free to circulate around the cylinder liners and leakage into the sump is prevented, due to a seal at their base. As with the dry interference fit, hairline cracks or **perforations** can lead to premature failure. If you have to remove the cylinder head to conduct repairs, the engine crankshaft must not be rotated unless the wet liners have been clamped into position. If the wet liners are moved while the cylinder head is removed, it is possible to damage the liner base seals and allow coolant to leak into the engine oil.
Piston damage 	Piston damage will normally occur for two main reasons: physical impact or overheating. The most common cause of physical impact is when valves strike pistons because of cam belt or timing chain failure. This impact damage may be minor (only small marks have been made in the piston crown) or major (the piston crown is damaged beyond repair). Whichever type of damage has occurred, it is important to carry out a proper visual inspection to make sure that other components have not been affected or damaged. Overheating of the piston can occur due to poor cooling, incorrect combustion or lack of lubrication. The most common cause of overheating is lack of lubrication and insufficient oil levels. Always check pressure and condition if piston damage has occurred.
Bent connecting rods 	The connecting rod joining the piston to the crankshaft is an extremely strong component. It is possible to bend this component if the engine has **hydro-locked**, caused by liquid entering the engine. Reasons why liquid may have entered the engine include: • severely leaking head gaskets • excessive over fuelling • water entering through the air intake. Water entering from the air intake is the most common reason, caused by driving through deep puddles, streams or floods. If water has entered through the air intake system, make sure you check the condition of the air filter once you have carried out the repair.

CONTINUED ▶

3 Diagnosis & rectification of light vehicle engine faults

Symptom	Possible fault
Engine block structural failure	This is where the **integrity** of the engine cylinder block has been compromised, normally from the inside out. Mechanical component failure may lead to physical block damage, cracks or holes, which will normally require the complete replacement of **short engine** components. Damage can also occur if the liquid cooling system is allowed to freeze. Internal pressures created as ice crystals form can produce enough force to crack the cylinder block. It is important to fully pressure test the cooling system following any repairs due to freezing.
Broken or seized piston rings	If an engine has overheated due to excessive friction, it is possible for the piston rings to be damaged. This damage may be individual compression of oil control rings, or it can affect all rings at once. The piston rings may get stuck in the ring grooves – if this happens, they need to be replaced. If you are fitting new rings to an old piston, you can use part of an old broken piston ring as a scraper to help clean out the grooves before fitting the new rings.
Crankshaft damage	Crankshaft damage can occur for a number of reasons: • Wear and tear caused by the operation of a cylinder power stroke can cause **ovality** or **barrelling** of the crank pins or main **journals**. This must be carefully measured and assessed during any repairs or overhaul. • Vibrations created by an imbalance of piston operation can cause the crankshaft to fracture, particularly if a weak point already exists due to manufacturing processes. • Excessive big end or main bearing wear can occur due to insufficient lubrication. This can cause scoring of the bearing surfaces. You should carefully assess the crankshaft if you are carrying out any repairs or overhaul.
Woodruff key failure	Many engine pulleys are attached to cam or crankshafts with bolts, and are located with woodruff keys to prevent rotational movement relative to the shaft. If the keys or keyways wear or break, the pulleys can spin on the shaft, causing excessive wear or valve timing issues. Always take care when assembling an engine following repairs to make sure that woodruff keys are correctly fitted.
Cam lobe/ follower wear	During the manufacturing process, many camshafts and followers are **case hardened**. Over time, it is possible for this case hardening to become worn away. Once the surface has gone, the metal underneath is much softer and wear is accelerated. Symptoms include tapping noises from the camshaft area and the need for continuous maintenance of the engine valve clearances. When you are overhauling engine components, it is very important to assess cam profiles and followers to ensure ongoing correct operation.
Cylinder head cracks	Over a period, the constant heating and cooling of the cylinder head in the combustion chamber area may lead to cracks in the material between the inlet and exhaust valves. If these cracks continue to grow, compression may be lost or leaks can allow combustion processes to enter the cooling system. If you remove a cylinder head, you must carefully check for cracks or perforation before reassembly.
Cam belt failure	The manufacturing processes of modern camshaft drive belts are so efficient that failure is normally caused by external influences. Reasons for cam belt failure include: • incorrect tension (too loose or too tight) • poor maintenance (exceeding recommended replacement schedules) • incorrect fitment (not following the manufacturer's recommended installation procedures) • contamination (oil, coolant or dirt allowed to come into contact with the belt) • pulley bearing failure (creating friction and heat) • fouling of engine components (items such as plastic covers rubbing against the belt) • worn sprockets (as crankshaft sprockets wear, their diameters are reduced, allowing the belt to drive through a narrower angle than designed).

CONTINUED ▶

Level 3 Light Vehicle Technology

Symptom	Possible fault
Bent valves	A sudden failure of the valve drive mechanism will normally lead to the collision of pistons and valves. Most valves will be damaged beyond repair and must be replaced to ensure the correct sealing of the combustion chamber. You should also examine the pistons for damage and replace them if necessary.
Valve return spring failure	The continuous opening and closing of inlet and exhaust valves can lead to fatigue of the valve return springs, eventually leading to fracture. If this happens, the valve may drop into the combustion chamber causing damage similar to that which occurs if the cam belt snaps.
Oil pump failure	Oil pump failure will lead to a sudden loss of oil pressure. Friction between engine mechanical components will create heat, leading to rapid seizure. If damage to an engine has occurred due to loss of oil pressure, you must examine the oil pump for wear following the manufacturer's instructions.

Key terms

Integrity – describes a system that is complete and unbroken.

Short engine – a fully reconditioned engine but without external parts such as head, sump, oil or fuel pump, etc.

Ovality – wear in a circular engine component that has caused it to become oval.

Barrelling – wear on a crankshaft journal that has caused it to become tapered towards the outer ends like a barrel.

Journal – part of a shaft which is supported by a bearing.

Case hardened – a metal that has been heat treated to make the surface hard and resistant to wear.

Engine restoration and repair

When undertaking engine restoration and repair, you must use a comprehensive diagnostic procedure that will lead you to the original fault. It is important to carry out as much of your diagnostic routine as possible before you start to strip the engine down. You should treat an engine failure like a crime scene: do as much of your investigation as possible without disturbing the 'evidence'. To do this, you will need to make use of various diagnostic tools and equipment.

Coolant pressure testing

The procedure for using an engine coolant pressure tester is described in *Level 2 Diploma in Light Vehicle Maintenance & Repair Candidate Handbook* (Chapter 3, page 248). You can diagnose various faults using an engine coolant pressure tester. You can find external coolant leaks quickly, but an internal engine coolant leak may require further investigation.

- If you suspect that coolant is leaking into an engine cylinder, remove the spark plugs or glow plugs and place the cooling system under pressure. If you then leave the engine for a period, especially overnight, with the cooling system pressure tester still attached, you can assess any leaks into the cylinder.
- If you suspect a cylinder head gasket failure, attach the cooling system pressure tester to the engine while the cooling system is cold and then start the engine. A rapid rise in pressure may indicate cylinder head gasket failure.

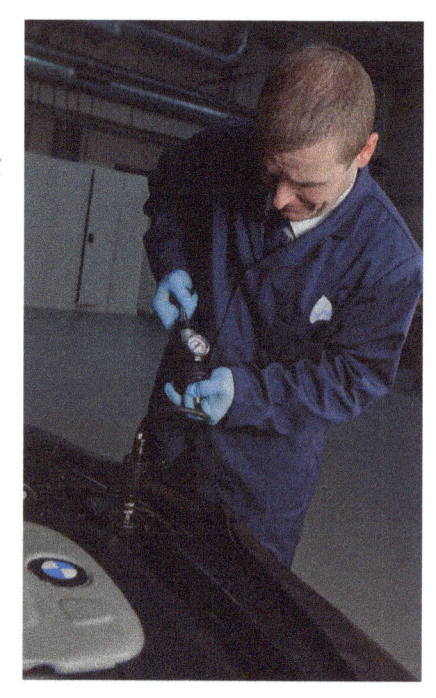

Figure 3.19 Cooling system pressure testing

3 Diagnosis & rectification of light vehicle engine faults

Oil pressure testing

The procedure for using an engine oil pressure tester is described in *Level 2 Diploma in Light Vehicle Maintenance & Repair Candidate Handbook* (Chapter 3, page 246).

If you find that the engine oil pressure is too high, it is possible that a blockage in the feed side of the lubrication system exists. Alternatively, the oil pressure relief valve is stuck in the closed position.

If you find that the oil pressure is too low, check for lubrication system faults such as:

- low quantities of engine oil
- worn or damaged oil pump
- a blocked strainer
- oil pressure relief valve stuck open.

Check for mechanical problems such as:

- worn crankshaft or main bearing journals.

Figure 3.20 Oil pressure testing

Cylinder balance testing

A cylinder balance check involves disabling engine cylinders individually and seeing how they affect the performance of the engine. Use an engine **tachometer** to assess the effect that disabling the cylinder has on engine idle. A dedicated piece of diagnostic equipment is needed to shut off individual cylinders without causing overfuelling or ignition problems which can damage the catalytic converter.

- If you see a large difference in the idle speed when the cylinder is disabled, this is usually an indication that the cylinder is working well.
- If you see a small difference or no difference in the idle speed when the cylinder is disabled, this is usually an indication that the cylinder is at fault.

Compression testing

You should always conduct an engine compression test if you suspect that a pressure leakage might have occurred within a cylinder.

> **Did you know?**
>
> An infrared laser thermometer can sometimes be used to help identify the cylinder that is not running very well. If you move the infrared thermometer slowly along the exhaust manifold, you can sometimes see a misfiring cylinder, as it will run colder than the rest.

Figure 3.21 Laser thermometer

Checklist			
PPE	VPE	Tools and equipment	Source information
• Steel toe-capped boots • Overalls • Latex gloves	• Wing covers • Steering wheel covers • Seat covers • Foot mat covers	• Compression gauge • Spark plug socket and ratchet • Torque wrench • Oil can	• Compression technical data • Spark plug torque setting information • Job card

1. Isolate the engine so that it cannot start.

2. Gain access to and remove all of the spark plugs.

3. You can now attach the compression tester to the engine, in place of the plugs.

4. Once the gauge is attached to a cylinder, crank the engine with the throttle held wide open. Continue to crank the engine for a short period, until the maximum reading is shown on the gauge. Record the reading and compare the figure with the manufacturer's specifications. Repeat the procedure for the remaining cylinders. (This is known as a **dry test**.)

5. Using an oil can, introduce a small amount of oil to the cylinder down the plug hole and repeat the compression test. Record the results and compare with the original readings. (This is known as a **wet test**.)

> **Key terms**
>
> **Tachometer** – a piece of diagnostic equipment that measures engine speed.
>
> **Wet test** – where oil is used during a compression test.
>
> **Dry test** – where no oil is used during a compression test.

During a wet compression test, the oil can often sit on top of the piston crown and form a temporary compression seal around the top piston ring.

- If the wet test shows an increase in compression pressure over a dry test, this can indicate that the leakage is occurring down past the piston and rings.
- If the compression pressure stays low, this can indicate that the leakage may be coming from the cylinder head or valve area.

Leakdown testing

A cylinder leakdown tester is a tool that you can use to help diagnose compression leakage within a cylinder. It consists of a pressure gauge, regulator and adapter which is screwed into the spark plug or injector hole in the engine.

3 Diagnosis & rectification of light vehicle engine faults

Checklist			
PPE	**VPE**	**Tools and equipment**	**Source information**
• Steel toe-capped boots • Overalls • Latex gloves	• Wing covers • Steering wheel covers • Seat covers • Foot mat covers	• Cylinder leakdown tester • Spark plug socket and ratchet • Sockets or spanners for injector removal • Torque wrench	• Engine technical data • Spark plug torque setting information • Job card

1. Strip out and remove the spark plugs or glow plugs.

2. Rotate the engine until the cylinder to be tested is on its compression stroke and the inlet and exhaust valve are both closed.

3. Lock the crankshaft to prevent it turning. You can do this physically or by placing the car in gear and applying the handbrake.

4. Screw the leakdown tester into a spark plug hole and connect the cylinder leakdown tester to a source of compressed air.

5. Introduce a regulated pressure into the cylinder.

6. The gauge can indicate if a significant difference in pressure exists in the cylinder when compared with the pressure introduced by the compressed air. If there is a pressure difference, then cylinder leakage may have occurred.

To help diagnose the location of the cylinder leakage you can:

- Listen for air leaking from the exhaust pipe: this is an indication that the exhaust valve is leaking.
- Listen for air leaking from the intake manifold (remove the air filter housing): this is an indication that the inlet valve is leaking.
- Listen for air leaking from the dipstick tube or the cam/rocker cover: this is an indication that pressure is leaking down past the piston.
- Remove the cooling system pressure cap, and check for bubbles in the engine coolant: this is an indication that pressure is leaking past the head gasket or cylinder head.

Smoke

A smoke generator is a tool that you can use to help diagnose engine system leakage. A chemical or oil is heated in the smoke generator, which then produces pressurised smoke from a pipe that can be connected to various engine systems.

- To check for an inlet system leak, remove a vacuum pipe from the inlet manifold and connect the smoke generator. Then block the air inlet using special bungs. Once the smoke has been introduced to the inlet system, you can then look for any signs of leakage that may cause engine running problems.
- To check for an exhaust system leak, place the pipe from the smoke generator in the exhaust tailpipe and seal it. Once the smoke has been introduced to the exhaust system, you can then look for any signs of leakage that may result in an engine emissions fault.

Cylinder block tester

A cylinder block tester is a tool which you can use to help diagnose compression leakage past the cylinder head gasket into the cooling system. You will need to fill a container with a special chemical and place it in the neck of the cooling system radiator or expansion tank. Once the engine has reached its normal operating temperature, use a hand pump to draw the fumes given off in the cooling system through the liquid. If the chemical liquid changes colour, it indicates the existence of hydrocarbons from exhaust gases. These hydrocarbons will normally have leaked from the combustion process inside an engine cylinder past the head gasket.

> **Did you know?**
>
> You can use an exhaust gas analyser to conduct a similar test to a cylinder block tester. With the engine at normal operating temperature, carefully remove the cooling system pressure cap and hold the probe from the exhaust gas analyser above the coolant to sample the fumes. You can then assess the amount of hydrocarbons, measured in parts per million, from the gas analyser screen. (Be careful not to dip the exhaust probe of the gas analyser into the coolant, as the liquid can be sucked up and damage the machine.)

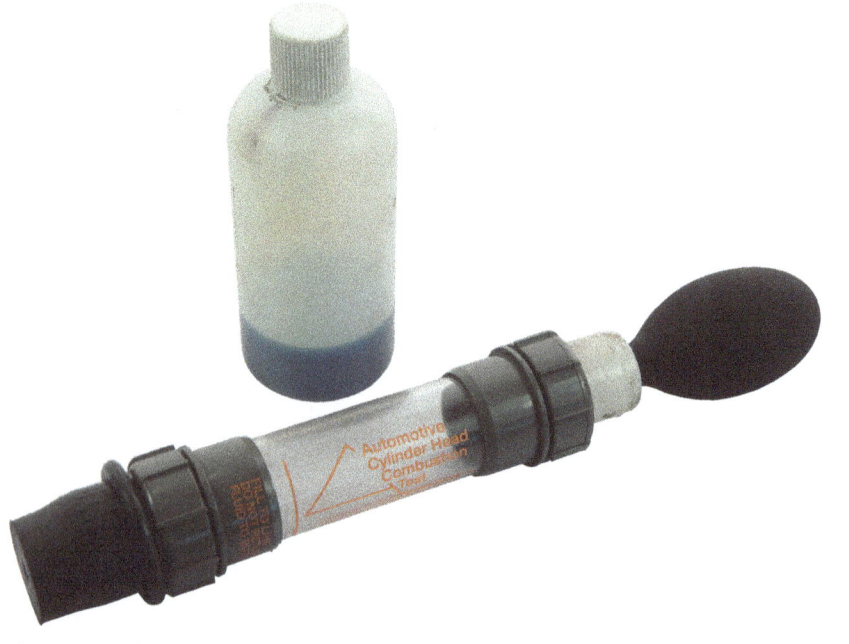

Figure 3.22 Block tester

3 Diagnosis & rectification of light vehicle engine faults

Endoscopy

Endoscopy normally involves inserting a small camera inside an engine to visually inspect for damage before the engine is stripped down. For example, you can put an illuminated probe with the camera attached down a spark plug hole, and use the image shown on the computer screen to check for physical engine damage.

Figure 3.23 Endoscope

Case study

Mr Byng has his car recovered to your garage on the back of a breakdown lorry. When he started the engine this morning, there was a loud tapping noise from inside the engine and it was misfiring. Your boss asks you to check it out.

Here's what you do:

- ✓ Listen to the customer's description of the fault.
- ✓ Question the customer carefully to find out the symptoms.
- ✓ Carry out a visual inspection, check fluid levels, etc.
- ✓ Get the customer's permission to start the engine briefly to confirm the symptoms described.
- ✓ Gather information from technical manuals (including engine data) before you start and take it to the vehicle.
- ✓ Devise a diagnostic strategy.
- ✓ Check the quick things first (10-minute rule) and conduct a visual inspection of engine components.
- ✓ Conduct as much diagnosis as possible without stripping down.
- ✓ Use a cylinder balance test to check which cylinder is misfiring.
- ✓ Conduct a compression test and cylinder leakdown test.
- ✓ Use the information from the tests to diagnose the fault(s).
- ✓ Use an endoscope to confirm your diagnosis.
- ✓ Contact the customer and explain the results of the diagnostic tests.
- ✓ Give an estimate for the cost of repair and gain authorisation to carry out the work.
- ✓ Strip down the engine and confirm the root cause of the fault.
- ✓ Keep the customer informed of progress and costs.
- ✓ Overhaul, remanufacture, repair and replace faulty engine components.
- ✓ Correctly reassemble any dismantled components/systems.
- ✓ Check all engine fluid levels.
- ✓ Thoroughly test the system to ensure correct function and operation.
- ✓ Inform the customer of any special precautions or 'running in' procedures.

Level 3 Light Vehicle Technology

> **Safe working**
>
> It is vital that the engine crankshaft and camshaft timing are accurately set before any engine strip-down or rebuild is attempted. This will ensure that the internal components (for example, valves and pistons) are in the correct positions and will not cause damage if the engine is rotated or started. Always follow the manufacturer's recommendations.

> **Key term**
>
> **Up-stand** – the amount that a cylinder liner protrudes above the level of the block.

> **Action**
>
> Using resources available to you, research how to use Plastigauge and write a short description of the procedure.

Engine measurement

It is important during engine overhaul and repair procedures to correctly assess component measurements. You can do this using various types of technical measuring equipment.

Table 3.5 provides a list of technical measurement tooling that you could use when testing and evaluating light vehicle engine mechanical systems. Due to the nature of engine wear, you will experience different requirements during your repair and overhaul routines and so you should adapt the list for your particular situation.

Table 3.5 Measurement tools and examples of their use

Measurement tool	Possible use
External micrometer	Measuring crank pin or main bearing journal wear
Internal micrometer	Measuring cylinder bore diameter
Depth micrometer	Measuring piston/liner **up-stand** at TDC
Vernier gauge	Measuring piston diameter
Dial test indicator	Measuring cam lobe profile lift
Feeler gauges	Used in conjunction with a straight edge to measure for cylinder head bow
Bore gauges	Used to measure engine cylinder wear
Plastigauge	Used to measure crankshaft shell bearing clearance

When you are measuring engine components, it is important to take measurements in a number of different positions so that you can conduct an accurate assessment of engine wear.

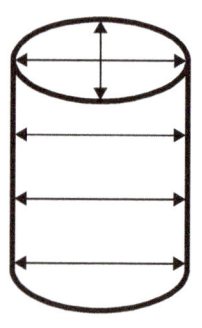

Figure 3.24 Measuring a cylinder in different positions

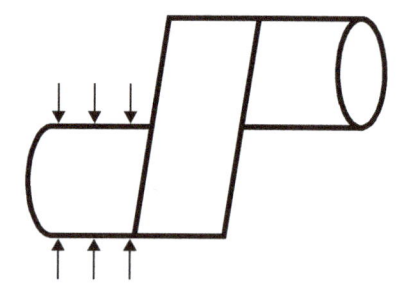

Figure 3.25 Measuring a crank pin in different positions

Restoration of engine components

Depending on the amount of wear or damage that has occurred to an engine, it is sometimes possible for an engineering workshop to machine components so that they are restored to a usable condition. Some examples of restoration are described in Table 3.6.

3 Diagnosis & rectification of light vehicle engine faults

Table 3.6 Examples of restoration

Restoration	Purpose
Cylinder head skimming	The cylinder head is machined flat after it has become warped, possibly due to overheating.
Cylinder block decking	The cylinder block is machined flat on its upper surface to restore its condition, possibly after corrosion.
Cylinder boring	The engine cylinder is machined oversize to restore it to an even diameter after excessive engine wear has created ovality. The machine shop will also need to supply a new set of oversize pistons and rings that are matched to the new size of the cylinder bores.
Cylinder honing	An accurate cutting tool is passed through the cylinder to restore the surface finish of the cylinder wall, which may have been damaged due to poor lubrication.
Cylinder glaze busting	An abrasive tool is rotated inside the cylinder bore to help remove the shiny glaze created as the piston moves up and down, and to restore the surface of the cylinder wall.
Crankshaft journal/crank pin grinding	The crank pins or journals can sometimes be ground down to restore their surface after damage caused by poor lubrication or wear. The machine shop will also need to supply a new set of oversize bearing shells that are matched to the new size of the crank pins or journals.
Cam lobe profiling	Depending on the materials that have been used during the manufacture of a camshaft, it is sometimes possible to grind the cam lobes and restore or change their profiles. This will normally have an effect on engine operation and performance, as changes to cam profile will alter power output, economy and emissions.
Valve seat re-cutting	If valve seats have become worn during operation, it is sometimes possible to re-cut their faces using a special grinder. The angle of the valve seat will vary according to the manufacturer, but is typically around 45°.

Skills for work

You have been asked to conduct your first full engine overhaul and rebuild. You are slightly unsure of the recommended procedures so you will have to seek the assistance and cooperation of some of your colleagues to successfully complete this task.

This situation requires that you use particular personal skills. Some examples of these skills are shown in Table 6.1 at the start of Chapter 6, on pages 304–305.

1. Using the examples given in Table 6.1, choose one skill from each of the following categories that you think you need to demonstrate in order to get the assistance and cooperation you require.
   - General employment skills
   - People skills
   - Self-reliance skills
   - Specialist skills
2. Now rank these skills in order of importance, starting with the one that it is most important for you to have in this situation.
3. Which of the skills chosen do you think you are good at?
4. Which of the skills chosen do you think you need to develop?
5. How can you develop these skills and what help might you need?

CHECK YOUR PROGRESS

1. List three engine mechanical faults.
2. What does the term ovality mean?
3. Name six tools that you could use to assess engine wear.

105

Level 3 Light Vehicle Technology

Did you know?

To ignite the air and fuel in a petrol engine, a very high voltage spark is required to initiate the burn.

With an engine speed of 6000 rpm the piston is travelling from **top dead centre (TDC)** to **bottom dead centre (BDC)** and back again 100 times a second. (6000 rpm divided by 60 seconds = 100; that's 200 strokes in one second!)

If a spark is required every fourth stroke, that's 50 sparks per second for each cylinder.

Key terms

Top dead centre (TDC) – when the piston is at the highest part of its stroke.

Bottom dead centre (BDC) – when the piston is at the lowest part of its stroke.

High tension (HT) – high voltage electricity.

Did you know?

Electricity and magnetism are very closely linked. If an electric current is passed through a copper conductor, such as a piece of wire, an invisible magnetic field is generated around that conductor. If the piece of copper wire is wound into a coil, then the magnetic field is intensified.

If a magnetic field is moved through a copper conductor, electric current is generated.

Ignition systems use magnetic flux to transform the 12V created by the battery into many thousands of volts that are required at the spark plug.

Ignition systems

The purpose of the ignition system is to initiate the combustion of the air/fuel mixture in the cylinder. In a petrol engine this is normally achieved by a high voltage spark created by a spark plug. An ignition system must take 12 volts from the battery and turn it into many thousands of volts, many times a second. A slight discrepancy in this process can lead to a misfire or the engine not running at all.

Ignition system components

An ignition system is made up of a number of components. These include:

- the battery
- the ignition switch
- the ignition coil
- a switching mechanism (such as an ECU)
- spark plugs.

Some older systems may include distributors and **high tension (HT)** leads, although many modern systems now use coil per cylinder, also known as coil on plug (COP).

Figure 3.26 shows a typical distributorless ignition system, which is described in full on pages 117–119.

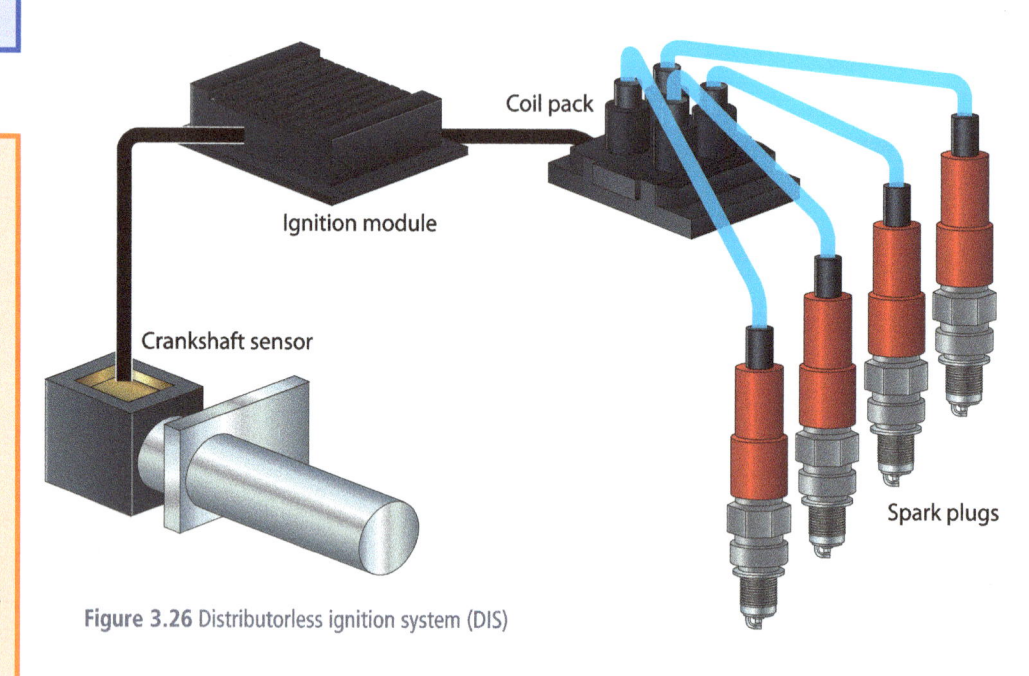

Figure 3.26 Distributorless ignition system (DIS)

3 Diagnosis & rectification of light vehicle engine faults

Diagnosing ignition system faults

When diagnosing ignition system faults, it is important to check for mechanical issues first before assuming that the fault lies with the electrics or electronics. Always check:

- ✓ compressions
- ✓ air/fuel mixture
- ✓ induction air leaks
- ✓ valve clearances
- ✓ valve timing
- ✓ spark plug condition
- ✓ spark plug gap.

Spark plugs

The purpose of the spark plugs in petrol engines is to ignite the fuel and start the rapid burn at precisely the correct time. The gap at the tip of the spark plug creates an open circuit, and no electric current can flow until a high enough voltage is produced to overcome the resistance of the air gap. Two electrodes are used to do this:

- a centre electrode, sometimes fed from a high tension ignition lead
- an earth electrode, normally manufactured as part of the spark plug shell.

The tips of these electrodes are machined to precise shapes.

A high tension spark finds it easier to jump from a sharp point on an electrode. The thinner and sharper an electrode can be made, the better it will function. Unfortunately, due to high temperatures and **spark erosion**, narrow pointy electrodes do not last very long. As a result, the electrodes found on a standard spark plug are machined flat with sharp edges, allowing sparks to be created at a number of points. Over time, the sharp edges will begin to erode and round off. This reduces the overall performance of the spark plugs.

To overcome this issue, a number of precious and rare metals, such as platinum and iridium, are used in the production of spark plugs. This means they are less prone to spark erosion and can be manufactured with smaller sharper electrodes.

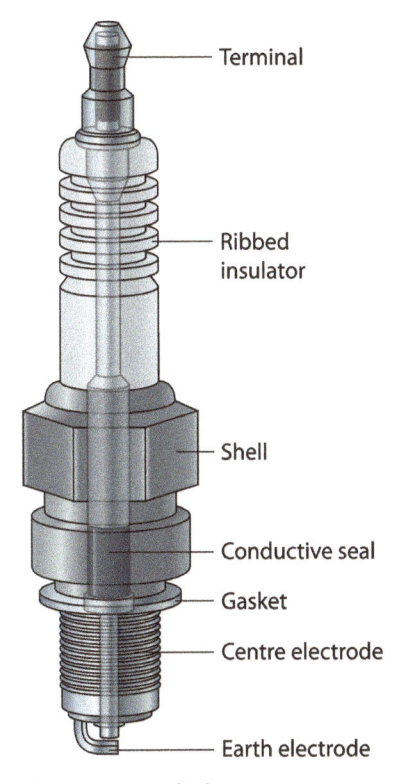

Figure 3.27 Spark plug

> **Key term**
>
> **Spark erosion** – the burning away of material caused by an electric spark.

Spark plug gap

The size of the air gap between the electrodes at the tip of the spark plugs is accurately calculated by the manufacturer during the engine design process. Each type of engine requires a spark plug with a particular size gap to produce the correct quality and timing of sparks used to ignite the petrol inside the combustion chamber.

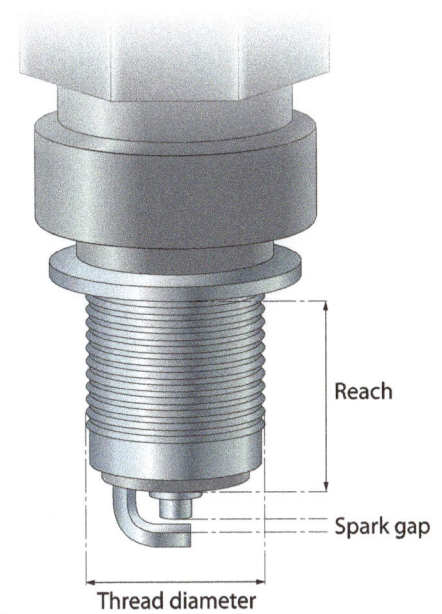

Figure 3.28 Spark plug gap

107

Level 3 Light Vehicle Technology

Did you know?

New spark plugs do not come ready-gapped from the manufacturer. As a result, they require setting before they are fitted to a particular engine.

When setting the gap on spark plug electrodes, always follow the manufacturer's recommendations and use feeler gauges to accurately measure the gap available.

Wire gauges are available to set spark plug gaps which have multiple earth electrodes.

Safe working

It is important to use spark plugs of the correct heat range during the maintenance of petrol engine systems. If you use the wrong ones, this can produce engine mechanical failure.

If spark plug gaps are incorrectly set, this may lead to poor running and even misfiring.

- Spark plug gaps that are larger than recommended by the manufacturer need higher firing voltages.
- Spark plug gaps that are smaller than recommended by the manufacturer need lower firing voltages.

If a multicylinder engine has spark plugs fitted with varying gap sizes, then each individual cylinder will receive a firing voltage of different proportions and timings, leading to uneven running.

Heat range

During normal operation, spark plugs are designed to operate within a certain heat range. An ideal range is between 500 and 850°C.

During the design process of new engines, manufacturers test different spark plug heat ranges to ensure that the correct spark plug is selected for the vehicle. This information is included in the vehicle's technical data.

- If the temperature is too low, carbon and combustion chamber deposits may not be burned off.
- If the temperature is too high, electrode and/or piston damage may occur (components could melt).

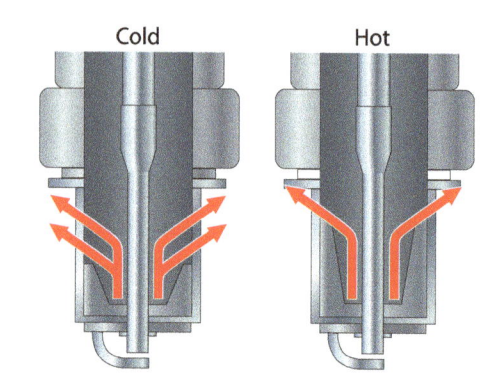

Figure 3.29 Hot and cold spark plugs

Other types of spark plug

Spark plug manufacturers are producing new designs to enable better ignition of the fuel and reduce the possibility of a misfire. Multiple earth electrodes are now commonplace. Although only one electrode is used at any one time, multiple earth paths mean that, as the electrodes erode due to spark ignition, alternative paths to ground may be found and the likelihood of a misfire is reduced.

Modern capacitor discharge ignition systems (CDI) may use surface discharge spark plugs. These have no earth electrode, but discharge directly to the shell of the spark plug casing. These types of spark plugs will produce a good spark even when fouled.

Ignition coil

The ignition coil is a step-up transformer – it takes battery voltage and turns it into the many thousands of volts required to produce a spark at the plug.

An ignition coil contains two sets of copper windings: a primary and a secondary winding, as shown in Figure 3.30.

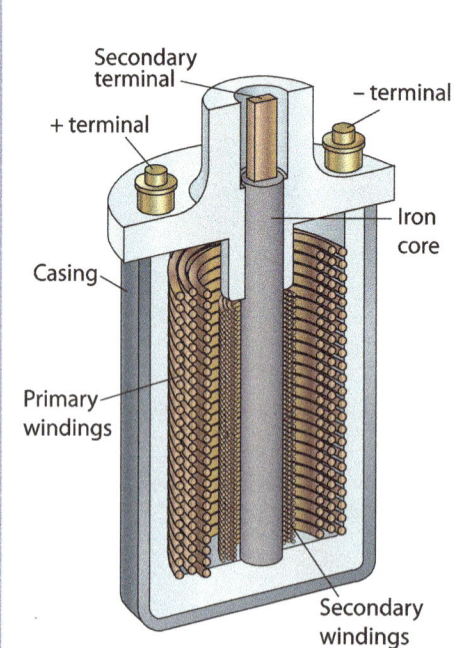

Figure 3.30 Ignition coil

3 Diagnosis & rectification of light vehicle engine faults

- The primary winding is connected to the low-tension (12V) circuit of the vehicle. It contains large copper wire wound loosely with several hundred turns.
- The secondary coil contains a very thin copper winding with many thousands of turns.
- A soft iron core is inserted between these two windings, as this helps increase the magnetic field produced when electric current flows.

When the low-tension (12V) circuit is switched on, current flows through the primary winding generating a magnetic field, which also covers the secondary winding and the soft iron core. When current is flowing through the primary circuit, the magnetic field is building/charging. This is called the **dwell period**.

When the current flow is switched off to the primary winding, a rapid collapse of the magnetic field occurs. This collapse of the magnetic field, as it passes through the secondary winding, induces a high tension voltage of many thousands of volts.

The voltage in the secondary winding builds until the pressure is great enough to bridge the air gap at the plug and a spark is produced.

Primary circuit testing (LED test lamp)

You can use an LED test lamp to check the switching of the primary coil circuit.

- Connect the clamp of the LED test light to a power source (i.e. battery) and probe to the negative coil wire.
- Crank the engine and see if the LED flashes.
- If not, the primary circuit is not being switched.

If the primary circuit is not switched, check the trigger signal with an oscilloscope to see if a wave form is being produced.

Primary circuit testing (oscilloscope)

You can check the correct function of the primary circuit using an oscilloscope.

- To connect an oscilloscope to the primary circuit, you need to select a single channel and connect the test probes to the appropriate sockets.
- You can now connect the black (ground) wire to a suitable earth and the coloured (signal) wire to the primary (negative) terminal.
- Now crank or start the engine.
- If you are unsure what voltage or timescale settings to use, start high and work down until the image is clearly produced on the screen.

The image should look something like the one shown in Figure 3.31.

What to look for:

- From the left-hand voltage axis, you should see a 12V trace.
- As the ignition coil begins its charge time (dwell period), the voltage will fall to zero. It will continue to charge until the primary circuit is broken.

> **Key term**
> **Dwell period** – the charge time of the ignition coil, when current is flowing in the primary circuit.

> **Safe working**
> Do not use a standard test lamp, as this will draw a high current and can damage electronic components.

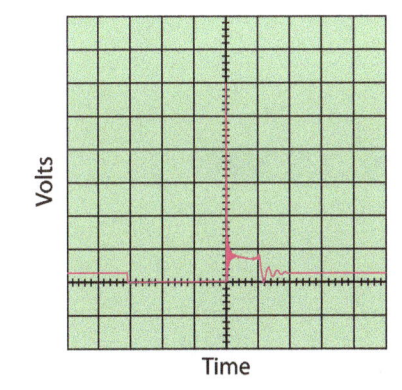

Figure 3.31 Primary wave form

> **Did you know?**
> If a trigger function is available on your oscilloscope, you can set it to hold the wave form in the centre of the screen.

109

- As the primary circuit is broken and the magnetic field collapses through the primary and secondary circuit, a high voltage discharge is produced. This is known as a back EMF.
- This high voltage discharge will be shown as a sudden spike on the wave form.
- After this initial spike, voltage will fall rapidly until it reaches a level where it can maintain a spark across the spark plug. This is known as the spark duration.
- At the end of the spark duration, you will see a small oscillation, showing that the back EMF is charging and discharging through the coils. This is similar to the bounce energy produced in a spring once it has been compressed and released.
- The wave form will now return to 12V before the primary circuit is once again connected and the dwell period is repeated.

Electronic ignition systems

The advantages of using an electronic system to switch the primary circuit of the ignition coil are:

- No mechanical wear is produced at the switching mechanism.
- Reaction time is far quicker than a conventional ignition system that uses **contact breakers**.
- Greater accuracy is available in the ignition timing.
- The system is more reliable.

Trigger signals

Many modern ignition systems use a low voltage signal, triggered by engine position, to switch the base of a **transistor** unit. This allows the large current to flow through the primary circuit of the ignition coil.

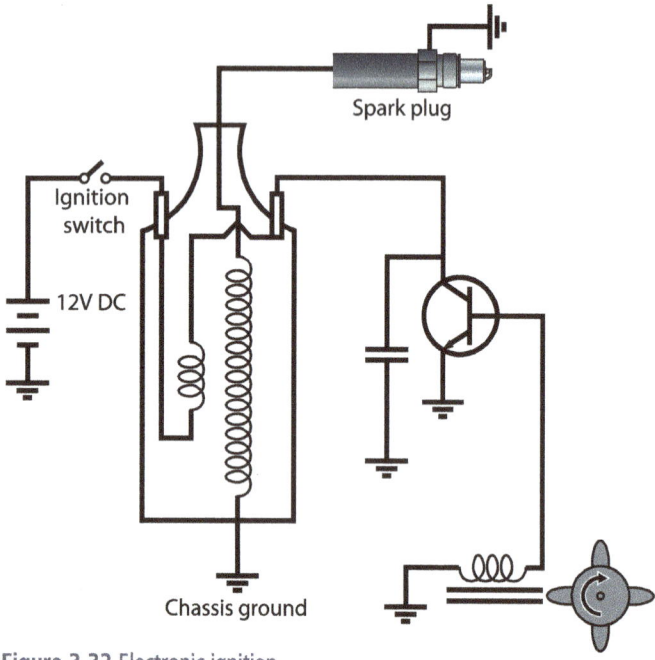

Figure 3.32 Electronic ignition

Inductive sensors

An inductive sensor is a component that generates a small electric current when its internal magnetic field is disrupted. This small electric current can be used to create signals that show engine position.

A **reluctor** ring is a rotating shaft (normally turning at half crankshaft speed) with a number of protrusions or fingers corresponding to the number of cylinders in an engine.

Mounted closely to this reluctor ring is the pickup – a permanent magnet with a small coil of copper wire wound around it.

As the reluctor ring rotates, and one of the fingers moves towards the pickup, the magnetic field produced by the permanent magnet is disrupted. When this magnetic field moves across the coil winding of the pickup, it creates a small voltage. As the reluctor finger comes level with

> **Key terms**
>
> **Contact breakers** – a mechanical switch used on early ignition systems to control current flow in the primary circuit. Contact breakers are also sometimes called 'points'.
>
> **Transistor** – an electronic component made from semi-conductive material. It can act like a switch with no moving parts.
>
> **Reluctor** – a small toothed wheel used in conjunction with an inductive pickup.

the pickup, disruption to the magnetic field falls and voltage within the coil of wire also falls to zero. When the reluctor finger moves away from the pickup, the magnetic field is once again disrupted, but this time in the opposite direction so a negative voltage is produced. This creates a small alternating current within the pickup.

The small alternating current is not sufficient to switch the ignition coil primary circuit on and off. Instead, the small signal is used to switch the base of the transistor inside the ignition amplifier unit. This then switches the primary circuit of the ignition coil, allowing a dwell or charge period and then a secondary high tension discharge to create the spark.

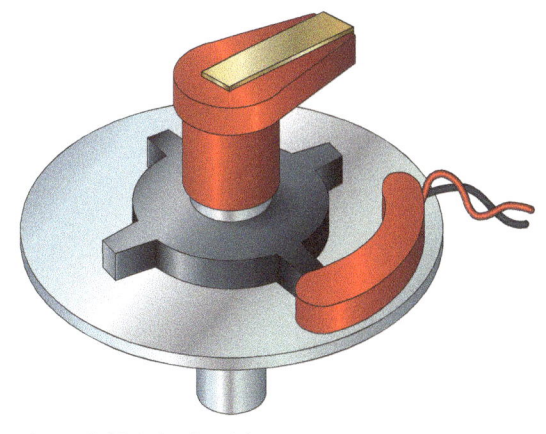

Figure 3.33 Inductive pickup

Testing an inductive sensor

A digital multimeter is too slow to give an accurate indication of correct operation when testing an inductive sensor. The display screen will not refresh quickly enough to show the alternating current. It is far better to use an oscilloscope.

- When using an oscilloscope to test an inductive sensor, select a single channel and connect the test probes to the correct sockets.
- As an inductive sensor creates its own voltage, you should connect the test probes to the two output wires from the pickup.
- With the engine being cranked or running, you should see an alternating wave form on the screen.
- If you are unsure of what voltage or timescale to use, start with a high setting and work down until a clear image is displayed on the screen of the oscilloscope.

Hall effect sensors

The Hall effect sensor is a small unit that produces a magnetic field when supplied with electric current. This magnetic field is sensed by a small **integrated circuit** and a signal is produced. A rotor drum with slots and vanes (doors and windows) rotates through the middle of the Hall effect sensor, interrupting the magnetic field. The integrated circuit is connected to a signal wire which produces an on and off pulse. This pulsing signal is used to switch the base of a transistor in the ignition amplifier in a similar way to an inductive sensor. The ignition amplifier is then able to switch the primary circuit of the ignition coil, creating a dwell period or charge time and then a high tension spark.

The number of slots in the rotor drum corresponds to the number of engine cylinders and therefore the number of sparks required.

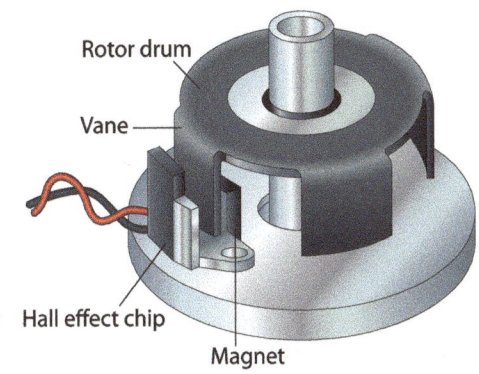

Figure 3.35 Hall effect sensor

> **Did you know?**
> A permanent magnet inductive sensor is described as 'active', as it will produce its own voltage in a similar way to a generator.

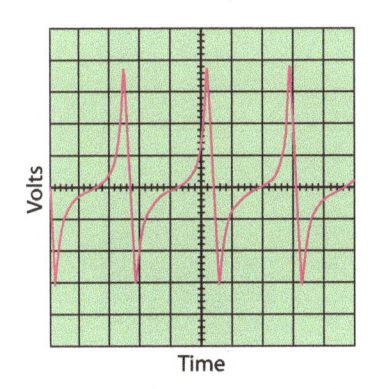

Figure 3.34 Inductive sensor wave form

> **Did you know?**
> It is sometimes possible for an engine management ECU to detect cylinder misfires by comparing the frequency of signals produced at cam and crankshaft sensors. As a cylinder misfire occurs, the speed of the crankshaft will slow slightly and then speed up again as the next cylinder begins its power stroke. This speed difference can be used by the ECU to recognise misfires and record a diagnostic trouble code (DTC).

111

Level 3 Light Vehicle Technology

Did you know?
The Hall effect sensor is an example of a passive sensor. It does not generate its own electrical current; it relies on an electric current normally supplied by the vehicle's ECU.

Did you know?
A transistor is a small semi-conductor, which is an electronic component that can operate like a switch. It has three wires to connect it to an electric circuit:
- a collector
- an emitter
- a base.

Electric current enters the transistor at the collector terminal but is blocked from leaving at the emitter until a small electrical current is supplied to the base. Because of this, a small electric current is able to switch a larger electric current and also act as an amplifier.

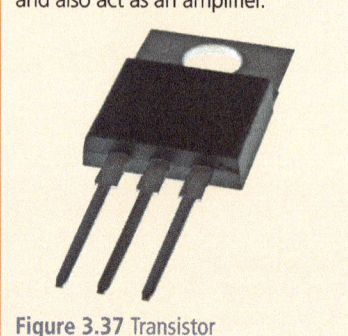

Figure 3.37 Transistor

Key terms
Integrated circuit – a small self-contained electronic circuit on the surface of a microchip.

Amplifier – an electronic device that increases the signals passing through it.

Ignition module – another name for an ignition amplifier.

Testing a Hall effect sensor
- When using an oscilloscope to test the Hall effect sensor, select a single channel and connect the test probes to the correct sockets.
- As a Hall effect sensor is a passive unit, it relies on the ECU to send it current. It will normally have three wires:
  - a positive feed
  - a negative earth
  - a signal wire.
- You need to connect the black probe of the oscilloscope to a suitable earth or ground. You can then use the coloured wire of the oscilloscope to test the correct function of the Hall effect sensor.
- It doesn't matter if you do not know which wire to probe.
- With the engine being cranked or running, one wire will produce a flat line at 0V, another will produce a flat line at 5V, and the third should produce an on and off square wave form (digital signal).

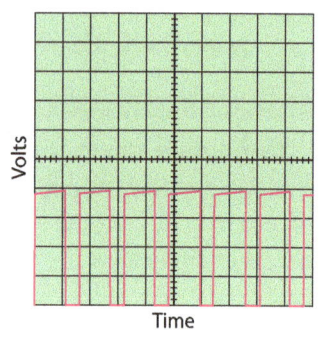

Figure 3.36 Hall effect wave form

Ignition amplifier
The small signals produced at the trigger sensors are not enough to switch the primary circuit of the ignition coil. To increase the signals, an ignition **amplifier** or **ignition module** is used.

The ignition amplifier effectively takes the place of the mechanical contact breakers that are found in a conventional ignition system. It contains transistors, which are electronic switches with no moving parts. When connected to an ignition circuit, the base of the transistor is normally joined to the trigger sensor, and the collector and emitter are joined to the ignition coil primary circuit. In this way, when a small trigger signal is produced, the primary circuit is switched on and off.

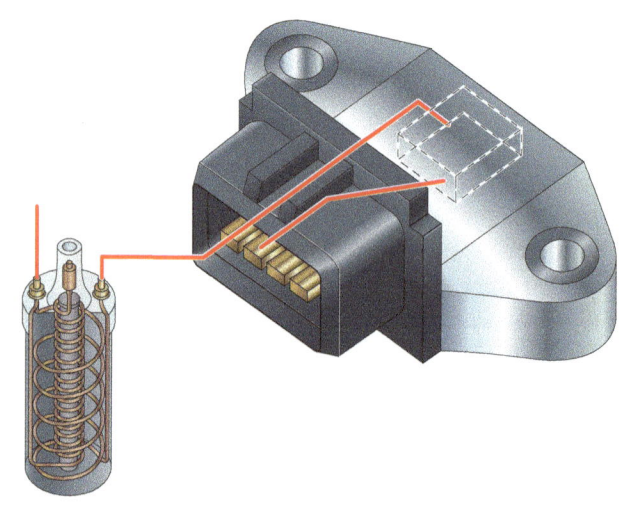

Figure 3.38 Ignition amplifier

3 Diagnosis & rectification of light vehicle engine faults

Safe working

Many ignition amplifiers are sealed in the factory during production, and you should never attempt to open these up. During the manufacturing process, dangerous chemicals such as hydrofluoric acid are used, which have the potential to cause serious harm.

Exposure to hydrofluoric acid can have harmful effects on health that may not be immediately apparent:

- **Inhalation:** It is severely corrosive to the respiratory tract and may cause sore throat, coughing, laboured breathing and lung congestion/inflammation.
- **Ingestion:** It may cause sore throat, abdominal pain, diarrhoea, vomiting, severe burns of the digestive tract and kidney dysfunction.
- **Skin contact:** This causes serious skin burns which may not be immediately apparent or painful. Symptoms may take eight hours or longer to be displayed. The fluoride ion readily penetrates the skin, causing destruction of deep tissue layers and even bone.
- **Eye contact:** Symptoms of redness, pain, blurred vision and permanent eye damage may occur.

Did you know?

Many ignition amplifiers are now incorporated inside the engine management ECU.

Action

Using a multimeter, check the primary coil resistance of four different vehicles in your workshop.

- Were they all the same?
- Which vehicle has the lowest impedance ignition coil?

Variable dwell angle

An advantage of electronic ignition is the opportunity to produce variable dwell periods.

With mechanical contact breakers, the dwell period of the ignition coil is set by the manufacturer's prescribed points gap. Once the engine is running, dwell variation will be minimal. This means that when designing a system, the manufacturer will set an average dwell angle so that, overall, the ignition coil performance is also average.

In a standard ignition system:

- If the dwell period is too long, at slow engine rpm, overheating of the ignition coil can occur. This may lead to eventual spark loss. It will produce a good spark at high engine rpm.
- If the dwell period is too short, a good spark may be produced at low engine rpm. But as the revs are increased, the spark will become weaker and weaker, eventually leading to a misfire.

The ignition amplifier can limit the current passing through the primary circuit of the ignition coil. This means that a low **impedance** (low resistance) primary winding may be used. This gives a short charge time of the ignition coil, but because current is limited by the ignition module, short or long dwell periods are no longer a problem. Having this ability allows a good-quality spark to be produced at any engine speed. It is commonly known as a constant energy ignition system.

Did you know?

On some constant energy ignition system wave forms, a small bump may occur during the dwell period. This shows current limiting. If engine revs are increased, it will normally move to the right and may disappear altogether.

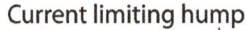

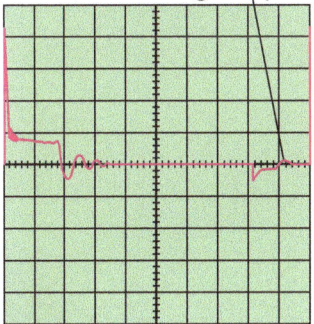

Figure 3.39 Constant energy current limiting

Level 3 Light Vehicle Technology

Key terms

Impedance – a form of electrical resistance.

Insulation – the material placed around electrical wiring to prevent the conduction of electricity.

Tracking – the short circuiting of high voltage ignition electricity to earth.

Did you know?

Producing a spark under normal atmospheric conditions is not very difficult. If a spark plug is removed, held to earth and the engine cranked, a spark can often be seen. However, when it is placed back into the engine, the harsh environment of the combustion chamber means that the spark may fail and a misfire might occur. Where possible, it is always best to test the spark plugs and high tension circuit when fitted to an engine under normal operating conditions.

Computer-controlled electronic ignition

The entire process that operates the primary circuit of the ignition coil may be contained within an engine management control unit. The switching mechanism for the primary circuit is normally still transistorised and engine position is calculated from crankshaft or camshaft sensors.

The advantage of putting ignition control into the hands of the engine management ECU is that the ECU can also take into account other sensor information, such as:

- engine temperature
- engine speed
- engine load
- knock sensing.

With ECU control, the timing of the spark can be adjusted to take into account all engine running conditions.

Secondary circuit

The secondary circuit is the part of the ignition system that contains very high voltage. It is often referred to as the HT (high tension) circuit.

Voltage requirements

The amount of voltage needed for the spark to jump the gap of the spark plug inside the cylinder of a running engine varies. A number of factors impact on this, including:

- cylinder compression
- air/fuel mixture strength.

Voltages in excess of 10,000 volts are normally required to overcome resistances within the high tension circuit, including the air gap at the plug. As a result, high levels of **insulation** are needed to stop this electrical voltage **tracking** directly to earth and avoiding the difficult route across the spark plug gap. The high tension ignition circuit should be kept very clean, as dirt will make an easy path to earth.

These high levels of installation can make it difficult to test secondary circuits with normal equipment. Specialist diagnostic equipment that uses inductive sensing of the magnetic fields produced by the secondary high tension voltage are available. Most of these give an indication of correct operation by flashing an LED (light emitting diode).

Figure 3.40 Inductive high tension tester

3 Diagnosis & rectification of light vehicle engine faults

Using an oscilloscope to test the secondary circuit on an ignition system

- Most oscilloscope manufacturers produce specialist wires for secondary ignition testing.
- These may take the form of clamps that go around the HT leads or inductive pickups that can be held against the HT circuit to produce a wave form.
- Always follow the manufacturer's recommendations when you connect the specialist leads to the oscilloscope.
- With these leads connected, crank or preferably run the engine and select a high voltage scale for a short time period.

Most oscilloscopes give the option of displaying a single cylinder or multiple cylinders on the screen at any one time. These options are described as:

- Single
- Parade (one after another)

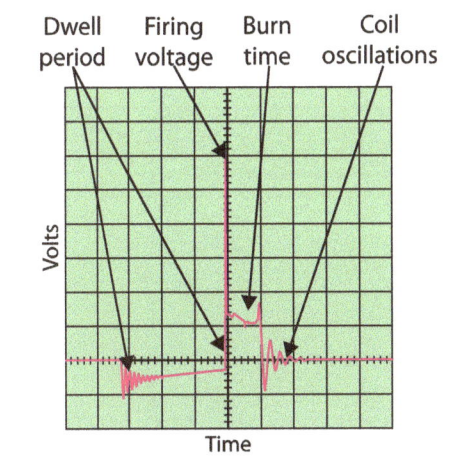

Figure 3.41 Secondary wave form

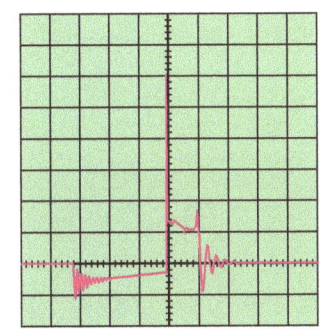

Figure 3.42 Single cylinder

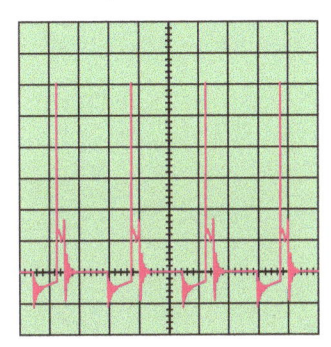

Figure 3.43 Parade

- Superimposed (laid over the top of each other)
- Raster (displayed one above another)

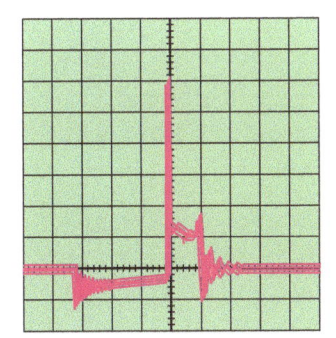

Figure 3.44 Superimposed

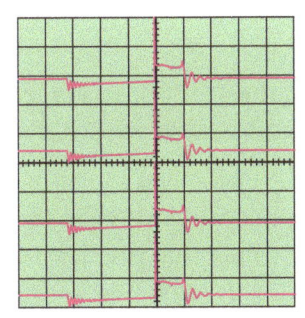

Figure 3.45 Raster

This gives you the option of comparing different cylinders all at once. If used in parade mode, the cylinders will be displayed in the firing order from left to right. If used in raster mode, the cylinders will normally be displayed in firing order from bottom to top.

115

Level 3 Light Vehicle Technology

When viewing a single cylinder, the wave form will normally conform to the following pattern:

A high voltage spike will be produced at the start of the firing pattern (measured in kilovolts on the vertical scale). This is a very good indication of cylinder comparison, showing the voltage needed for the spark to jump the gap of the plug. If cylinders are compared, a high firing voltage may indicate such issues as:

- wide spark plug gaps
- large rotor arm to distributor cap gap
- break in a plug lead
- break in the king lead
- worn spark plugs
- large reluctor air gap (on the inductive pickups)
- high cylinder compression
- a weak air/fuel mixture.

If the cylinders are compared, a low firing voltage may indicate such issues as:

- small spark plug gap
- incorrect ignition timing
- voltage tracking to earth
- fouled spark plugs
- low compression
- a rich air/fuel mixture.

Once the initial firing voltage has been produced, less effort is now required to maintain the spark at the plug gap, and voltage will fall. You will now see a short period of wave form, showing the spark duration. This spark duration should be roughly the same for all cylinders and should remain relatively even. If the spark duration voltage starts high and rapidly falls (an angled line moving sharply downwards from left to right), this may indicate a high resistance in the high tension circuit or spark plug. If this occurs on all cylinders, it may indicate a high resistance in the secondary feed circuit (king/coil lead).

Following the spark duration, you should see a small amount of oscillations. These represent the coil or condenser back EMF being produced.

Once these oscillations have smoothed out, the voltage will then fall to zero, showing that the discharge process has finished and the dwell period of the ignition coil has once again started.

At the end of the ignition coil dwell period, you will once again see a high tension discharge spike and the process repeats.

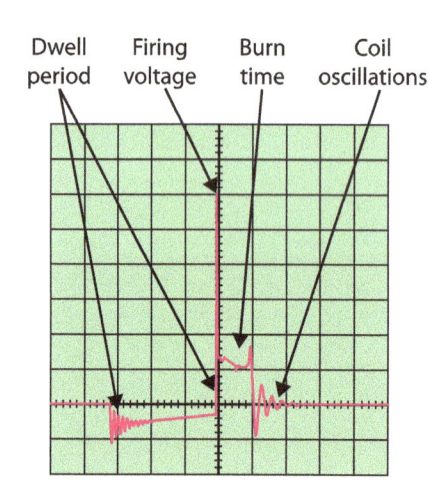

Figure 3.46 Secondary wave form description

Key term

Pinking – a small metallic tapping sound created in the engine by over-advanced ignition timing.

Knock sensors

If the air/fuel mixture inside the combustion chamber ignites early, power will be lost and a small knocking noise called **pinking** may occur. Over long periods, this can lead to engine damage.

3 Diagnosis & rectification of light vehicle engine faults

To overcome this problem, many engines are fitted with knock sensors. The knock sensor will be mounted high in the cylinder block or on the cylinder head, normally between two cylinders.

If the burn begins too early and pinking occurs, the knocking noise can be interpreted as a small vibration from the engine. **Piezoelectric crystals** are contained inside the knock sensor, and when pinking occurs, the corresponding voltage output can be translated by the ECU. The ECU will then normally retard ignition timing by two degrees. If pinking continues, it will be retarded a further two degrees, and so on until pinking stops. When pinking no longer exists and signals from the knock sensor stop, the ECU will start to advance the ignition timing back up one degree at a time until pinking reoccurs and the whole process repeats itself.

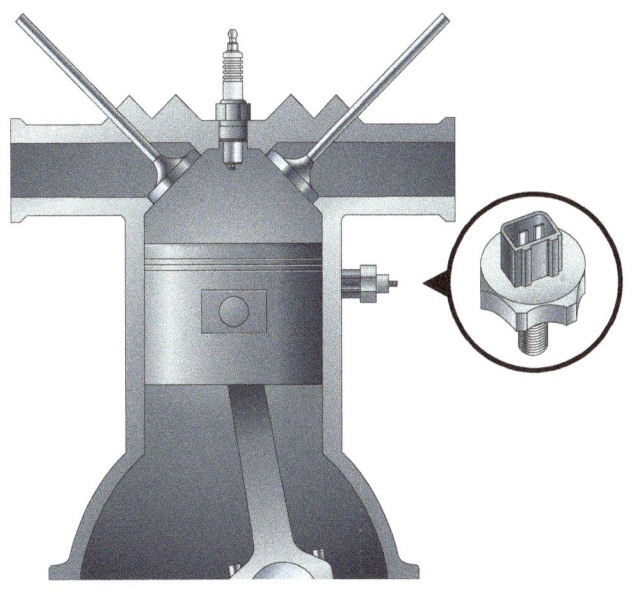

Figure 3.47 Knock sensor

Testing a knock sensor with an oscilloscope

- Knock sensors are active and produce their own voltage.
- They can be attached to an oscilloscope and an analogue signal will be produced when knocking occurs.
- Once you have connected the probes to the oscilloscope and selected a single channel, if the knock sensor has two terminals, you should connect the probes of the oscilloscope here.
- The engine may be running or stationary, and you should lightly tap the cylinder block near the knock sensor with a metallic implement.
- A wave form should now appear on the screen, as shown in Figure 3.48.

Wasted spark and distributorless ignition systems (DIS)

A distributorless ignition system (DIS), as the name suggests, is one that no longer requires a distributor in order to send the spark to the correct spark plug at the appropriate time.

The advantages of the system are:

- fewer moving parts
- no mechanical timing
- less maintenance
- no mechanical load on the engine
- increased coil saturation (charge).

As most engines have an even number of cylinders, two cylinders normally operate in pairs (go up and down together). As the pair of companion cylinders move upwards, one will be on its compression stroke while the other will be on its exhaust stroke. The cylinder on the compression stroke will require a spark, while the one on the exhaust stroke will not. It is normally the job of the distributor to choose which cylinder to supply with a high tension spark.

> **Key term**
>
> **Piezoelectric crystals** – a type of crystal that creates a small electrical current when flexed.

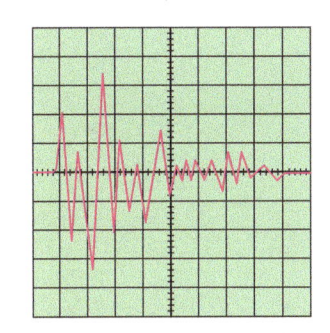

Figure 3.48 Knock sensor wave form

> **Did you know?**
>
> You can use this quick and easy diagnostic test to check if a knock sensor is working:
>
> With an ignition timing strobe light attached to the engine and shining on the timing marks, use a small metallic implement to tap the engine block near to the knock sensor. If the knock sensor is functioning correctly, you should be able to see that the ignition timing marks retard themselves accordingly.

Level 3 Light Vehicle Technology

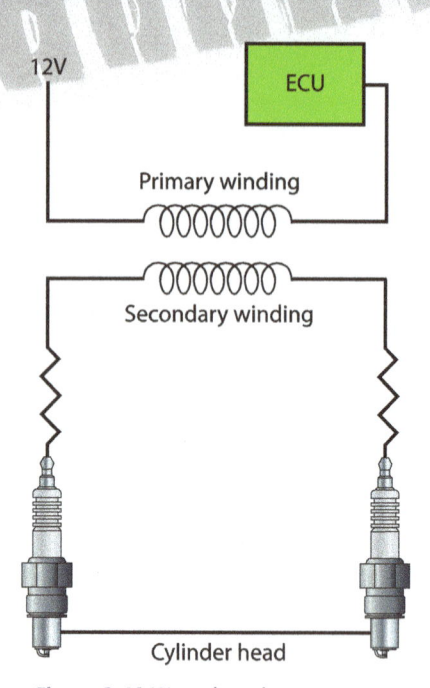

Figure 3.49 Wasted spark

Wasted spark works on the principle that if a spark is provided to both companion cylinders at the same time, the one on its compression stroke will ignite the fuel and start the power stroke, while the one on the exhaust stroke will be wasted. The next time around, after 360° of crankshaft revolution, the other cylinder will be on its compression stroke and ignite the air/fuel mixture, while its companion will be on its exhaust stroke and the spark will be wasted. By using this process, the timing of a pair of companion cylinders can be reduced by half, removing the need for a distributor and associated components. Normally in this type of system, a paired ignition coil is used to fire a set of companion cylinders.

Operation of wasted spark ignition

As both spark plugs fire at the same time, the system does not need to know which plug needs the spark – it just needs to know when.

A spark plug is connected to each end of the secondary winding on a pair of companion cylinders. The cylinder head acts to complete the circuit between the two spark plugs. When the primary circuit of the ignition coil is supplied with power, it charges. This creates an invisible magnetic field that cuts across the secondary circuit of these paired spark plugs. When the primary circuit is switched off, the magnetic field collapses, which induces a high voltage in the secondary circuit of the companion cylinders. High voltage flows instantly through the secondary winding, spark plugs and cylinder head, producing the spark.

Because current flows in one direction, one spark plug will receive a positive spark (jumping from the centre to the earth electrode) and the other will receive a negative spark (jumping from the earth electrode to the centre electrode) to complete the circuit. This phenomenon can lead to spark erosion and wear at the tips of the spark plug, which may be different. Also, if one spark plug fails (becomes an open circuit), the other spark plug may also fail.

Testing wasted spark ignition

When using an oscilloscope to test a wasted spark system, connect it in the same way as you would for a standard secondary ignition system (see page 115). Depending on the type of oscilloscope used, the wave form will be the same as that described on page 116, although as one spark travels from the earth electrode to the centre electrode, its wave form may appear inverted (upside down). Most oscilloscopes that are designed for automotive use will automatically turn this wave form the other way up so it appears on the screen the same as the rest.

- The cylinder on the compression stroke is known as the 'event' cylinder. As it is under compression, has air/fuel mixture, etc., its firing voltage will be high.
- The cylinder on its exhaust stroke is known as the 'wasted' cylinder. As it is not under compression, and is expelling exhaust gas through the exhaust valve, its firing voltage will be low.

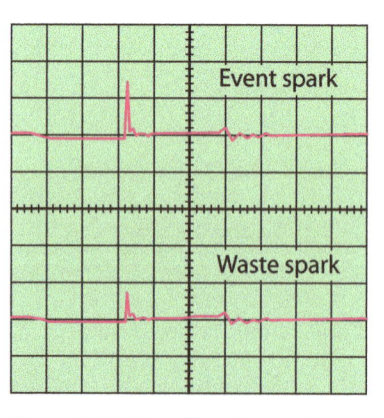

Figure 3.50 Wasted spark wave form

After 360° of crankshaft revolution, the event and waste will swap over, and so will the firing voltages. Although the firing voltages will be constantly rising and falling, you can analyse the patterns in the same way as described for secondary wave forms (see page 116).

Direct ignition systems

A direct ignition system uses individual coils on the top of each spark plug, often known as 'coil on plug' (COP). Although normally triggered in a wasted spark pattern, as each spark plug has its own ignition coil, firing voltages will be positive, with the spark jumping from the centre electrode to the earth electrode. Unlike a standard wasted spark system, if one spark plug fails, its companion should continue to operate as normal.

The advantages of this type of system include:

- The ignition coil does not have to work as hard as it would do if it had to fire more than one cylinder by itself.
- The system is under less strain, and the coil is given more time to become fully charged and therefore produce a much more reliable spark.
- As the high tension voltage is produced right on top of the spark plug, there is no need for spark plug leads, etc. This reduces overall circuit resistance and creates far less radio interference. (As high voltage travels along a plug lead, it gives off an electromagnetic wave that will affect radio reception.)
- If one coil fails, they are normally replaceable as individual items.

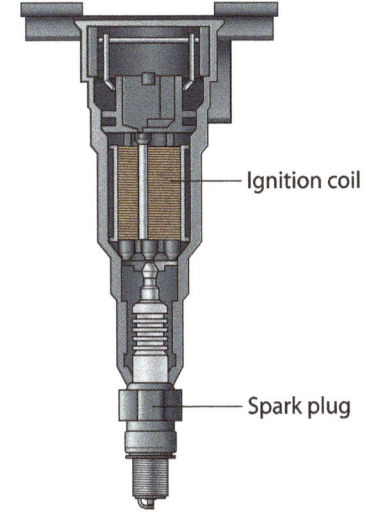

Figure 3.51 Coil on plug

Testing coil on plug

As plug leads are no longer used in this type of system, you cannot use an inductive clamp. Instead, inductive testers have been produced that can be applied directly to the coil on plug and which pick up a signal to produce a wave form. If an inductive tester is unavailable, it is sometimes possible to connect an adapter plug lead between the coil and spark plug that enables the use of a standard inductive clamp.

> **Key term**
>
> **Capacitor** – a temporary storage for electric charge.

Capacitor discharge ignition (CDI)

Capacitor discharge ignition systems may work with single ignition coils (standard systems or wasted spark) or individual coils (coil on plug).

In a non-CDI system, 12V in the primary circuit is converted into many thousands of volts in the secondary circuit. In a CDI system, 12V is initially stepped up to around 400V and this is used to charge a capacitor. When the spark is required, the capacitor discharges through the

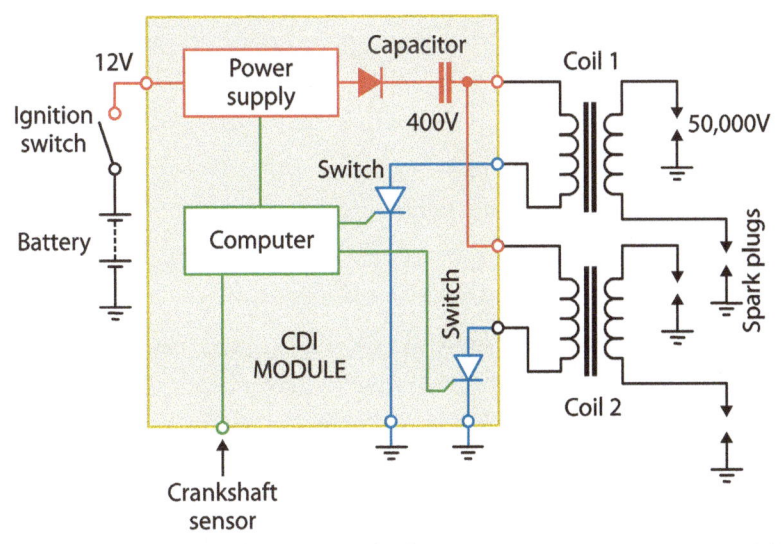

Figure 3.52 Capacitor discharge ignition (CDI)

Level 3 Light Vehicle Technology

> **Action**
> Connect an oscilloscope to an ignition system component and retrieve an operating wave form.

primary winding of the ignition coil, and the 400V is stepped up to a much higher constant voltage.

Although cold-starting can sometimes be an issue with a CDI system, some systems are able to produce multiple sparks in the same time period as a normal ignition system, thereby reducing the possibility of a misfire. The charging and discharging of the system is normally controlled by an ECU.

Testing DIS and CDI

Due to the compact nature of DIS and CDI and the fact that lots of the components are sealed at the factory, testing can be difficult.

The use of on-board diagnostics (OBD) may be required to check for faults. Feedback from the system can generate diagnostic trouble codes (DTCs), which help to point you in the right direction. Unfortunately, OBD will not recognise mechanical faults, so always check these if a fault does not generate a DTC.

Faults and symptoms in ignition systems

Table 3.7 gives an indication of some of the symptoms that you may experience when working on an ignition system. This list is not exhaustive, and you should conduct a thorough diagnostic routine to ensure that you have correctly located the fault.

Table 3.7 Ignition system faults

Symptom	Possible fault
No spark	Crankshaft position sensor open circuit; no engine speed signal being produced
Misfiring	High resistance ignition coil windings, causing it to break down under load
Backfiring	Incorrect ignition timing, causing fuel to ignite in the inlet manifold
Cold/hot starting problems	Fouled spark plugs, caused by incorrect operating heat range
Poor performance	High resistance plug leads, giving a weak spark under acceleration
Pre-ignition	Spark plug earth electrode overheating, igniting the fuel before the spark, caused by incorrect spark plug being used
Detonation	Petrol spontaneously igniting caused by using fuel with a lower octane rating than recommended
Excessive exhaust emissions	Retarded ignition timing, causing combustion temperatures to rise and creating excessive oxides of nitrogen
High fuel consumption	Incorrect spark plug gaps, causing uneven running and leading to excessive fuel consumption
Lack of power	Advanced ignition timing, causing pressure build-up when the connecting rod is vertical
Unstable idle speed	Loose reluctor ring, causing variable engine speed signals

> **Skills for work**
>
> Following a routine service, the car that you have been working on has developed a misfire. Although this is probably a straightforward fault to diagnose and rectify, you need to inform your supervisor and the customer that you will need extra time to finish the job.
>
> This situation requires that you use particular personal skills. Some examples of these skills are shown in Table 6.1 at the start of Chapter 6, on pages 304–305.
>
> 1. Using the examples given in Table 6.1, choose one skill from each of the following categories that you think you need to demonstrate in order to inform your supervisor and customer about this situation.
>   - General employment skills
>   - Self-reliance skills
>   - People skills
>   - Customer service skills
>   - Specialist skills
> 2. Now rank these skills in order of importance, starting with the one that it is most important for you to have in this situation.
> 3. Which of the skills chosen do you think you are good at?
> 4. Which of the skills chosen do you think you need to develop?
> 5. How can you develop these skills and what help might you need?

> **CHECK YOUR PROGRESS**
>
> 1. Name two precious metals that are used in the construction of spark plugs.
> 2. Give two possible faults that could cause a high plug firing voltage and two possible faults that could cause a low firing voltage.
> 3. What does CDI stand for?

Combustion

Modern engines require high levels of fuelling and combustion controls to enable them to comply with strict emission regulations while maintaining fuel economy and performance. It has been discovered that these three outcomes (emission controls, fuel economy and performance) are nearly achieved if:

- the air/fuel ratio is maintained at 14.7:1 by mass
- ignition timing is advanced or retarded, as required.

Unfortunately, there is always a trade-off between performance, economy and emissions, so perfect engine operation is rarely accomplished.

Emissions are tightly controlled by regulations. As a result, they tend to be regarded as more important than fuel economy and performance. The main emissions produced by a spark ignition engine are:

- carbon monoxide (CO)
- hydrocarbons (HC)
- oxides of nitrogen (NOx)
- oxygen (O_2)
- carbon dioxide (CO_2)

If an engine is accurately controlled, and a **stoichiometric** combustion process is produced, many of these emissions can be

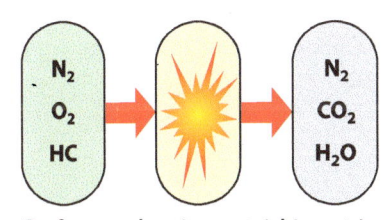

Perfect combustion – stoichiometric

Figure 3.53 Ideal combustion

> **Key term**
>
> **Stoichiometric** – a balanced chemical reaction often used to describe the ideal air/fuel ratio.

Level 3 Light Vehicle Technology

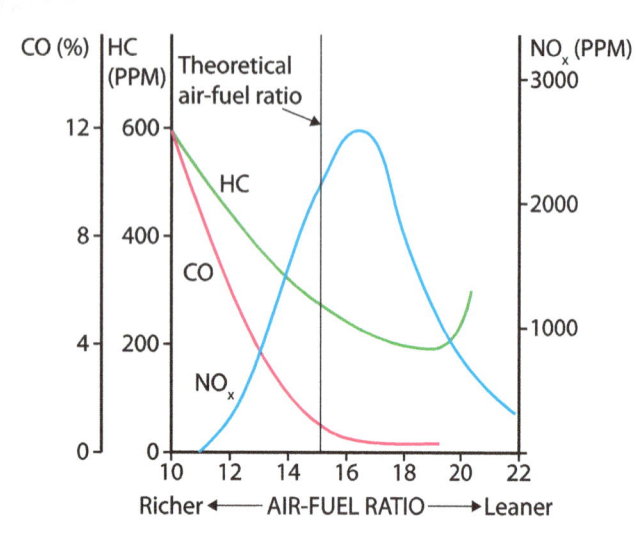

Figure 3.54 Exhaust emissions graph

reduced considerably. As the air/fuel ratio of a petrol engine approaches the ideal of 14.7 to 1, carbon monoxide and hydrocarbons fall considerably, but unfortunately carbon dioxide and oxides of nitrogen rise (see Figure 3.54).

If you burn a fossil fuel, one by-product is carbon dioxide. Although carbon dioxide is not considered a poisonous gas, it has an environmental impact as it is believed to contribute to global warming. The better the combustion process, the higher the output of carbon dioxide.

Oxides of nitrogen also rise as the ideal air/fuel ratio is reached. Oxides of nitrogen are produced by the high combustion temperatures and must be dealt with in another way (for example, by exhaust gas recirculation (EGR), which is described on page 128).

Emissions diagnosis

The correct use of an exhaust gas analyser can provide some very useful diagnostic information. Table 3.8 shows some examples of exhaust gas diagnostic readings and their possible faults.

Table 3.8 Examples of exhaust gas diagnostic readings and their possible faults

Exhaust gas	Approximate recommended values	Fault	Possible cause
Hydrocarbons (HC)	Less than 1200 ppm (parts per million)	Too high	Engine misfire, allowing unburnt fuel into the exhaust system.
Carbon monoxide (CO)	Less than 3 per cent pre catalytic converter	Too low	Excess oxygen in combustion process (weak mixture). Possible induction air leak.
	Less than 0.3 per cent post catalytic converter	Too high	Not enough oxygen in combustion process (rich mixture). Possible leaking injector.
Carbon dioxide (CO_2)	More than 14 per cent	Too low	Engine mechanical condition/efficiency fault. Possible low compressions.
Oxygen (O_2)	Around 0.5 per cent	Too low	Engine running rich. Possible blockage in the air intake.
		Too high	Too much oxygen entering the intake or exhaust (leaks).

3 Diagnosis & rectification of light vehicle engine faults

Common terms

Table 3.9 explains some common terms associated with fuelling, combustion and exhaust emissions.

Table 3.9 Fuelling, combustion and exhaust emission terms

Term	Meaning
Flame travel	This is the way that the flame spreads in the burning air/fuel mixture within the combustion chamber. To provide good performance and efficient combustion, the engine is designed to ignite the fuel at a particular position inside the combustion chamber. This is normally around the tip of a spark plug or fuel injector. The flame created by the burning mixture should then spread out evenly away from this point, giving a smooth but rapid pressure rise within the cylinder. This pressure will react correctly against the piston crown and obtain maximum effort to turn the crankshaft.
Pre-ignition	Pre-ignition normally occurs on petrol engine vehicles. This is a condition where the air/fuel mixture ignites before the operation of the spark plug. Pre-ignition is normally a result of a combustion chamber deposit (carbon build-up) which is burning/glowing red hot and will act as an alternative source of ignition.
Detonation	Detonation is a condition that exists when the fuel in the combustion chamber spontaneously ignites due to a rapid rise in heat, normally created by compression pressures. This can occur in both petrol and diesel engines. Fuel droplets begin to burn in a number of different places at once, and the collision of flame fronts as they spread out creates a distinctive knocking noise; this is particularly apparent in diesel engines.
Pinking	Pinking is a condition that occurs when the ignition of the air/fuel mixture happens too early in the combustion cycle. The most common reason for this is over-advanced ignition timing on petrol engines or over advanced injection on diesel engines. This results in the pressure build-up occurring before the piston and the connecting rod have reached the required position after top dead centre. If the piston is at top dead centre and the connecting rod is vertical, turning effort to the crankshaft is wasted and performance is lost. The condition can sometimes be identified by a light pinging or metallic tapping noise from the engine when accelerating or putting the engine under load.
Octane rating	The octane rating of a fuel refers to petrol. It is a measurement of petrol's resistance to detonation, i.e. how stable the fuel is under compression pressures. The higher the octane rating, the more stable the fuel is and therefore the more suitable it is for high performance or high compression engines. Fuel is classified with a research octane number (RON), which helps to identify its properties.
Cetane rating	The cetane rating of a fuel refers to diesel. It is a measurement of diesel's delay period, i.e. how quickly it will ignite and burn after injection. The higher the cetane rating, the shorter the delay period and the faster the ignition of the fuel. A diesel fuel with a high cetane rating is desirable for good combustion and performance, as it gives a smooth and even flame spread within the combustion chamber.
Volatility	Volatility refers to a fuel's ability to vaporise at certain temperatures. Petrol is a volatile fuel which turns into a vapour readily at room temperature. Diesel is less volatile and needs its temperature to be raised before it will vaporise. The volatility of a fuel is important as it is the vapour of a fuel that burns, not the liquid.

CONTINUED ▶

Term	Meaning
Calorific value	The calorific value of a fuel is a measurement of the heating power it contains. It refers to the amount of energy released when the fuel is burned under specified conditions. The amount of energy stored in a fuel depends on its composition and, as a result, different fuels have different values.
Flash point	Flash point is the lowest temperature at which a fuel will vaporise to form an ignitable mixture in air. Because of the volatility of petrol, it has a lower flash point than diesel.
Hydrocarbon content	Whether a fuel is petrol or diesel depends on how much hydrogen and carbon the fuel contains. This reflects the number of carbon molecules in the hydrocarbon chain. The amounts can differ slightly depending on how the fuel is refined, but as a general rule: • When the chain has between five and nine carbons, the hydrocarbon is petrol. • When the chain has about twelve carbons, the hydrocarbon is diesel. Depending on how diesel and petrol are refined, approximate percentages of carbon and hydrogen are 84% carbon and 14% hydrogen (plus 2% other elements).
Composition of air	During the induction process, air is mixed with fuel to form an ignitable mixture – because oxygen is required for combustion. At sea level, air is composed of approximately 21% oxygen and 78% nitrogen (plus 1% other gases).
Air/fuel ratio	Air/fuel ratio refers to the quantities of air and fuel in the mixture created for combustion. An ideal air/fuel ratio for correct combustion is 14.7:1 by mass (weight). This means that the amount of air in the combustion mixture will weigh 14.7 times more than the amount of fuel.
Stoichiometric ratio	Stoichiometric ratio refers to a balanced chemical reaction where all the component elements are used up during a reaction process. This is often used to describe the correct air/fuel ratio of 14.7:1 for ideal combustion.
Homogenous charge	A method of evenly mixing the air/fuel charge in the engine cylinder, used in direct injection spark ignition engines (also known as GDI or gasoline direct injection). For more information on GDI see page 145.
Stratified charge	A method of injecting fuel towards the spark plug in the engine cylinder to form a concentrated layer when lean burn operation is in use. This form of injection is used in direct injection spark ignition engines (also known as GDI or gasoline direct injection). For more information on GDI see page 145.
Lambda window	The term lambda window is sometimes used to describe the point when exhaust emissions fall within a desired tolerance for correct engine operation. Engine management is designed to use various sensors to monitor and control engine functions to maintain this lambda window.
Carbon monoxide	If during the combustion process, the burning fuel goes out, perhaps due to lack of oxygen or rapid cooling, carbon monoxide (CO) is produced. Carbon monoxide is a product of incomplete combustion and it is harmful to health. It is colourless, odourless and tasteless, but if inhaled it replaces oxygen in the blood and starves the organs of required oxygen. Measurements of carbon monoxide in exhaust gases can give an indication of air/fuel ratios. Modern engines operate with approximately 3% CO. A figure higher than this indicates a rich mixture. A figure lower than this may indicate a weak mixture. (To get an accurate measurement of CO, you need to take a sample before the catalytic converter.)

CONTINUED ▶

3 Diagnosis & rectification of light vehicle engine faults

Term	Meaning
Carbon dioxide	With effective combustion, the chemical elements carbon and oxygen combine to form carbon dioxide (CO_2). This should normally be higher than 14% by volume. Carbon dioxide can be used as an effective diagnostic gas and can give a good indication of engine mechanical operation and condition. If its percentage by volume is low, you should assess the engine condition and function. Carbon dioxide is a greenhouse gas and is considered an environmental pollutant.
Hydrocarbons	If fuel passes through the combustion process with no chemical change, hydrocarbons (HC) are given off. These should be kept as low as possible, usually under 200 parts per million at idle. High values of hydrocarbons normally give an indication of an engine misfire. Hydrocarbons are considered harmful to health and may cause lung damage or cancer.
Oxides of nitrogen	Most of the combustion process takes place at temperatures between 2000 and 2500°C. In this extreme heat, the oxygen and nitrogen in the incoming air are combined to produce oxides of nitrogen (NOx), which are pollutant gases. Unfortunately, the better the combustion process is, the more oxides of nitrogen are produced. Because of this, specific methods of control are required, such as exhaust gas recirculation (EGR) – see page 128. Oxides of nitrogen can damage the lungs and may also harm plant life and reduce visibility. Most four gas exhaust analysers are unable to measure oxides of nitrogen.
Oxygen	The oxygen found in exhaust gas is not considered harmful to health or an environmental pollutant. If the stoichiometric ratio is correct, nearly all of the available oxygen in the inducted air (21%) will be used up during the combustion process, leaving around 0.5%. High levels of oxygen in the exhaust gases can indicate an air leak in either the induction or exhaust system. Low levels of oxygen in the exhaust gases may indicate an over-rich mixture caused by excessive fuelling.
Lambda	Lambda is often used to represent the ideal air/fuel ratio of 14.7:1 by mass. Most exhaust gas analysers are unable to measure true lambda, but instead use a mathematical calculation created from the measured gases (CO, CO_2, HC and O_2) to work out the air/fuel ratio. Any discrepancy in one of the measured gases can lead to an incorrect value of lambda being displayed on the analyser. If lambda values are incorrect, your diagnostic routine should focus on any of the four measured gases, which may be out of tolerance.

MOT requirements for exhaust emissions

During the annual MOT inspection, light vehicle exhaust emissions are tested to make sure they conform to strict limits. Petrol vehicles are tested using an exhaust gas analyser that measures the specific quantities of gaseous emissions at the tail pipe.

During an MOT test, diesel engine vehicles are tested for the amount of smoke or **particulate matter** produced at the exhaust tail pipe. A smoke meter is attached to the car at the tail pipe and the engine is revved up. An **opacity meter** in the smoke tester measures the exhaust gas to see if it falls within preset limits.

> **Key terms**
>
> **Particulate matter** – dirt and soot produced in the exhaust system.
>
> **Opacity meter** – a tool that measures how much light is absorbed by diesel exhaust smoke.

> **Action**
>
> Using sources of data available to you, find out the emission limits of a current basic emissions test.

Level 3 Light Vehicle Technology

> **Did you know?**
>
> A BET (basic emissions test) will confirm if the vehicle complies with general emission standards. If the vehicle fails to meet the standards of the BET, then vehicle specific data is used to conduct a further test to ensure it conforms to the manufacturer's specifications.

Exhaust emission standards

Exhaust emission standards are continually being updated and revised, with Euro 4 being the most common standard currently. The stricter Euro 5 regulations are now compulsory for all new cars currently on sale in the UK.

The European emission standards are set out in Table 3.10. They show the limits set in grams per kilometre (g/km).

Table 3.10 European emission standards

Standard	Commencing	Carbon monoxide limits	Total hydrocarbons	Non-methane hydrocarbons	Oxides of nitrogen	Hydrocarbons plus oxides of nitrogen	Particulate matter (soot)
Limits for diesel engine cars							
Euro 1	July 1992	2.72 (3.16 COP)	N/A	N/A	N/A	0.97 (1.13 COP)	0.14 (0.18 COP)
Euro 2	January 1996	1.0	N/A	N/A	N/A	0.7	0.08
Euro 3	January 2000	0.64	N/A	N/A	0.50	0.56	0.05
Euro 4	January 2005	0.50	N/A	N/A	0.25	0.30	0.025
Euro 5	September 2009	0.500	N/A	N/A	0.180	0.230	0.005
Euro 6	September 2014	0.500	N/A	N/A	0.080	0.170	0.0025
Limits for petrol engine cars							
Euro 1	July 1992	2.72 (3.16 COP)	N/A	N/A	N/A	0.97 (1.13)	N/A
Euro 2	January 1996	2.2	N/A	N/A	N/A	0.5	N/A
Euro 3	January 2000	2.3	0.20	N/A	0.15	N/A	N/A
Euro 4	January 2005	1.0	0.10	N/A	0.08	N/A	N/A
Euro 5	September 2009	1.000	0.100	0.068	0.060	N/A	0.005
Euro 6	September 2014	1.000	0.100	0.068	0.060	N/A	0.005

COP = Conformity of Production

Exhaust emission control

Advances in engine management, and the control of fuel injection and ignition systems, have made it possible to limit most exhaust emissions, but further methods are needed in order to meet all of the strict limits imposed. Other methods to control exhaust emissions are shown in Table 3.11.

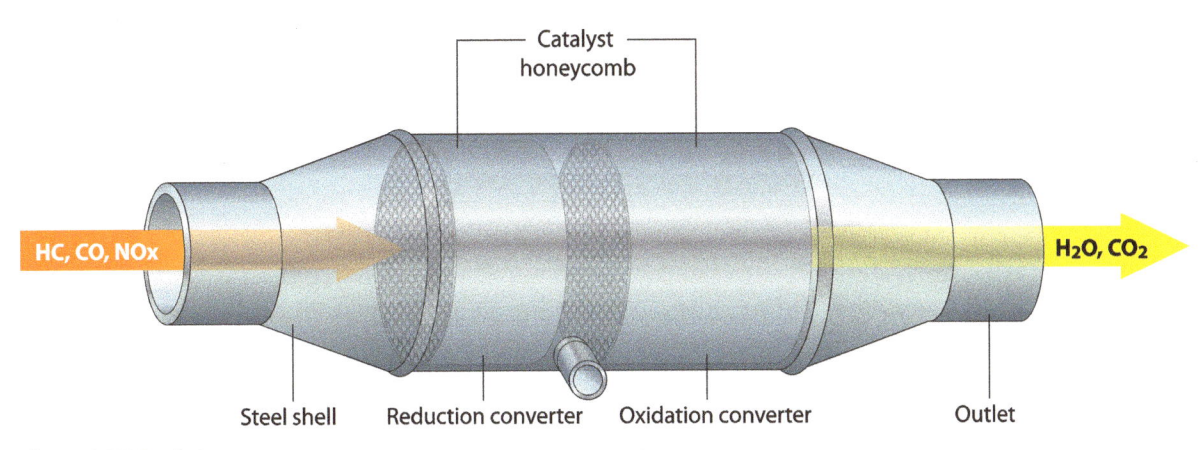

3 Diagnosis & rectification of light vehicle engine faults

Figure 3.55 Catalytic converter

Table 3.11 Methods of controlling exhaust emissions

Catalytic converters	Catalytic converters are fitted to some vehicles to help reduce the amount of harmful pollutants present in exhaust gases. A catalytic converter is normally mounted in the exhaust system very close to the inlet manifold where the exhaust gases exit the cylinder head. (The closer the catalytic converter is to the exiting exhaust gases, the quicker it heats up and begins to operate in an efficient manner.) From the outside, a catalytic converter may look similar to an exhaust system silencer. On the inside, it is normally made up of a matrix-style mesh which contains precious metals such as platinum and rhodium.
	The mesh-style matrix increases the surface area exposed to the exiting exhaust gases. When the gases pass over the precious metals, a chemical/catalytic reaction takes place (a catalyst is anything that makes a situation or substance change).
	The exhaust pollutants are not removed from the exhaust gases, but are converted to far less harmful pollutants.
	When the exhaust gas comes into contact with the reduction catalyst (see Figure 3.55 – reduction converter) a chemical reaction takes place that strips the nitrogen molecules from any oxides of nitrogen and retains them in the catalyst, while allowing the oxygen to continue.
	When the exhaust gases come into contact with the oxidation catalyst (see Figure 3.55 – oxidisation converter) a chemical reaction takes place which recombines hydrocarbons and carbon monoxide with some of the remaining oxygen to produce carbon dioxide and water.
	This way, the amount of harmful emissions produced by the exhaust system are reduced.
	To work efficiently, catalytic converters require the engine to operate within the ideal stoichiometric values. This means that accurate engine management is essential.
Selective catalytic regeneration (SCR)	On some three-way catalytic converters, oxides of nitrogen are removed using a chemical reaction. As a result of the chemical reaction, the nitrogen molecules are retained inside the catalytic converter and must periodically be cleaned out. Some engine management systems achieve this by changing their fuel injection program to make the system run rich – this alters the chemical processes going on inside the catalytic converter, helping to remove the nitrogen molecules. Other systems have a reservoir containing a special chemical, normally based on urea. This is injected into the exhaust gas before the catalytic converter and alters the processes going on inside the catalytic converter, helping to remove the nitrogen molecules. This chemical additive is a service item that must be topped up according to the manufacturer's maintenance schedule.

CONTINUED ▶

Secondary air injection	A secondary air injection system is sometimes fitted to vehicles to help with emission-related issues. A small air pump, controlled by the car's computer (ECU), can blow air into the exhaust manifold close to where it exits the cylinder head. By introducing extra oxygen to the hot exhaust gases, unburnt hydrocarbons can continue the combustion process and be burnt away. As with blowing on the embers of a fire, introduced air helps heat the exhaust gases. If the vehicle is fitted with a catalytic converter, this will assist with the converter's warm-up period, enabling it to operate efficiently sooner.
Diesel particulate filters (DPF)	A diesel particulate filter can be mounted in the exhaust system of a compression ignition powered car to help remove soot and other small contaminate particles that are produced during the combustion cycle. A car fitted with a DPF in good working order should show no visible smoke from the tailpipe when running. Unlike a catalytic converter, a DPF is a filter – the exhaust gases are passed through the filter and particulate matter is retained. This means that a DPF requires maintenance to prevent it from becoming blocked. Some diesel particulate filters are a replaceable service item, while others can be regenerated in a similar manner to SCR. Pressure and temperature sensors are often mounted in the exhaust system so that the engine management system is able to monitor and maintain correct function of the DPF. To work efficiently, DPFs need to operate within a certain heat range. Frequent slow driving and short journeys will not allow the DPF to reach its correct operating temperature and this may lead to early failure due to blockages.
Exhaust gas recirculation (EGR) – petrol	Exhaust gas recirculation (EGR) can be used as an emission control system to reduce the production of oxides of nitrogen (NOx). EGR is a method for taking some of the burnt exhaust gases and directing them through pipes or galleries back into the inlet manifold. By introducing exhaust gas back into the inlet manifold of a petrol engine, the amount of incoming air and fuel that reaches the cylinder is reduced. This means that the performance output of the engine is reduced, and so engine temperatures will be lower. This helps prevent the production of NOx. It is important that performance is not affected during most of the operating conditions under which a car is driven. A device called an EGR lift valve is fitted so that exhaust gas is only recirculated when outright performance or acceleration are not required. For example, when cruising on a motorway at a steady speed with your foot resting only lightly on the accelerator pedal, exhaust gas can be recirculated and the production of NOx reduced.
Exhaust gas recirculation (EGR) – diesel	In a diesel engine, it is very difficult to achieve ideal air/fuel ratios of 14.7:1 because diesel engines draw the same amount of air every time, and it is the amount of fuel that is injected that dictates how fast the engine will run. This means that when a standard diesel engine is operating under normal conditions, the air/fuel ratio will be extremely weak at low revs and idle, and rich during acceleration. These extremes of air/fuel ratio create emission and pollution problems. By introducing a quantity of burnt exhaust gases back into the intake manifold of a diesel engine, the amount of fresh air entering the cylinder can be more accurately controlled, which results in better air/fuel ratios. Exhaust gas will still compress and raise the temperature in the cylinder for ignition of the fuel.

CONTINUED ▶

Evaporative emission control (EVAP)	Petrol fumes are a pollutant and because of this, fuel tanks are sealed to reduce the possibility of fumes escaping into the atmosphere. If the tank has nowhere to store these evaporated fumes, high pressures could build up in the tank, causing leaks or damage. The vapours must be captured, stored and disposed of in an environmentally friendly way. Fuel emission vapours are stored in a charcoal canister. When the engine temperature is above 55°C the fumes are vented and burnt along with the rest of the air/fuel mixture. This enables all of the hydrocarbon (HC) vapours to go through the combustion process and eliminate these polluting emissions. This system is called an evaporative emission system (EVAP).
Positive crankcase ventilation (PCV)	Positive crankcase ventilation (PCV) was introduced in the 1960s and marked the beginning of evaporative emission control systems in the cars to help control pollution. The initial purpose of the PCV system was to capture crankcase vapours and prevent them from being passed into the atmosphere through atmospheric breather pipes. Combustion chamber gasses are able to escape into the crankcase through a process known as 'blow-by'. Blow-by occurs when the compressed fuel/air mixture is forced past the seal created between the piston ring and the cylinder wall. When this happens these gasses become trapped in the crankcase where hydrocarbons may then leak to the atmosphere. A PCV valve meters the return of the crankcase vapours to the engine's intake manifold. The vapours then mix with the engine's intake air and/or fuel/air mixture and re-enter the combustion chamber to be burnt. This makes sure that all of the hydrocarbon (HC) vapours go through the combustion process and eliminate HC emissions from the crankcase to the atmosphere.

CHECK YOUR PROGRESS

1. List five gases emitted from a petrol engine exhaust pipe.
2. With reference to fuel, what does the term volatility mean?
3. List three emission control devices.

SI fuel systems (petrol)

Although fuel injection systems have been around for a long time, early systems were mechanical. They relied on fuel distributed to the correct injectors under pressure, similar to the method used in standard diesel engines (although the fuel was injected into the intake manifold, rather than directly into the cylinder). These early mechanical fuel injection systems were an improvement on carburettors, but were still inaccurate in many ways.

Single point or throttle body injection

Early developments in electronic fuel injection included single point or throttle body fuel injection. A throttle body of similar size, shape and position to a carburettor was mounted in the air inlet tract. A single large fuel injector was mounted above the throttle butterfly. Intake air was regulated by the throttle butterfly, and a solenoid type fuel injector

Level 3 Light Vehicle Technology

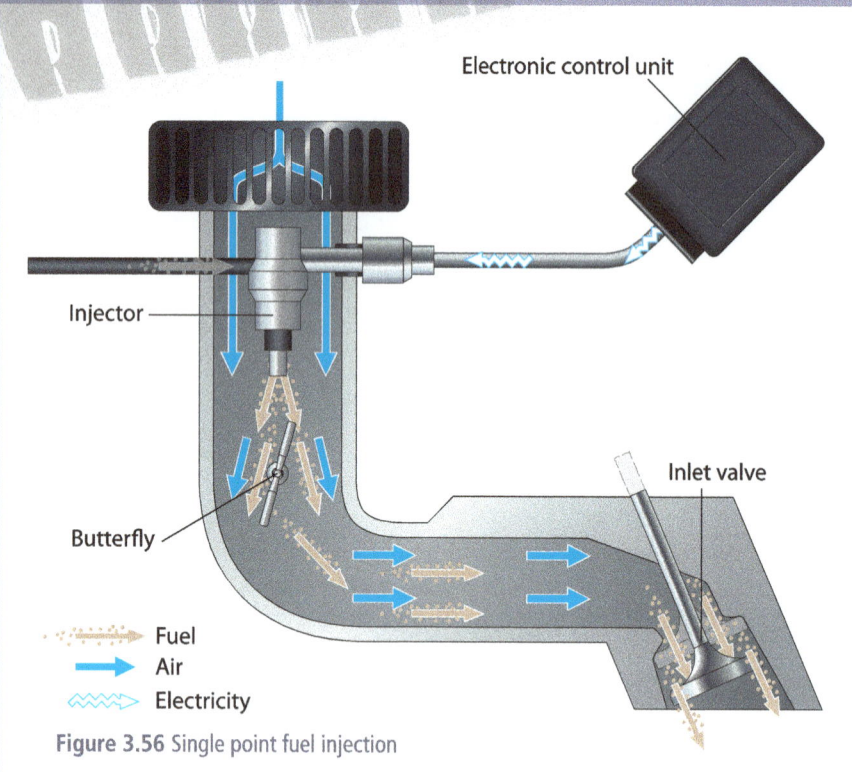

was used to atomise varying quantities of fuel into the air stream. This mixture of air and fuel travelled along the intake manifold and into the cylinder.

As emissions regulations became tighter and tighter, and the operation of catalytic converters became more reliant on precise air/fuel metering, single point fuel injection became unsuitable.

Electronic fuel injection (EFI) – multipoint

Multipoint fuel injection is where each cylinder has its own fuel injector mounted just before the intake valve. The air intake tract only draws air, and fuel is injected at the last moment before it enters into the cylinder. In this way, the metering of fuel and its accuracy can be adequately controlled.

Figure 3.56 Single point fuel injection

Three main systems of injector operation can be used with multipoint electronic fuel injection.

- **Simultaneous** – this is where the injectors are not timed to any particular cylinder and all injectors operate at the same time, controlled by a common electronic driver circuit.

- **Grouped** – this is where the injectors are grouped in combinations (normally pairs) and the grouped injectors are operated at the same time by a common electronic driver circuit for each pair.

- **Sequential** – this is where each individual injector works independently of the others and is timed to operate with the cylinder's induction stroke in the same pattern as the engine firing order.

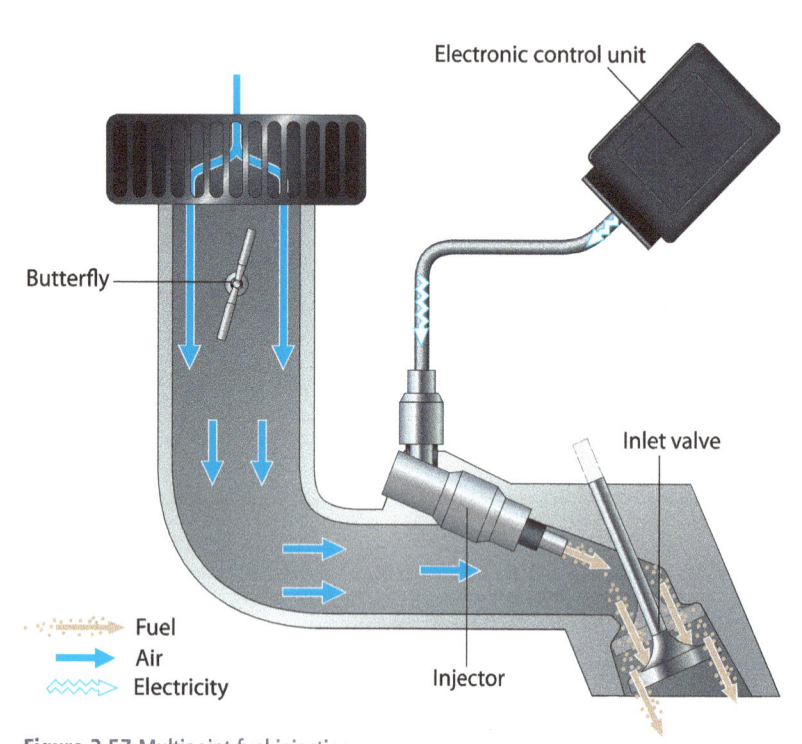

Figure 3.57 Multipoint fuel injection

3 Diagnosis & rectification of light vehicle engine faults

Fuel supply

The main components making up the fuel supply circuit are:

- fuel tank
- fuel pump
- fuel filter
- fuel rail
- fuel injectors
- fuel pressure regulator
- overflow return to tank.

Fuel tank

The fuel tank is the main reservoir and supply of cool, clean fuel for the vehicle's fuel injection system. Fuel tanks are sealed to the atmosphere, to reduce the possibility of hydrocarbon emission due to evaporation.

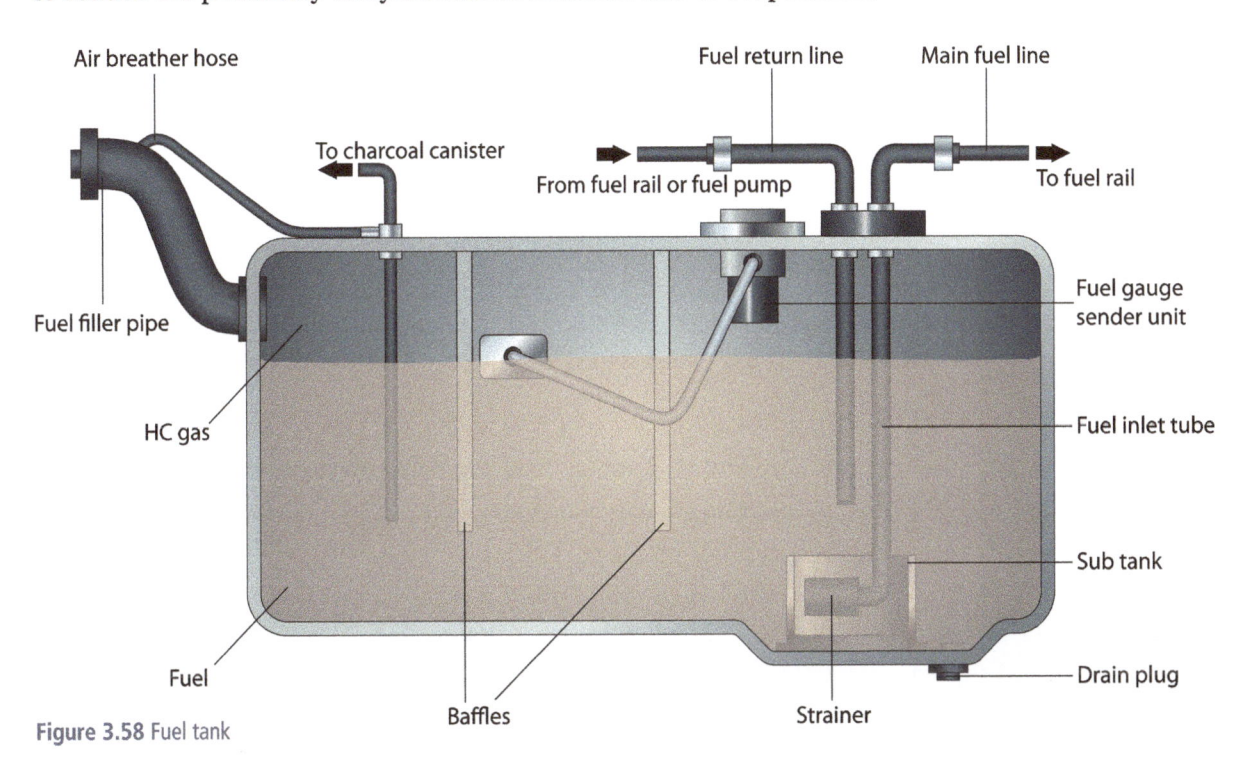

Figure 3.58 Fuel tank

Fuel pump

An electronic fuel injection system needs a large quantity of fuel, supplied under pressure to function correctly. This is the job of the fuel pump.

Many fuel pumps are of the roller cell type.

The fuel pump can be mounted in the fuel line under the vehicle, or submerged within the fuel tank itself. Many modern vehicle manufacturers are choosing to submerge the fuel pump inside the tank, as this assists with cooling and noise suppression. Even if the fuel pump is mounted in line under the vehicle, fuel will flow through the middle of the electric motor, and this will assist with cooling.

Level 3 Light Vehicle Technology

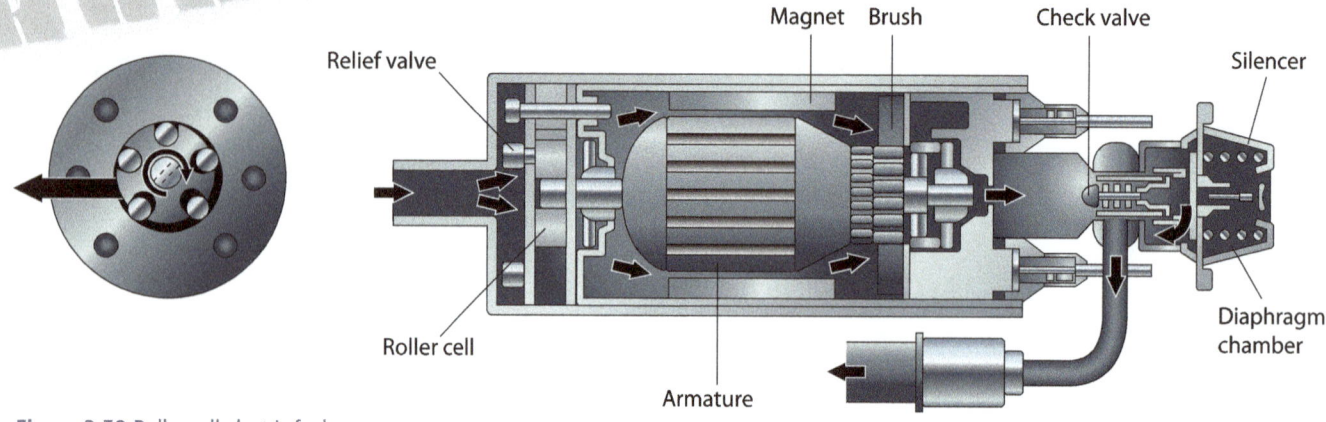

Figure 3.59 Roller cell electric fuel pump

As with all electric motors, fuel pumps create sparks during normal operation. The risk of ignition is very low, due to the fact that fuel vapour (which is the part that burns) is minimal. Also, any liquid fuel passing through the pump will quench any sparks. Because of the lack of air no fumes can be produced and therefore the fuel should not ignite.

A standard electronic fuel injection fuel pump is capable of supplying 4 to 5 litres of fuel per minute at a pressure of approximately 8 bar (120psi). Inside the pump, there is a pressure relief valve which is designed to lift off its seat if system pressure exceeds 8 bar. This is in case a blockage occurs (e.g. of the fuel filter). On the outlet end of the fuel pump there is a non-return valve, so that when the fuel pump, is switched off, it closes and maintains pressure within the system.

Testing pressure and volume

Fuel pressure and volume are vital to vehicle performance.

- You need to check fuel pressure using a suitable pressure gauge, as close as possible to the fuel rail.
- With the engine running, use the manufacturer's data to assess if pressure is within the required limits.
- You should also check system holding pressure, as this will help diagnose leaking fuel injectors.
- When you have checked engine running pressure, switch off the vehicle. When the non-return valve closes, you need to record the holding pressure for a certain time (normally for about 10 minutes).
- If pressure falls during the holding time, something in the system is allowing pressure to leak away. This could be a leaky injector, a faulty fuel rail pressure regulator or a faulty fuel pump non-return valve.

It is very important to check fuel delivery volume. This test is often overlooked by vehicle technicians. Volume and pressure are not the same thing – it is quite possible for a fuel injection system to have the correct pressure but not deliver the correct quantity of fuel, and this may lead to **starvation**.

Figure 3.60 Checking fuel pressure

Key term

Starvation – symptom where the engine does not operate fully/holds back, normally caused by a lack of fuel.

3 Diagnosis & rectification of light vehicle engine faults

Diagnostic tooling manufacturers have produced a number of tools to check flow volume. If you do not have access to some of the specialist tools, you should remove the fuel delivery hose and place it in a **graduated container**. The fuel pump relay may need to be bypassed, so that the fuel pump will run for a set period. You can calculate the fuel delivery rate per minute by:

- measuring the quantity of fuel delivered over a 10-second period
- multiplying this figure by six.

> **Safe working**
>
> Because of the volatile nature of petrol, you must take special precautions when testing fuel pressure and volume.
>
> - Always inform others of what you are doing.
> - Work in a well-ventilated area.
> - Make sure that the engine is cold.
> - Remove all sources of ignition.
> - Have a suitable fire extinguisher handy.
> - Clear up any spills immediately.

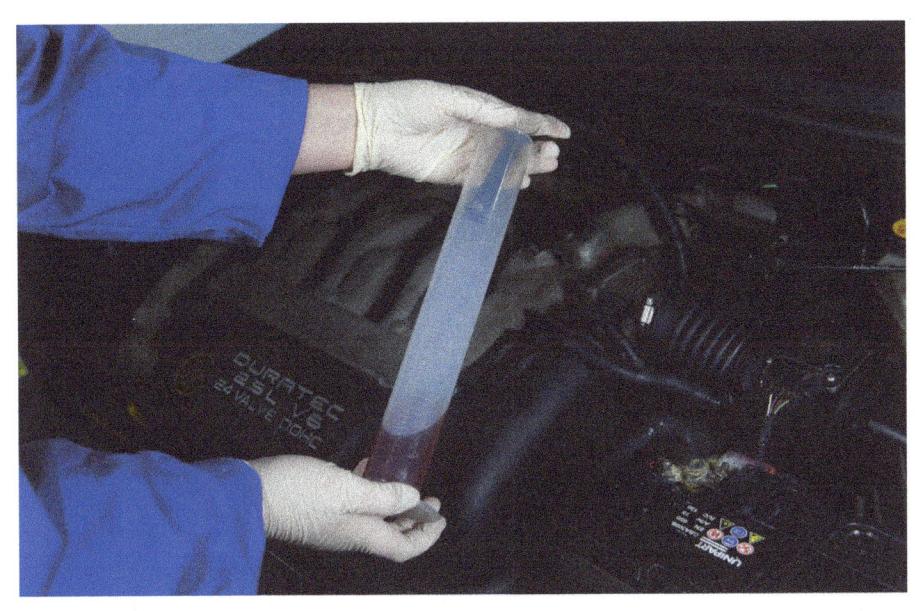

Figure 3.61 Checking fuel volume

Testing a fuel pump in situ

It is not always necessary to remove the fuel pump to test its operation. Amperage or current draw can give you a good indication of fuel pump function and condition. By using an inductive amps clamp or connecting an ammeter in series on the fuel pump feed wire, you can record an amperage reading. Under a normal operating pressure of around 2 bar (30psi), current flow will be around 3 to 5 amps (Bosch KE-Jetronic systems may be around 8 amps).

- If current draw is higher than this, the fuel pump is struggling. This may be due to a blockage or a worn fuel pump.
- If the current draw is lower than this, then the fuel pump is finding it too easy. This may be due to a faulty fuel pressure regulator, or in some cases a leakage.

The **commutator** is the electrical connection in the back of the fuel pump. It is the segmented copper component to which the **brushes** are connected. These commutator sections allow electricity to be transmitted to the field windings of the motor, making it operate.

It is common for commutator sections to fail as the fuel pump gets older. This will normally lead to intermittent starting problems.

Figure 3.62 Inductive amps clamp

> **Key terms**
>
> **Graduated container** – a measuring jug, or similar container.
>
> **Commutator** – a segmented electrical contact mounted on the end of a motor armature. It is designed to change electrical polarity as the motor turns.
>
> **Brushes** – spring-loaded electrical contacts that transfer current to the rotating armature of a motor.

Level 3 Light Vehicle Technology

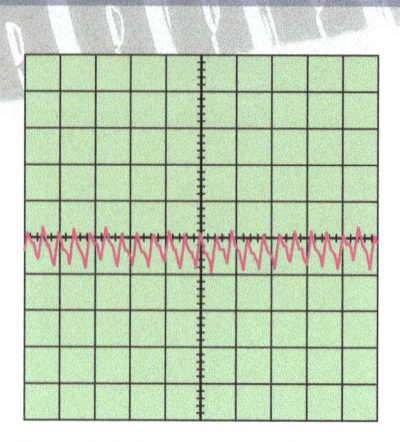

Figure 3.63 Commutator ripple

Depending on where the fuel pump stops when it is switched off, it may come to rest at a dead spot on the commutator. If the commutator section is not functioning correctly, the motor will not start to spin unless shaken or tapped.

Using an oscilloscope will allow you to test the commutator without removing it from the fuel tank.

- Connect an inductive amps clamp to the oscilloscope in the appropriate sockets.
- Gain access to the feed wiring of the fuel pump. Although it is possible to connect the inductive amps clamp at the fuse box, for this test it is normally recommended to get as close to the fuel pump as possible.
- With the inductive amps clamp around the feed wire to the fuel pump, switch on the system and select an appropriate scale on the oscilloscope.

A typical fuel pump commutator ripple is shown in Figure 3.63.

As each commutator section connects with a brush, current will build in a **field winding**, creating a slope on the wave form. This slope will drop sharply as the commutator moves away from the brush. Any faulty commutator sections will now be displayed as flat lines or voltage spikes (these may be positive or negative spikes).

If your oscilloscope has the facility to add cursors that measure **delta time**, you can also calculate the speed of the fuel pump without removing it from the fuel tank. Most fuel pump motors have six or eight commutator sections. This means that if you freeze the pattern on the screen, you should be able to see a repeat ripple showing that the pump has made one complete revolution. If you add the cursors to the peaks of these two ripples, you can measure delta time. If you divide the delta time into 60,000, this will give you the rpm of the pump. (Between 3000 and 7000 rpm are approximate figures for a petrol pump.)

The fuel pump is normally connected to a power source via a relay. This relay will only operate when it receives an earth via the ECU. The ECU will only operate the relay when a speed signal is received in the engine showing that it is cranking or running. In this way, if an accident were to occur and the fuel lines split: the engine would cut out (no longer rotating); the ECU would not receive a cranking signal and would cut the earth to the fuel pump relay, stopping the fuel pump.

> **Key terms**
>
> **Field winding** – the electromagnetic copper wiring inside a motor.
>
> **Delta time** – the difference in duration between two points on an oscilloscope wave form.

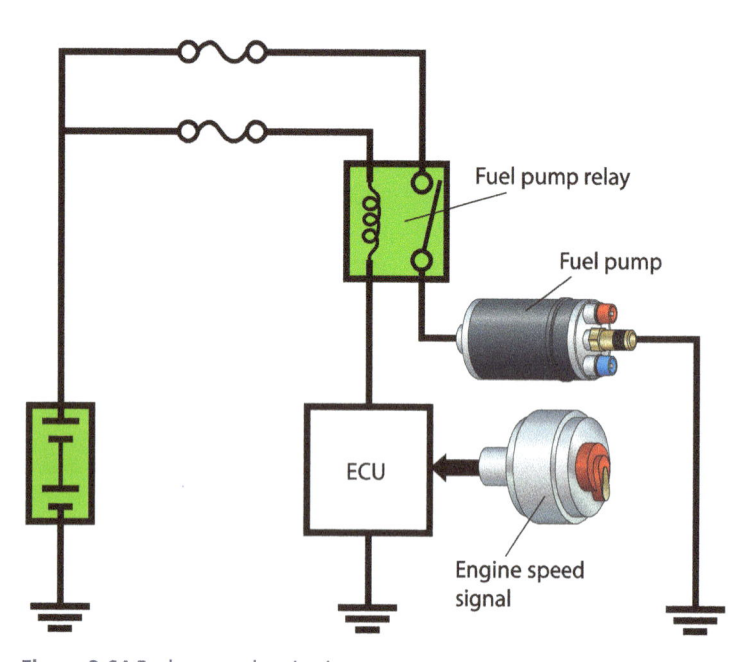

Figure 3.64 Fuel pump relay circuit

3 Diagnosis & rectification of light vehicle engine faults

Checking the fuel pump electrical supply circuit

If the fuel pump fails to operate you should you should check the system supply voltage (for more information see Chapter 4, page 187) at the following points:

- the fuel pump fuse
- the fuel pump relay
- the fuel pump feed connection
- the fuel pump earth connection.

When testing always start with the parts of the circuit that are easiest to access first. Voltage should be available all the way to the fuel pump feed connection but should not be seen on the earth connection.

Figure 3.65 Fuel filter

Fuel filters

Because of the very small tolerances within an electronic fuel injection system, small amounts of dirt would easily block fuel injectors. As a result, an inline fuel filter should be fitted. These are service items, but are often overlooked when diagnosing fuel starvation problems. If services are missed, then these fuel filters may not be replaced for some considerable time and may become blocked. If current draw on the fuel pump is high, you should suspect the condition of the fuel filter.

Fuel pressure regulators

A known pressure in the fuel pressure rail is vital to the ECU's calculation of how much fuel is being delivered to the engine. Very few electronic fuel injection systems have methods for measuring fuel flow. Maintaining a known pressure in the fuel rail is the job of the fuel pressure regulator.

If a known pressure of fuel exists in the rail, then the ECU does not have to measure the flow or quantity of petrol, but merely needs to know how long to hold the fuel injector open to calculate the amount of fuel being delivered.

Fuel pumps are designed to deliver high volumes of fuel, at a far higher pressure than required by the fuel injection system. In this way, a surplus of fuel should always be available under all engine running conditions.

> **Did you know?**
>
> Some fuel injection systems are returnless, which means that fuel flow and pressure are regulated in the tank. If a vehicle uses a returnless system, you will not see a pressure regulator at the fuel rail.

Pressure regulator operation

Fuel from the pump is delivered to one end of the fuel rail from which the injectors feed. At the other end of the fuel rail is a return system to the petrol tank. How much fuel returns to the tank is controlled by the fuel pressure regulator. A spring-loaded valve is fitted inside the fuel pressure regulator that will lift off its seat when a predetermined pressure is reached (approximately 3 bar). If, for example, 10 bar of fuel pressure was delivered to the rail, the regulator would maintain 3 bar in the rail and allow 7 bar to return to the tank.

Level 3 Light Vehicle Technology

Atmospheric pressure outside the fuel injection system is not always the same as the pressure inside the intake manifold. For example, at low speeds with the throttle butterfly restricting the airflow, manifold pressure is usually negative. Because of this, pressures at the nozzle of the injector vary with engine load. This will have an effect on the amount of fuel being delivered into the intake manifold.

For example:

- At low engine speeds, large vacuums exist inside the manifold, effectively sucking extra fuel through the injector and delivering a rich mixture.
- At high engine speeds, manifold pressure is almost atmospheric, so less fuel is being sucked from the nozzle of the injector and the system may deliver a weak mixture.

To overcome these problems, many fuel pressure regulators are adaptable to engine speed and load. A spring-loaded valve inside the pressure regulator is connected to a diaphragm. One side of the diaphragm is connected to the intake manifold via a vacuum pipe. With the engine running, the vacuum in the manifold acts directly on the diaphragm of the fuel pressure regulator. Manifold pressure will now alter the value of fuel pressure that is needed to lift the valve off its seat.

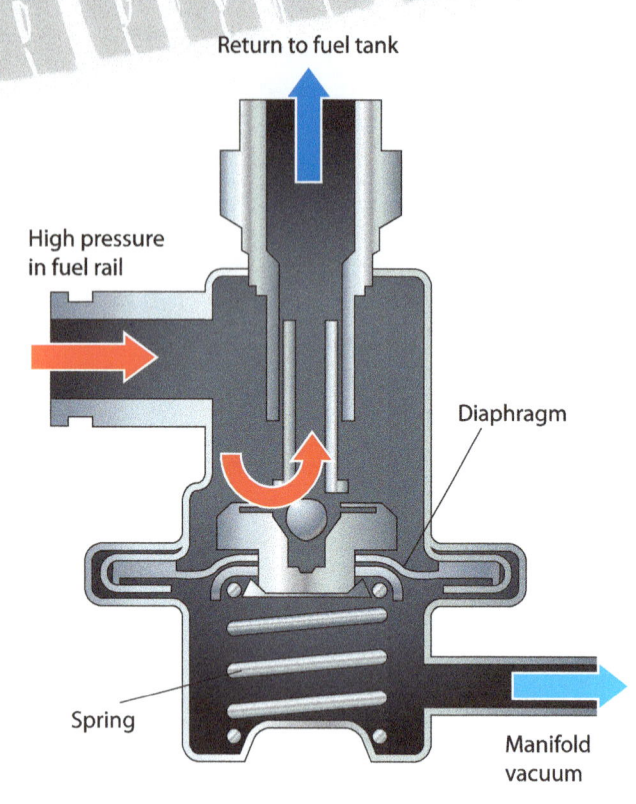

Figure 3.66 Vacuum-operated fuel pressure regulator

For example:

- Low engine speed, with a large vacuum in the manifold, draws down on the spring-loaded valve, allowing it to lift off its seat earlier so pressure in the fuel rail is reduced. A low pressure in the fuel rail will help to cancel out the vacuum at the tip of the fuel injector and maintain the correct quantity of fuel delivered.

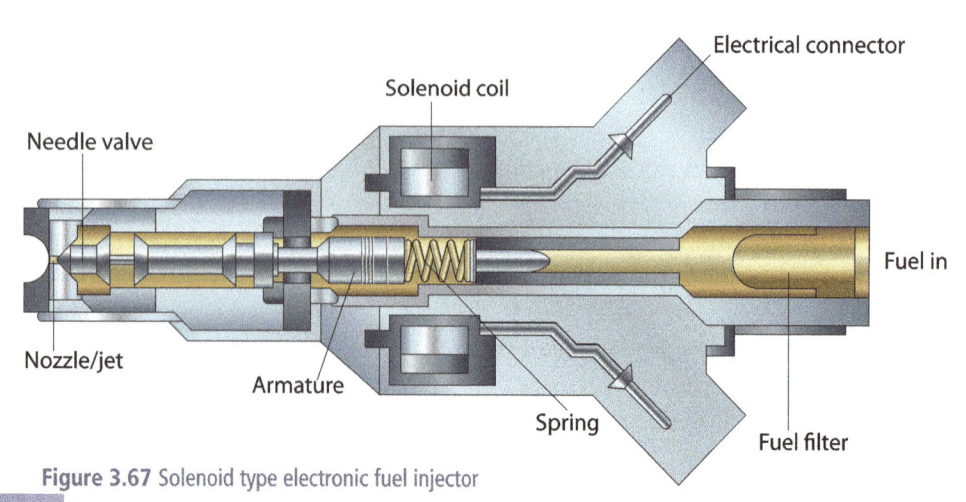

Fuel injectors

Fuel injectors are of the solenoid type. A small coil of wire is wound around a movable armature. When current is applied to this coil, a magnetic field is created, drawing the armature through the middle of the winding. When current is removed from the coil, the magnetic field collapses, and a return spring is used to move the armature back to its original position.

Figure 3.67 Solenoid type electronic fuel injector

The armature forms a needle valve at one end and fuel pressure is supplied to the needle valve. When the winding of the solenoid is supplied with current and the needle of the armature lifts from its seat, fuel can be sprayed under high pressure into the intake manifold.

Air induction

In a petrol engine, speed and load are controlled by the position of the throttle butterfly. As the throttle butterfly is opened, more air is allowed into the engine, and so the engine speeds up. The throttle butterfly can be mechanically operated by an accelerator cable or, more usually on modern vehicles, it can be electronically controlled. A sensor picks up the position of the accelerator pedal, and motors drive the throttle butterfly into the correct position with no direct connection. So that the fuel injection system ECU can calculate engine demands (i.e. load and speed), throttle position and air quantity must be measured.

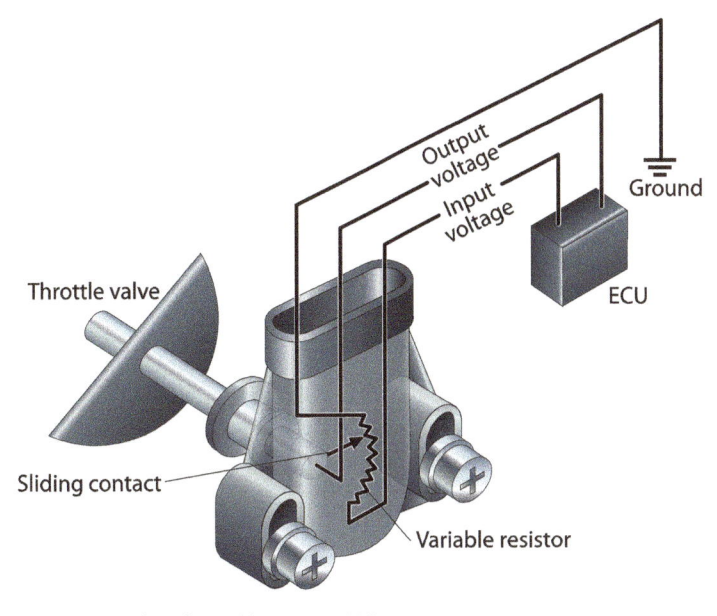

Figure 3.68 Throttle position sensor TPS

Throttle position sensors

A potentiometer (variable resistor) can be mounted on the throttle body. The potentiometer is able to accurately measure the position of the throttle butterfly (how far the butterfly is open). For information on how to test a throttle position sensor, see page 141.

Auxiliary air valves

Engine demands can vary, even at a tickover. As electrical loads, power steering and air-conditioning are operated, idle speed can fall to a level where it can't be maintained and the engine will stall. If small quantities of air are allowed to bypass the throttle butterfly, tickover speed can be increased when needed. In many cases, this air bypass is controlled by a small valve or solenoid actuated by the engine management ECU. By regulating the amount of air that passes the throttle butterfly, correct idle can be maintained. This is sometimes called idle air control (IAC).

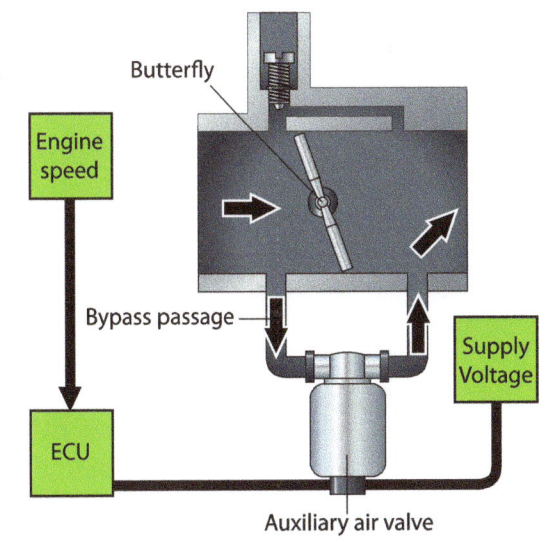

Figure 3.69 Auxiliary air valve

Plenum chambers

The length of the air inlet tract (from intake to valve) will vary from manufacturer to manufacturer. The ideal length of the air inlet tract depends on engine load and speed demands.

Air induction is not a smooth process. In a multi-cylinder engine, each piston will be on a different stroke, which creates

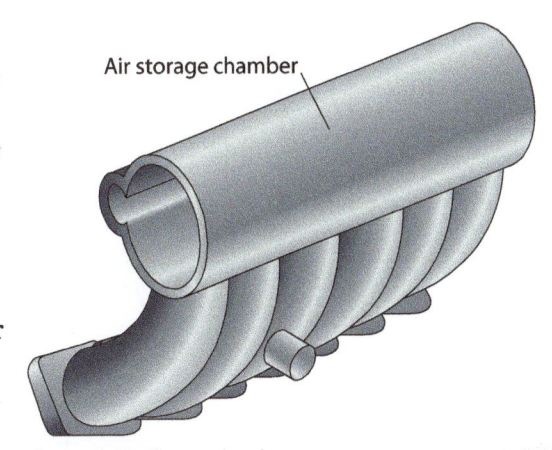

Figure 3.70 Plenum chamber

137

Level 3 Light Vehicle Technology

Figure 3.71 Airflow meter

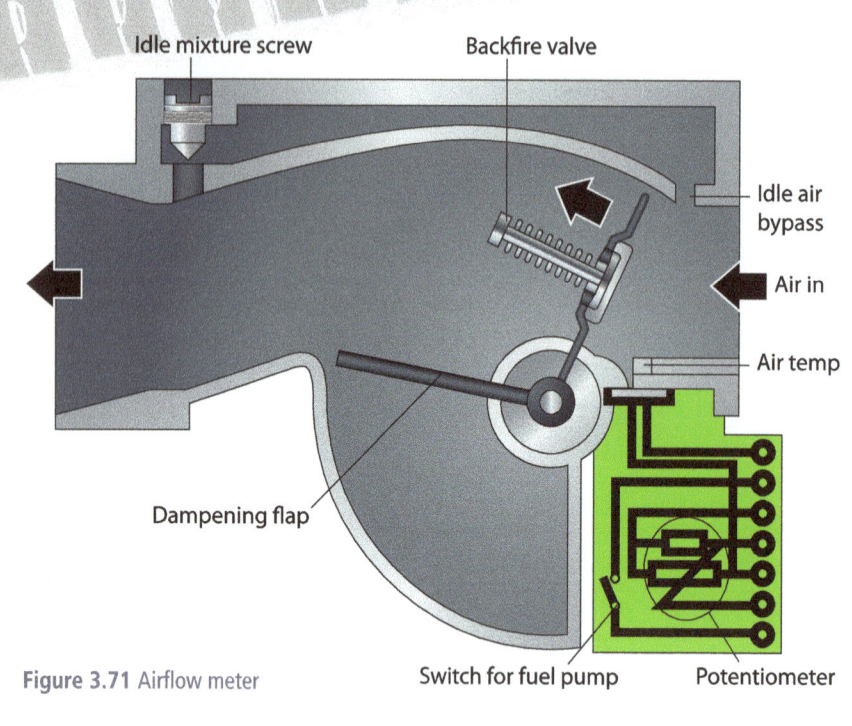

Labels: Idle mixture screw, Backfire valve, Idle air bypass, Air in, Air temp, Potentiometer, Switch for fuel pump, Dampening flap

a pulsing of the inducted air. The pulsing induction places different requirements for quantities of air and fuel and the distance it must travel along the air intake tract.

To help overcome this, some manufacturers use a plenum chamber in the design of their intake manifolds. This is a chamber that can be used as a reservoir of air from which the cylinders can draw. This means the intake air does not have to travel the entire length of the intake tract.

Air quantity measurement

To maintain the correct air/fuel ratio, the engine management system needs to know the amount of air being drawn into the cylinders; this is the job of air measurement sensors.

Airflow meters

Early electronic fuel injection systems used airflow meters to measure the quantity of air entering the engine.

An airflow meter consists of a small spring-loaded flap that is pushed open by the incoming air. On top of the spring-loaded flap is a potentiometer (variable resistor) that can accurately measure position, giving an indication of the amount of air entering.

Another flap is attached at right angles to the first, and during operation it is moved within a damping chamber. Induction pulses act evenly on both flaps at the same time, reducing fluctuation and preventing incorrect signals being sent to the ECU.

Testing the airflow meter with an oscilloscope

- The airflow meter is a passive sensor, which means that it is fed with voltage from the ECU.
- Three wires are normally needed to operate the airflow meter. These wires will be feed, earth and a signal wire.
- First, connect the probes to the oscilloscope and select a single channel. Then connect the black (ground) wire from the oscilloscope to a good earth.
- You can now use the coloured probe to test for a signal.
- If you don't know which one is the signal wire, you can probe all three.
  - The earth wire will give a flat line at a 0V.
  - The ECU feed will give a flat line at approximately 5V.
  - The signal wire will give a signal proportionate to how far the flap is opened.

Did you know?

Flap type airflow meters are inefficient and as a result they have been replaced with other forms of air measurement on modern fuel injection systems.

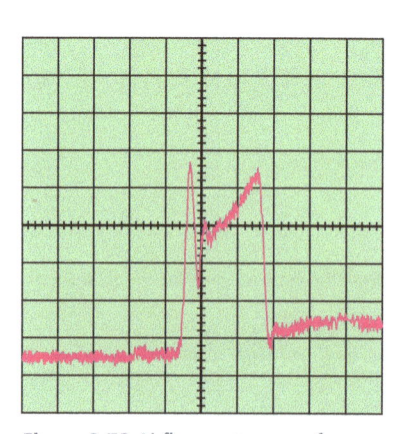

Figure 3.72 Airflow meter wave form

3 Diagnosis & rectification of light vehicle engine faults

You can perform this test with the engine running or just with the ignition on.

- With the engine running, slowly increase the revs, allow them to settle and then release. The wave form created on the screen should have a smooth curve up and down (this will show that the potentiometer is operating correctly with no dead spots).
- With the ignition on, you can operate the flap inside the airflow meter manually (with your finger or a screwdriver). As you carefully open and close the flap, a smooth curve up and down should appear on the wave form.

Manifold absolute pressure sensors (MAP)

A manifold absolute pressure sensor can also be used to calculate air intake for the engine, and will also give a good indication of engine load. For any given size of manifold, if the amount of vacuum or low pressure can be measured, the ECU can calculate roughly how much air is being drawn.

The manifold absolute pressure sensor is connected to the intake manifold, either directly or by a vacuum pipe. The low pressure (depression) created in the manifold will be sensed by a diaphragm, converted into a signal, and sent to the ECU for calculation.

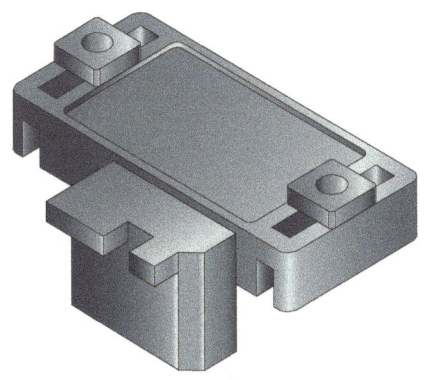

Figure 3.73 Manifold absolute pressure sensor (MAP)

Testing a manifold absolute pressure sensor with an oscilloscope

Two types of manifold absolute pressure sensor are in common use: analogue and digital. The testing procedure is the same for both types, although the wave form created on the screen of the oscilloscope will be different.

- First, connect the probes to the appropriate ports on the oscilloscope. Then connect the black probe to a good source of earth.
- The manifold absolute pressure sensor will have three wires:
  - feed from the ECU
  - earth
  - a signal wire.
- Select a single channel on the oscilloscope then probe these wires. If you don't know which one is the signal wire:
  - The earth wire will give a flat line at 0V.
  - The ECU feed will give a flat line at approximately 5V.
  - The signal wire will give a varying voltage proportional to the amount of vacuum being drawn inside the manifold.

If it is an analogue sensor, a rising and falling sine wave will be created that will look very similar to the one produced by the airflow meter – see Figure 3.74.

If it is a digital sensor, a small integrated circuit will convert the signal into an on and off square wave form. The frequency of this wave form will change in proportion to the amount of depression created in the manifold – see Figure 3.75.

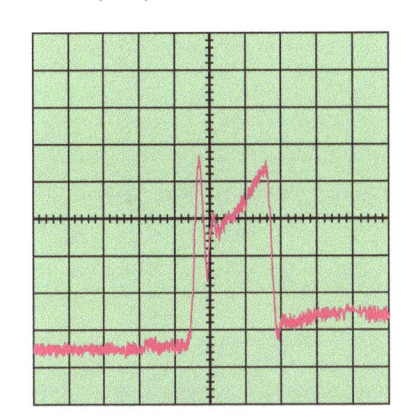

Figure 3.74 Manifold absolute pressure sensor wave form (analogue)

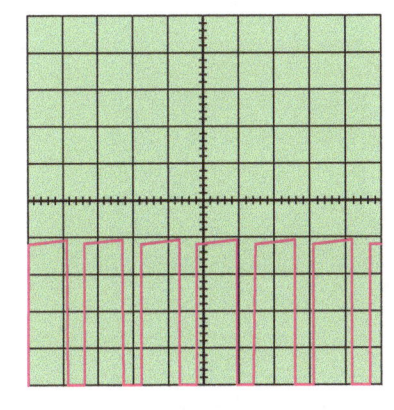

Figure 3.75 Manifold absolute pressure sensor wave form (digital)

Level 3 Light Vehicle Technology

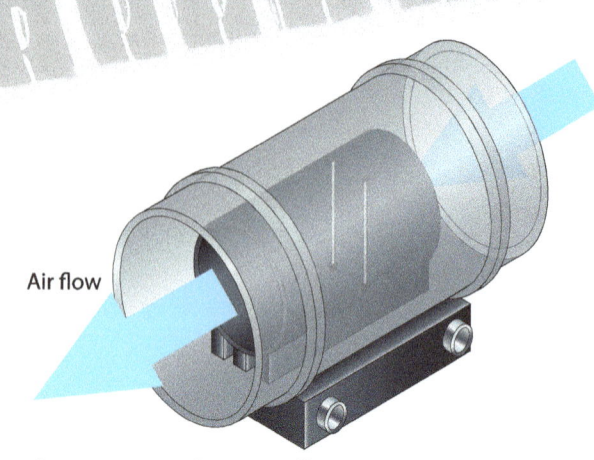

Figure 3.76 Hot wire mass airflow sensor MAF

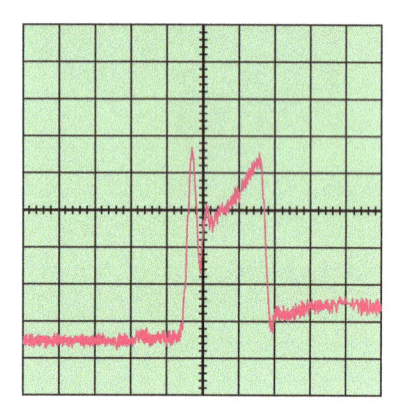

Figure 3.77 Hot wire mass airflow sensor MAF wave form

Key term

Transducer – an electrical device that converts one form of energy into another.

Did you know?

It is often possible to tell the difference between a throttle position switch and a throttle position sensor by listening very carefully. You can sometimes hear a small click as the throttle is allowed to return to the idle position or opened fully to the wide open throttle position (this is the sound of the switch opening and closing).

Hot wire mass airflow sensors (MAF)

A hot wire mass airflow sensor is a very accurate way of measuring the amount of air entering an engine.

Normally mounted before the throttle butterfly, the mass airflow sensor contains wires or heating elements, sometimes called hot film. When supplied with an electric current, they will get hot. As the intake air passes over these wires or elements, they are cooled. To maintain a constant heat, the ECU must supply more electric current. As a result, the amount of current required to keep the wires at a constant temperature is proportional to the mass of air entering the engine.

Testing a hot wire mass airflow sensor with an oscilloscope

As with flap type airflow sensors and MAP sensors, an MAF should have three wires.

- First, connect the oscilloscope probe wires to the appropriate sockets and select a single channel. Then connect the black probe to a good source of earth.
- With the engine running, use the coloured probe to check the output signal from the mass airflow sensor.
- If you don't know which wire is the signal wire, you can probe all three:
  – The earth wire will give a flat line at 0V.
  – The feed wire will give a flat line at approximately 5V.
  – The signal wire will produce a trace on the screen that rises and falls with airflow.

Barometric pressure sensors

Some engine management systems make allowances for air pressure due to weather conditions and altitude. In this way, the density of the oxygen in the air can be calculated. A small pressure **transducer** can be mounted in the air intake system, normally before the throttle butterfly, and works in a similar way to a manifold absolute pressure sensor. You can use live data from a scan tool to help check the correct function and operation of a barometric pressure sensor.

Throttle position switches

Early electronic fuel injection systems used throttle position switches to help provide the ECU with information. They normally have three connections, which could include: feed in, feed out at idle and feed out at a wide open throttle.

Voltages applied at the feed out wires can be detected by the ECU when the throttle is closed, as the switch will make the idle circuit live. It can also detect when the throttle is wide open, as this circuit then becomes live.

3 Diagnosis & rectification of light vehicle engine faults

Testing the throttle position switch

You can use a multimeter to check a throttle position switch.

- Connect the red and the black probes to the appropriate sockets and turn the selector dial to 20V DC. Then connect the black probe to a good source of earth.
- Switch the ignition on and test the three wires of the throttle position switch.
- By probing the wires, you should find one with a live feed in (usually around 5V).
- When the throttle is in the closed position, one wire on the idle circuit should also become live.
- With a wide open throttle, the final wire should become live.

Figure 3.78 Adjusting a throttle position switch

Some throttle position switches are adjustable, with slotted mounting screws. With the multimeter connected to the idle circuit and the throttle butterfly closed, you can slowly adjust the switch so that it is just in the closed position and the ECU receives the correct signal.

Throttle position sensors

Throttle position sensors use potentiometers (variable resistors). A sliding contact moves along a variable resistor inside the potentiometer and sends the ECU information about throttle position. Many engine management systems are now 'drive by wire', which means that there is no throttle cable. Potentiometers are used to sense and dictate the position of the throttle butterfly.

Testing a throttle position sensor with an oscilloscope

As with many of the other sensors described, throttle position sensors are passive. They normally have three wires (unless this is a multiple track potentiometer):

- a feed in (usually around 5V)
- an earth wire
- a signal wire (there may be more than one).

- Connect the probes to the correct sockets on the oscilloscope then select a single channel.
- Connect the black probe to a good source of earth.
- Use the coloured probe to find the signal wire. (If you don't know which one the signal wire is, you can probe all three wires.)
- The earth wire will give a flat line at 0V, the feed wire will give a flat line at approximately 5V, and the signal wire will give a voltage proportional to the amount that the throttle is being opened.
- As the throttle is operated, you should see a pattern on the screen that rises and falls smoothly in relation to throttle operation.

Did you know?

As a failsafe method, it is now common to find multiple tracks (resistors) within a throttle position sensor. This means that you may find more wires than are described in the testing instructions below. You should test each wire found on the sensor connector plug individually to ensure correct function and operation.

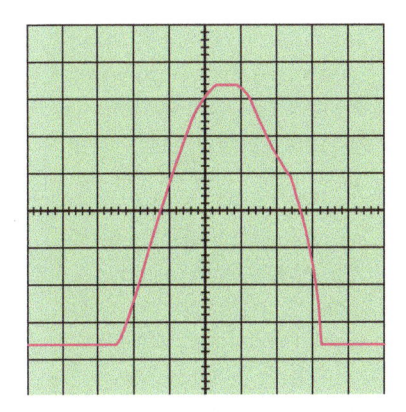

Figure 3.79 Throttle position sensor wave form

Engine coolant temperature sensors (ECT)

As fuelling demands vary according to engine temperature (for example, a cold engine requires more fuel, similar to applying a choke on a carburettor), the engine's ECU needs to measure temperature in order to control fuelling as required.

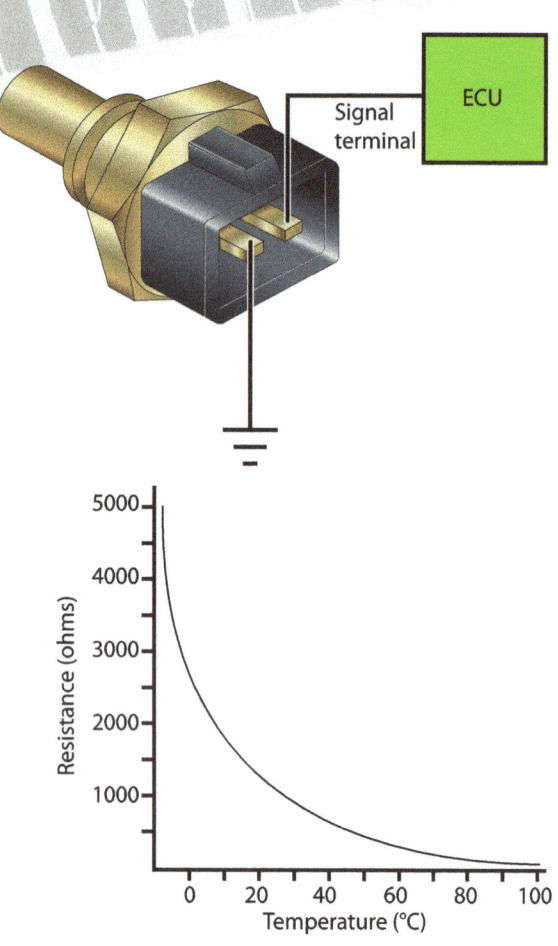

The engine coolant temperature sensor (ECT) is a heat sensitive variable resistor called a **thermistor**.

Two types of ECT are in common use:

- negative temperature coefficient NTC
- positive temperature coefficient PTC.

Testing engine coolant temperature sensors

Testing using a multimeter

As an ECT is a temperature sensitive resistor, you can use a multimeter on the ohms scale to determine whether its resistance changes with heat.

- Connect the multimeter probes to the appropriate sockets and select a low ohms scale on the multimeter dial.
- Then calibrate the ohmmeter by touching the probes together to see if the display shows zero resistance.
- With the wiring harness plug disconnected, you can warm up the ECT in a cup of hot water.

The resistance values should change (check the manufacturer's specifications):

- NTC thermistor: as temperature rises, resistance should fall.
- PTC thermistor: as temperature rises, resistance should increase.

Engine coolant temperature sensors are normally one or two wire passive sensors. A voltage is applied down the signal wire from the ECU, and as the engine warms up, the voltage on the signal wire will rise and fall due to the resistance of the thermistor.

You can test the voltage change using a voltmeter:

- Set the multimeter to 20V DC and connect the probes to the appropriate sockets.
- Then connect the black (common) probe on the multimeter to a good source of earth.
- With the wiring harness still connected, probe the signal wire of the thermistor and start the engine.
- As the engine warms up, voltage will either rise or fall depending on the type of thermistor fitted:
  - NTC thermistor: voltage should fall as the engine warms up.
  - PTC thermistor: voltage should rise as the engine warms up.

Figure 3.80 Engine coolant temperature sensor ECT

Key term

Thermistor – a temperature sensitive resistor.

Did you know?

A negative temperature coefficient thermistor is the most common type, although positive temperature coefficient sensors are sometimes used and may look identical externally. Many coolant temperature sensors are colour-coded, and the colour may also match the wiring harness plug. If a new coolant temperature sensor is purchased, and is supplied as a different colour from the original, it is worth checking that it is the correct one for the vehicle. It may be a modification, or it may be the wrong type of sensor.

3 Diagnosis & rectification of light vehicle engine faults

Testing using an oscilloscope

- Connect the oscilloscope probes to the appropriate sockets and select a single channel.
- Connect the black probe to a good source of earth and use the coloured probe to test the signal wire.
- Select a fairly low voltage scale and use the slowest timescale.
- When the engine is started and run, the wave form on the screen should rise or fall as the engine warms up:
  - NTC wave form will fall.
  - PTC wave form will rise.

Air temperature sensors

As the density of oxygen changes due to air temperature, the ECU must be able to measure this so that it can adjust the amount of fuel injected. On a cold day, the air charge will be denser and contain more oxygen. If this is not compensated for, stoichiometric values may be inaccurate.

Air temperature sensors may be NTC or PTC thermistors. You can test function and operation in exactly the same way as for an engine coolant temperature sensor (ECT) (see page 142).

Engine speed sensors

A number of methods are used to sense engine speed, but they are mainly confined to crankshaft position sensing or ignition primary trigger signals.

The ignition pulses or analogue waves produced by the sensors can be converted into digital frequency signals that are translated into engine speed at the ECU. You can test the signals using the same process as described for electronic ignition inductive sensors (see page 111).

Lambda sensors (pre- and post-catalytic converter)

To make sure that the correct air/fuel ratio is achieved from the engine management system, an oxygen sensor (also called a lambda sensor) is fitted in the exhaust system before the catalytic converter. The lambda sensor measures the oxygen content of the exhaust gases and instructs the ECU if the engine is running too rich or too weak.

- Too much oxygen in the exhaust gas and the engine is running weak.
- Too little oxygen in the exhaust gas and the engine is running rich.

Many lambda sensors use a **zirconia** ceramic to measure the oxygen content. One end of the sensor is inserted into the exhaust gas and the outer end is exposed to the outside air. The difference in oxygen between these two points causes a corresponding resistance in the zirconia ceramic and signals the ECU, which can now calculate the amount of oxygen in the exhaust gas. The resistance in the sensor will continuously rise and fall, creating an average figure that can be used to maintain an ideal air/fuel ratio.

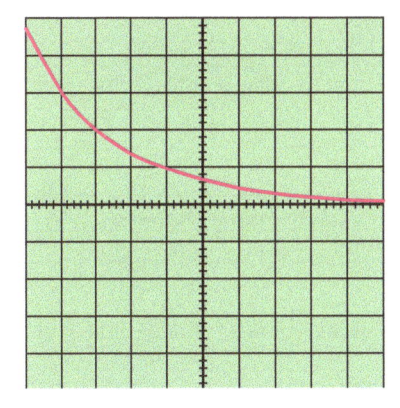

Figure 3.81 Engine coolant temperature sensor wave form

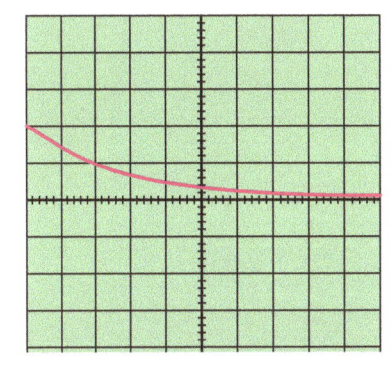

Figure 3.82 Air temperature sensor wave form

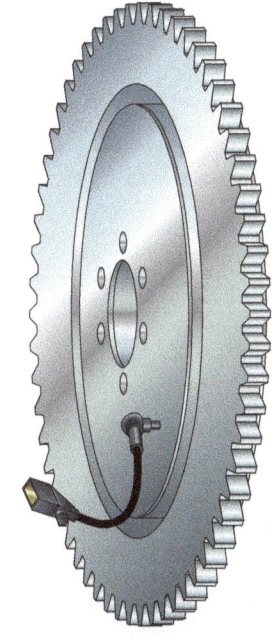

Figure 3.83 Crankshaft position sensor (CPS)

Level 3 Light Vehicle Technology

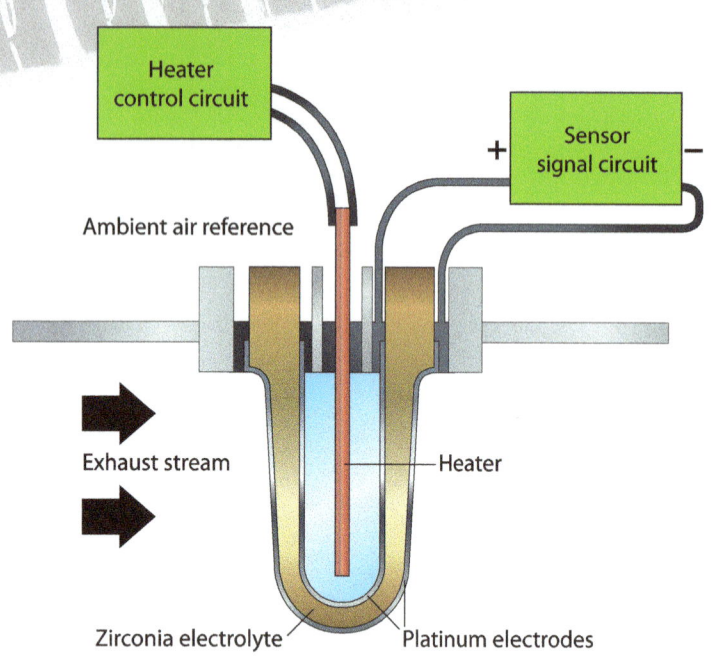

Figure 3.84 Lambda sensor

Key terms

Zirconia – a steel-grey strong metallic element with a high melting point that is used especially in combination with other metals.

Open loop – when the engine management ECU operates on pre-programmed values from its memory.

Closed loop – when the engine management ECU operates with signals supplied by sensors.

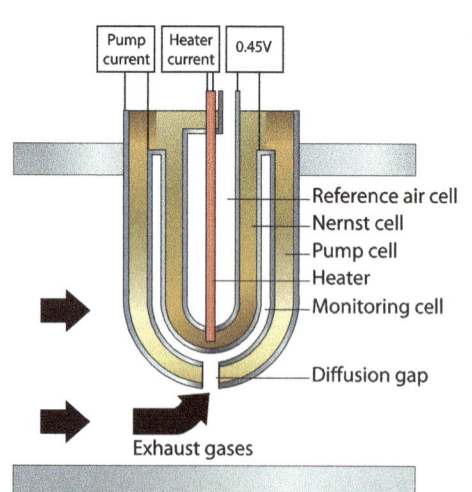

Figure 3.86 Wideband lambda sensor

A lambda sensor will only begin to operate correctly once it has reached an operating temperature above 300°C. Because of this, it normally contains an additional heating element to get it up to operating temperature as quickly as possible. This additional heating will also help prolong the lifespan of the sensor.

- During the warm-up period of the catalytic converter and lambda sensor, signals from the oxygen sensor to the ECU are ignored. When this happens, the engine management system is running **open loop**.
- As soon as the catalytic converter and lambda sensor are up to the required operating temperature, the signals produced are used to correct fuel injection for the ideal air/fuel ratio. When this happens, the engine management system is running **closed loop**.

To make sure that the catalytic converter is operating correctly, some manufacturers include a second lambda sensor mounted after the catalytic converter. The signals produced by the two lambda sensors (pre-cat and post-cat) are compared. If they are found to be similar, no chemical reaction is taking place within the converter and therefore the catalyst has failed. The engine malfunction indicator lamp (MIL) will be illuminated and a diagnostic trouble code stored.

Figure 3.85 Pre- and post-catalytic converter lambda sensors

The failure of a lambda sensor will normally illuminate the engine management malfunction indicator lamp (MIL). A failure will also affect the efficiency of the injection system and catalytic converter operation, leading to excessive exhaust emissions.

If the MIL is illuminated, the engine management system should be checked for any stored diagnostic trouble codes using a scan tool as described in Chapter 4, page 196.

If a fault with the heater circuit is indicated this can be tested using the following method:

1. Make sure the exhaust is cool.
2. Disconnect the sensor's wiring harness and set your multimeter to read 'Ohms'.

3. Connect the ohmmeter to the two heater wires.
4. The resistance should be between 1 and 20 ohms depending on type of lambda sensor.
5. Normal failure condition is a burnt-out heater leading to a very high or open-circuit reading.
6. If the resistance is high or it is an open circuit the sensor should be replaced.

If a fault with the oxygen sensor circuit is indicated it can be tested using an oscilloscope using the following method:

1. First connect the probes to the oscilloscope and select a single channel and then connect the black (ground) wire from the oscilloscope to a good earth.
2. You can now use the coloured probe to test for a signal.
3. Start the engine and let it warm up to normal operating temperature.
4. Adjust the voltage and timescales so that an image appears on the screen of the oscilloscope.
5. Raise and hold the engine revs to around 2000 rpm and, if the lambda sensor is working correctly, a rising and falling waveform should be seen.

Specialist lambda sensor testers are also available. They consist of a row of eight or ten LEDs which, when connected to an operating sensor, will light up progressively according to the sensor voltage output.

> **Action**
>
> Using a vehicle from your workshop, identify the following sensors (if fitted):
>
> 1. Mass air flow sensor
> 2. Manifold absolute pressure sensor
> 3. Throttle position sensor
> 4. Intake air temperature sensor
> 5. Engine coolant temperature sensor
> 6. Camshaft position sensor
> 7. Crankshaft position sensor
> 8. Knock sensor
> 9. Oxygen (lambda) sensors (pre and post catalytic converter).
>
> How many did you find/identify?

NEW TECH

Wideband oxygen sensors

To improve the efficiency of the emission control system, some manufacturers use wideband oxygen sensors. This design of lambda sensor does not rely solely on the change of resistance in a zirconia ceramic. Instead, two chambers are created: one containing exhaust gas and the other open to air for reference (see Figure 3.86).

A component called an oxygen pump is embedded in the wideband sensor, and through an electrochemical process, it tries to maintain a stable oxygen quantity in one chamber. The quantity of oxygen in this chamber is measured by the zirconia ceramic. The amount of current supplied to the oxygen pump to maintain the correct oxygen content is proportional to the amount of oxygen in the exhaust gas. From this information, the engine management ECU is able to maintain a very stable air/fuel ratio.

Gasoline direct injection (GDI)

Advances in fuel delivery processes have allowed manufacturers to design petrol engines with direct injection into the combustion chamber. Many of the sensors discussed in the section on multipoint fuel injection are still used with this system, but the operation of the fuel injectors is different.

Level 3 Light Vehicle Technology

In this system, fuel is delivered to a common fuel rail from an engine driven high pressure pump. Instead of being mounted in the intake manifold, the solenoid type fuel injectors are located in the cylinder head, with the injector tips inside the combustion chamber. When the engine is operating, only air is drawn in through the intake valve and fuel can be injected in varying quantities directly into the combustion chamber. The design of the combustion chamber, injector position and shape of the piston crown allows the engine management system to inject fuel and mix it with the incoming air in various modes.

The direct injection of petrol gives the opportunity to create 'lean burn' technology for economy and low emissions. Two modes of injection can be created:

- **Homogeneous** – the air and fuel are evenly mixed in the cylinder to provide a smooth, even delivery of power when required with an air fuel ratio of 14.7:1 by mass.
- **Stratified** – small quantities of petrol are injected in a concentrated layer around the area of the spark plug. This will provide a very lean mixture of around 40:1 by mass for driving situations where performance is not required and economy and emissions are desirable.

Actuators (output signals)

Once sensor information has been processed by the engine management system, signals will be sent out to operate controlling components known as actuators.

Electronic fuel injectors

An electronic fuel injector is a solenoid operated valve. Depending on design, fuel may enter at the top of the injector or around the body, normally through a small mesh gauze filter.

The fuel injector contains a small solenoid coil, which when energised draws an armature needle from its seat, allowing the high pressure fuel from the rail to be **atomised** into the inlet manifold. When current is removed from the solenoid coil, a return spring closes the armature needle and fuel injection stops.

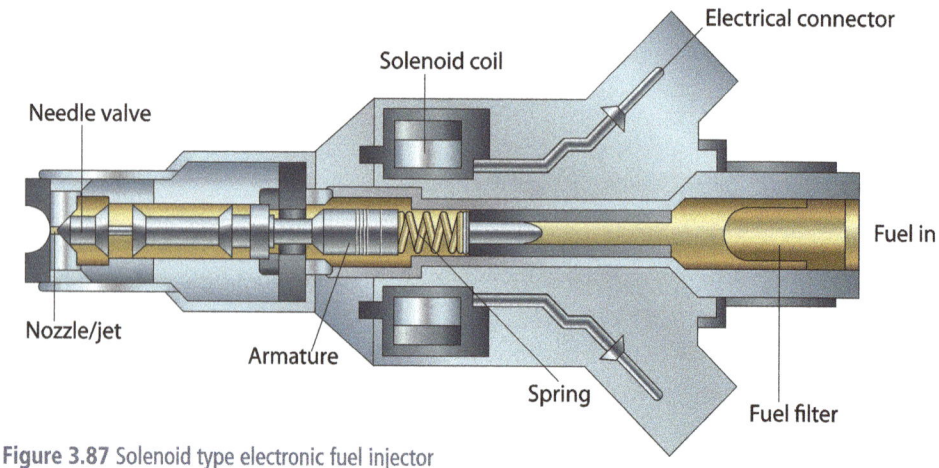

Did you know?

Lean burn technology can create large quantities of the exhaust pollutant oxides of nitrogen (NOx). To help monitor oxides of nitrogen and regulate injection, GDI systems include a NOx sensor after the catalytic converter, in a similar position to a post-cat lambda sensor.

Key term

Atomised – broken up into very small particles or a fine spray.

Figure 3.87 Solenoid type electronic fuel injector

3 Diagnosis & rectification of light vehicle engine faults

Testing electronic fuel injectors

Noid lights

Noid lights are small light emitting diode (LED) testers. They are normally specific to manufacturers, and can be purchased in sets to enable testing of a wide range of fuel injectors. When connected to the wiring harness at the fuel injectors, with the engine cranking or running, they will flash if an operating signal is being correctly produced.

Multimeter testing

You can test the resistance of the solenoid coil in an electronic fuel injector with an ohmmeter.

- Switch off the fuel injector and disconnect the wiring harness.
- Connect the multimeter probes to the appropriate sockets and select a low ohms scale setting.
- Calibrate the ohmmeter by holding the two wires together: a reading of 0 ohms resistance should be recorded. If there is any other reading on the ohmmeter screen, you will need to add or subtract this value from your final reading.
- Now connect the two probes of the ohmmeter across the fuel injector terminals and compare the results with the manufacturer's specifications.

Figure 3.88 Noid lights

As the switching of the fuel injectors is too quick for a voltmeter to cope with, best results are obtained by using an oscilloscope.

Oscilloscope testing

- Connect the probes to the appropriate sockets on the oscilloscope then select a single channel.
- Connect the black probe to a good source of earth.
- Use the coloured wire to probe the signal wire from the fuel injector; this is usually the earth side.
- If you do not know which wire to probe, you can test both until a wave form appears on the screen.
- If the feed wire is probed, you will see a flat line of approximately battery voltage.
- If the signal wire is probed, you should see a pattern.

Level 3 Light Vehicle Technology

Wave form pattern description:

- With a single injection pulse, the wave form should start at approximately battery voltage.
- When the injector is switched on by the ECU (earthed), the wave form should drop to zero and be held open for a period; this is called the **pulse width**.
- When the injector is switched off (the earth circuit is broken), the magnetic field created in the solenoid coil will collapse. This induces a high voltage back electromotive force (EMF), which will produce a large voltage spike, somewhere in the region of 50V.
- As the back EMF dissipates, the wave form will fall in a sloping pattern until it once again returns to battery voltage.
- The process then starts over again for the next injection pulse.

> **Key term**
>
> **Pulse width** – the period when the injector is switched on.

> **Did you know?**
>
> After the high voltage spike produced by the back EMF, as the voltage returns to battery voltage, you can sometimes see a small disruption somewhere on the curve of the wave form. This tiny blip usually indicates the mechanical closing of the armature needle. This is due to the magnetic field being disrupted as the armature moves. Although the oscilloscope is measuring the electrical switching of the fuel injector, this doesn't mean that mechanical movement cannot be seen.

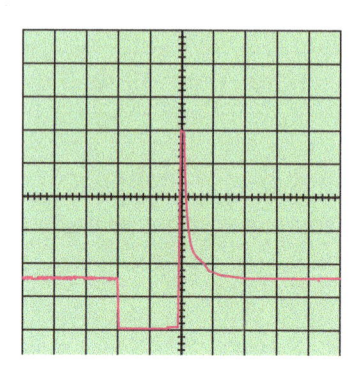

Figure 3.89 Injector voltage pulse width wave form

Amperage testing with an oscilloscope

You can also test the amperage used by the fuel injector using an oscilloscope.

- With an inductive amps clamp connected to the oscilloscope and around one of the wires of the fuel injector, select an appropriate scale.
- With the engine running, a wave form will be produced on the screen showing a rising curve as amperage builds up within the solenoid coil of the fuel injector.
- This curve should be relatively smooth and not too sharp, and when switched off should drop rapidly back to zero.
- If the amperage rises too quickly, resistance in the solenoid coil is too low.
- If amperage rises too slowly, resistance in the solenoid coil is too high.

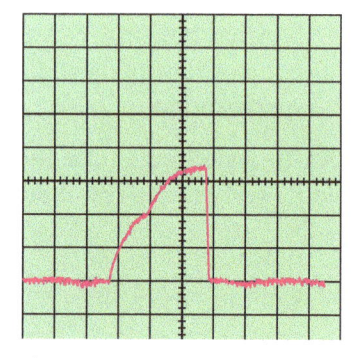

Figure 3.90 Injector amp ramp wave form

> **Did you know?**
>
> On close inspection, you can sometimes see a small blip on the amperage curve as current builds up. This is the disruption to the magnetic field in the solenoid coil produced by the opening of the injector armature.
>
> If the oscilloscope has two channels that operate at once, you can use one for voltage pulse width and the other for amperage ramp testing. The two wave forms can be superimposed. This will allow you to see true injector needle opening times by comparing the two small disruptions in the magnetic field when opening on the amps curve and closing on the volts curve.

Pump in tank returnless systems

In a standard petrol fuel injection system, heat created by the running engine is transferred to the fuel rail by conduction. This heat creates unpredictable pressures in the fuel rail and causes evaporation and pollution as petrol returns to the tank. Both of these conditions lead to inaccurate injection and unacceptable emissions.

Manufacturers are now producing returnless fuel injection systems. In this type of fuel injection system, the pressure regulator and filter have been moved inside the fuel tank and a pressure sensor is mounted on the fuel rail instead. Pressure is regulated in the tank by an internal return system before the fuel is sent to the rail. This stops engine heat being transferred into the fuel returned to the tank, and reduces evaporation.

> **Action**
> Connect an oscilloscope to a fuel injection system component and retrieve an operating wave form.

CHECK YOUR PROGRESS

1 Explain the terms homogeneous and stratified.
2 What is a noid light?
3 What is the purpose of the post-catalytic converter lambda sensor?

CI fuel systems (diesel)

Electronic diesel control (EDC)

Advances in electronic control have allowed standard diesel engines to be managed more accurately. Fuel injectors have been developed with needle motion sensors. Drive by wire systems are used, where a throttle potentiometer controls the injection pump via an ECU, instead of a throttle cable.

EDC has a number of advantages over a purely mechanical system:

- better control over the amount of fuel injected
- stable idle speed control
- the easy addition of cruise control
- electronic data gathering
- EGR control.

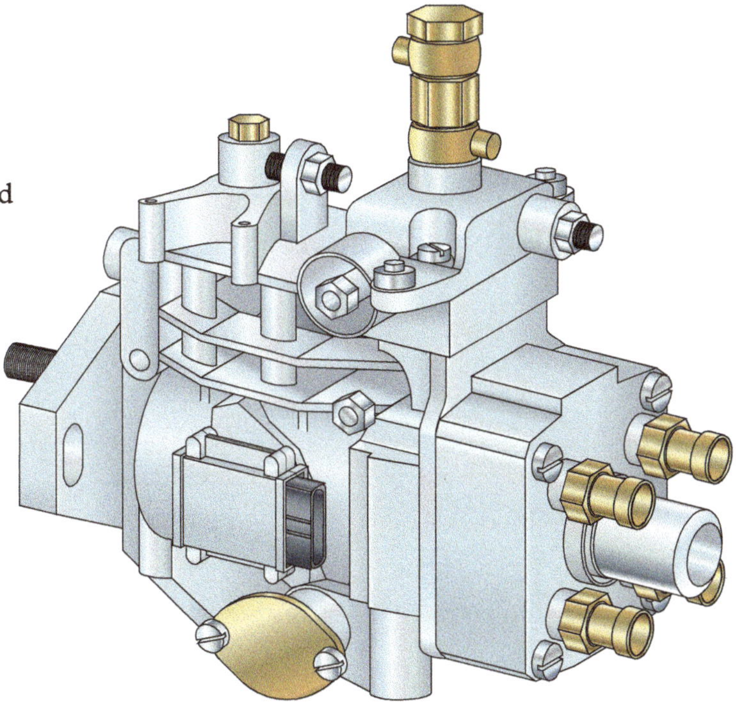

Figure 3.91 EDC injection pump

Common rail direct injection diesel

Modern diesel engines have moved away from mechanical pumps and injection systems and now operate in a very similar manner to electronic fuel injection in a petrol engine. A low pressure electric pump is used to transfer fuel from the tank to the engine through a fuel filter. An engine-driven high pressure pump is then used to raise the pressure of the fuel to around 1800 bar, at which point it is stored in a pressure accumulator or fuel rail that feeds all fuel injectors. As a result, this type of system no longer relies on engine speed to control injection pressures. (An early system with a mechanical pump provided poor pressure at low speeds or tickover.)

A number of sensors are used to monitor engine requirements, including load, speed and temperature. This information is then processed by the ECU and used to control electronic solenoid or piezoelectric fuel injectors, which atomise diesel directly into the superheated compressed air of the combustion chamber.

> **Safe working**
> Due to the extremely high pressures involved in common rail diesel injections systems, precautions need to be taken when working on these components. If the fuel system is not correctly depressurised before work is started, diesel can be released at high pressure causing severe injury or death. Always follow the manufacturer's instructions and recommendations.

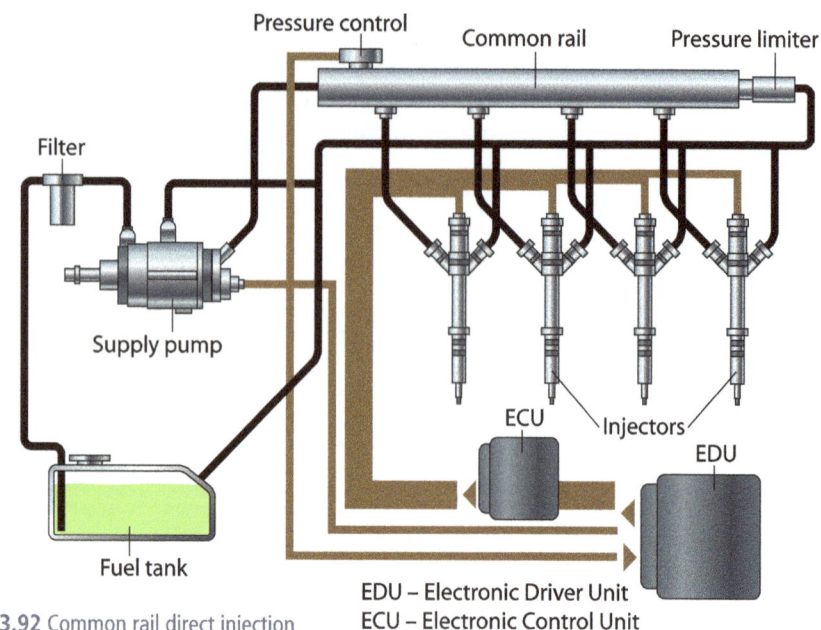

Figure 3.92 Common rail direct injection

EDU – Electronic Driver Unit
ECU – Electronic Control Unit

Fuel supply

The main components making up the fuel supply circuit are:

- fuel tank
- low pressure fuel pump
- fuel filter
- high pressure fuel pump
- fuel rail
- fuel pressure regulator and return system
- fuel injectors.

Fuel tank

The fuel tank is the main reservoir used for supply of clean fuel for the common rail injection system.

Low pressure fuel pump

A common rail fuel injection system needs a method of transferring fuel from the tank to the engine. This is the job of the low pressure fuel pump.

Many fuel pumps are of the roller cell type and operate in a similar manner to those found on petrol fuel injection systems.

Fuel filters

Due to the very small tolerances found within a common rail fuel injection system, small amounts of dirt would easily block fuel injectors or damage high pressure pumps. As a result a fuel filter is fitted between the low pressure pump and the high pressure pump.

High pressure fuel pump

A common rail fuel injection system needs extremely high fuel pressures in order to correctly atomise the fuel from the injectors in the combustion chamber. A mechanically driven pump is operated by the engine that is able to raise system pressure to around 1800 bar, regardless of engine speed.

Fuel rail and pressure regulator

The fuel rail is a common reservoir of diesel held at a constant high pressure from which all the injectors are fed. The pressure in the rail is monitored from an electronic sensor and a solenoid valve is able to allow any excess pressure and fuel to return to the tank.

Fuel injectors

There are two main types of fuel injector used with common rail systems:

- solenoid type
- piezoelectric.

Solenoid fuel injectors

In a solenoid type fuel injector a small coil of wire is wound around a movable armature. When current is applied to this coil a magnetic field is created drawing the armature through the middle of the winding. When current is removed from the coil the magnetic field collapses and a strong return spring is used to move the armature back into its original position.

The armature forms a needle valve at one end and fuel pressure is supplied to the needle valve. When the winding of the solenoid is supplied with current and the needle of the armature lifts from its seat, fuel can be sprayed under extremely high pressure into the combustion chamber.

Piezoelectric injectors

To improve the speed and accuracy of fuel injectors many manufacturers now use piezoelectric injectors. Instead of using solenoids to open the injectors, a series of piezoelectric crystals are stacked above the needle. When supplied with an electric current from the ECU the crystals expand in the electric field. As the crystals expand, the injector is opened extremely fast and with accurate control. As the amount of expansion of the crystals in each fuel injector may be slightly different due to its construction, each injector will often have to be calibrated.

To calibrate the injectors, a method known as injector coding is often used (see Chapter 4, page 196). Each injector will have a serial number marked on the body, which must be programmed into the fuel systems ECU using a scan tool. The serial number contains information about the function of the individual injector and allows the engine management system to operate it accurately.

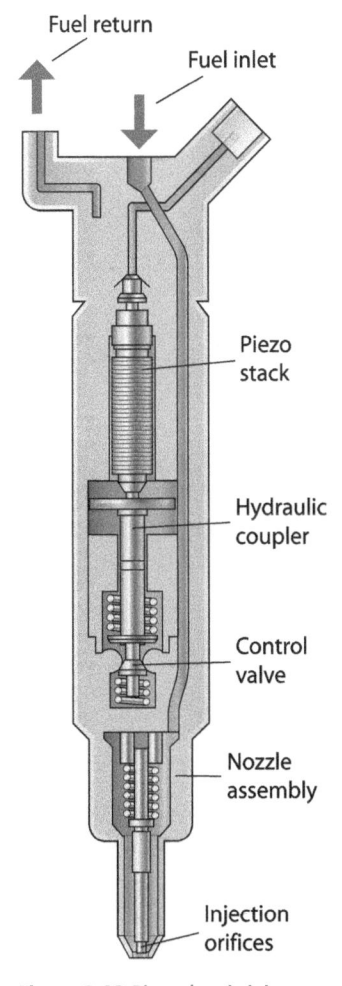

Figure 3.93 Piezoelectric injector

Level 3 Light Vehicle Technology

> **Key term**
>
> **Delay period** – the time between when the fuel is injected and when it starts to burn.

Fuel delivery

The advanced nature of common rail systems allows diesel fuel to be introduced by the injectors in distinct steps:

- **Pilot injection:** A small amount of fuel is injected prior to the main injection. The purpose of this is to initiate combustion and reduce the **delay period** normally associated with compression ignition. As a small amount of combustion is already underway when the main injection takes place, the flame spread is much more controlled. Detonation is reduced, performance is improved and diesel knock/noise is lessened.

- **Main injection:** Most of the fuel is atomised in the combustion chamber for power delivery. As the fuel is initially injected, the combustion chamber temperature and pressure falls slightly, leading to a pause before the fuel begins to burn. This is known as the delay period. As the fuel ignites, pressure and temperature rapidly rise as the flame spreads out in the combustion chamber, forcing the piston down on its power stroke. As the piston moves down, pressure drops and the combustion process ends with complete burn.

- **Post injection:** A small amount of fuel is injected into the combustion chamber after the main injection. The purpose of this is to help complete the combustion process and reduce the amount of hydrocarbon emissions. If the emission control system also incorporates a diesel oxidation catalyst and a catalysed soot filter DPF, the additional fuel injected into the cylinder is used for the purpose of increasing the DPF inlet temperature during the soot regeneration.

Sensors

The common rail direct injection system uses many of the same sensors found in petrol electronic fuel injection, including the following:

- mass air flow sensor
- throttle/pedal position sensor
- intake air temperature sensor
- engine coolant temperature sensor
- camshaft position sensor
- crankshaft position sensor.

Information on the function and operation of these sensors can be found in the section on electronic fuel injection, pages 137 to 143.

Glow plug operation on common rail systems

As with standard diesel engines, glow plugs are often used to pre-heat the air in the combustion chamber to assist with cold starting. Many plugs are left on for a short period after the engine has started and this is normally referred to as 'post glow'. They run at a reduced current to help prevent them from burning out, and the additional heat supplied helps to lower start-up emissions. For information on how to test glow plugs see Chapter 4, page 192.

> **Did you know?**
>
> The monitoring of air mass in a diesel engine is often used by the engine management system to control the amount of exhaust gas recirculation EGR for performance and emissions purposes. (See Table 3.11 on pages 124–125.)

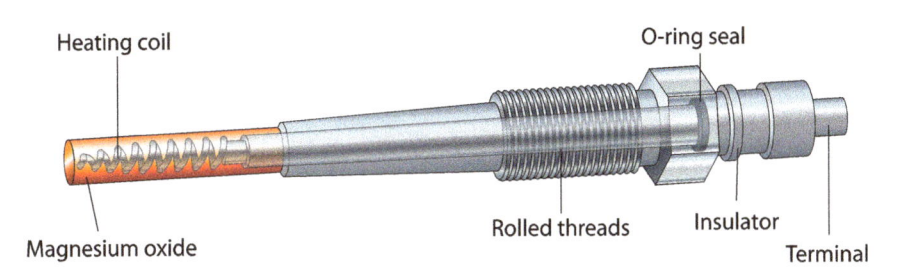

3   Diagnosis & rectification of light vehicle engine faults

Figure 3.94 Glow plug

Faults and symptoms in petrol and diesel fuel injection systems

Table 3.12 lists some of the symptoms that you may experience when working on petrol or diesel injection systems. This list is not exhaustive and you should conduct a thorough diagnostic routine to ensure that you have correctly located the fault.

Table 3.12 Faults and symptoms in petrol and diesel fuel injection systems

Symptom	Possible fault
Cold or hot starting problems	Coolant temperature sensor signal incorrect, affecting fuel injection quantity Faulty glow plugs on a diesel engine
Poor performance	EGR valve stuck open, reducing volumetric efficiency and performance
Excessive exhaust emissions	Fuel pressure regulator stuck closed, raising injection pressure and making the engine run rich
High fuel consumption	Incorrectly adjusted throttle position sensor, meaning that engine load is miscalculated and the engine is over-fuelled
Erratic running	Open circuit mass airflow sensor, leading to incorrect air quantity measurement
Low power	Blocked fuel filter, reducing the amount of fuel available for injection and reducing power because of starvation
Unstable idle speed	Induction air leak, allowing unmetered air to enter the engine, which causes surging

CHECK YOUR PROGRESS

1   What does the term EDC mean?
2   List the three steps of fuel injection used in a common rail system.
3   What is meant by the term post glow when describing glow plug operation?

153

Level 3 Light Vehicle Technology

Engine management

The introduction of computerised engine management systems has greatly enhanced the operation and control of light vehicle propulsion. Engine management improves performance and fuel economy and helps to reduce emissions.

Electronic control principles

To manage an engine, the operation involves:

Input (from sensors) → Processing (in an ECU) → Output (to actuators)

Electronic control is achieved using computers. Different manufacturers call these computers different things:

- electronic control unit (ECU)
- electronic control module (ECM)
- powertrain control module (PCM).

Despite their different names, essentially they all function in a very similar manner. For all their complexity, they can be considered as a large earth switch connected to a calculator.

Ignition control principles

To ensure a spark arrives at the correct spark plug at precisely the right time, regardless of speed and load, an ignition management system can be used.

- CPS or crankshaft position sensors provide information on engine speed and top dead centre position of cylinder number one.
- Distributor pickup or camshaft position sensors provide information for cylinder recognition, so that firing order can be determined.
- Engine coolant temperature sensors (ECT) provide information so that ignition timing can be advanced or retarded in response to engine temperature and fuelling requirements.
- Manifold absolute pressure sensors (MAP) provide information on engine load. This information may also be used to advance or retard the ignition timing.
- Battery voltage is often sensed, so that the rise time (or dwell period) of the ignition coil can be precisely calculated. This ensures that a good spark is available under all running conditions.

The combination of the sensor signals can be processed by the ECU. The ECU will send out signals to the ignition coil, switching off the primary circuit. The resulting collapse in magnetic field will induce a high voltage in the secondary winding, producing a spark at the plug.

Fuel system control principles

A fuel management system may be used to ensure that the correct quantity of fuel is injected by the correct fuel injector at precisely the right time, regardless of speed and load.

Figure 3.95 Electronic control unit (ECU)

Did you know?

The electronic control principles used to manage vehicle engines work in a similar way to the human brain operating different functions in the body. Information is received, that information is processed, and an action is taken.

Analogy: If you touch something hot, nerves in your fingertips sense this and send an electrical signal to your brain. Your brain processes this information and sends an electrical signal out to the muscles in your arms which (actuate) contract and move your fingers away from the hot surface.

- Airflow meters (AFM), mass airflow meters (MAF) or manifold absolute pressure sensors (MAP) can be used to provide information on the volume of airflow in the intake manifold. This means that the correct amount of fuel can be injected for stoichiometric operation.
- Engine coolant temperature sensors (ECT) provide information so that the quantity of fuel injected may be increased or decreased due to engine temperature.
- Throttle position sensors (TPS) provide information to the ECU on throttle butterfly position. When used in conjunction with airflow volume or mass, this can help the ECU calculate engine load.
- Distributor pickup or camshaft position sensors provide information for cylinder recognition, so that **sequential fuel injection** may be used. This is where the injectors operate in the firing order of the engine.
- Battery voltage is often sensed so that the pulse width of the fuel injector can be altered in accordance with the rising and falling voltage found when an engine is running.
    - If voltage is high, the fuel injector will open early and close late.
    - If voltage is low, the fuel injector will open late and close early.

This will have an effect on the amount of fuel injected.

> **Key term**
>
> **Sequential fuel injection** – when fuel is injected in the same sequence as engine firing order.

Networked systems

Engine management is designed to bring these two separate systems (ignition and injection) under one umbrella, combining them to provide a system with the lowest emission output, better fuel economy and strong all-round performance. Many early engine management systems had one ECU to control both the fuel injection system and the electronic ignition system. With the introduction of network systems (see Chapter 4, pages 198–201), many ECUs are able to share information. This means that separate processing systems have been developed to work in harmony with each other in order to provide overall engine control.

A modern engine management system can use many sensors and actuators to provide overall control. These include:

- Engine temperature sensors (ECT)
- Oxygen sensors (sometimes called lambda sensors)
- Knock sensors
- Airflow sensors
- Mass airflow sensors (MAF)
- Manifold absolute pressure sensors (MAP)
- EVAP vacuum controllers
- EGR valve
- Canister purge valve
- Canister air valve
- Throttle position sensors (TPS)
- Air temperature sensors
- Idle actuators
- Fuel pressure regulators
- Camshaft position sensors
- Crankshaft position sensors (CPS)
- Ignition coils
- Fuel injectors
- Malfunction indicator lamps (MIL).

Level 3 Light Vehicle Technology

> **Did you know?**
>
> With keep alive memory (see Table 3.13 below), information is not lost when power is disconnected.
>
> Adaptive strategy within the keep alive memory system of the ECU can cause issues if a technician's diagnostic strategy involves diagnosis by substitution. For example, if the technician does not fully understand the operation of an engine management system, and an engine is running badly, he or she may substitute components to see if it makes a difference to the running of the engine.
>
> Every time a component is substituted, adaption can occur inside the ECU's keep alive memory, making any running issues worse. The technician is basically reprogramming the ECU.
>
> As the situation becomes worse, the technician may resort to substituting the ECU with a new one from a supplier. Once connected, the running problems on the vehicle may reprogram the ECU's adaption strategy. This means that if the ECU is removed from the vehicle and sent back to the supplier, they now have a faulty ECU. If not reset to original values, the next person to use this ECU will suffer engine management operation issues.

This list is not exhaustive, and new types of sensors and actuators are continually being developed by vehicle manufacturers to improve sensing and engine control.

Engine management power supply

In order to process information coming from engine sensors, ECUs rely on voltage and frequency patterns. Unfortunately, vehicle system voltages are unstable and battery voltage rises and falls in accordance with engine operating conditions. For example:

- When the engine is cranked, the potential difference created at the battery may make system voltage fall to around 8V.
- When the engine starts and the alternator charges, system voltage will rise to around 14.5V.

These rising and falling voltages will have a large impact on ECU processing capabilities – basically they confuse the ECU because the input signals are not standardised.

As a result, many systems use a resistor to reduce the processing voltages from the sensors to around 5V. This means that, regardless of engine operation, cranking or charging, signal voltages will be stabilised to between 0V and 5V at all times. This ensures correct interpretation under all engine running conditions.

Table 3.13 lists some of the processing operations that take place within an ECU.

Table 3.13 Processing operations inside an ECU

Read only memory (ROM)	The read only memory contains a program to control the ECU, and therefore engine management. The program is vehicle generic and not restricted to one vehicle type. The information in this part of the ECU cannot be changed.
Random access memory (RAM)	This is a temporary memory inside the ECU. It is used by the central processing unit to store and retrieve information. With random access memory, information is lost when power is disconnected.
Programmable read only memory/erasable programmable read only memory (PROM/EPROM)	The programmable read only memory contains information which is vehicle specific, and is sometimes known as the engine management map. This information is updateable; this is sometimes called remapping. Some systems allow remapping to be done via the serial port or data link connection, where a new program or map is substituted for the original. This information tends to be read only.
Keep alive memory (KAM)	The keep alive memory system of a vehicle's ECU allows it to adapt to different running conditions. This is known as adaptive strategy, or adaption. As engine operation changes from that found in the ECU map (due to wear, faults or driving style), changes are substituted to the calculations carried out by the central processing unit. This allows fuel injection and ignition systems to adapt. Because of this, a stable engine operation is achieved, and driveability remains constant regardless of the situation (within reasonable limits).

CONTINUED ▶

3 Diagnosis & rectification of light vehicle engine faults

Limited operating strategy (LOS)	If a fault occurs in a sensor or actuator, default information is supplied and may bypass incoming and outgoing data to the ECU. The engine is now running on pre-programmed values and maintains a relatively stable engine operation. This is known as limited operating strategy (LOS), or sometimes 'limp home'.
Open and closed loop operation	For the ECU to accurately measure and control the engine, certain operating conditions must be met. (The engine should be at normal operating temperature, for example.) Until these conditions have been met, the ECU controls the engine using pre-programmed values and ignores sensor inputs – this is known as open loop. When the operating conditions have been achieved, sensor information is used to monitor and control engine operation – this is known as closed loop.

Case study

Mr Theophilus brings his car into your garage and complains that the engine management warning light on the dashboard stays on and the car lacks power. Your boss asks you to check it out.

When you turn on the ignition and start the car, the engine management light illuminates and stays on. You think you may need to road test the car to investigate the problem fully, so you go back to the reception to check that it's okay to do so.

Here's what you do:

- ✓ Listen to the customer's description of the fault.
- ✓ Question the customer carefully to find out the symptoms.
- ✓ Road test the car to confirm the symptoms described by the customer.
- ✓ Gather information from technical manuals (including how to retrieve fault codes) before you start and take it to the vehicle.
- ✓ Devise a diagnostic strategy.
- ✓ Check the quick things first (10-minute rule) and conduct a visual inspection of engine components.
- ✓ Ask the customer for authorisation if further testing is required.
- ✓ Conduct as much diagnosis as possible without stripping down.
- ✓ Using a scan tool, read the diagnostic trouble codes (P0404 Exhaust Gas Recirculation circuit range/performance and P0170 Fuel trim malfunction (bank 1)).
- ✓ Record the diagnostic trouble codes and clear them from the memory.
- ✓ Road test the car and re-scan to check that the code has returned. (Only code P0404 Exhaust Gas Recirculation circuit range/performance has returned.)
- ✓ Concentrate your diagnostic procedure on the EGR circuit.
- ✓ Locate the root cause of the problem (EGR valve staying open).
- ✓ Test the component and circuit.
- ✓ Connect an oscilloscope to the EGR valve and check for correct actuator signal.
- ✓ Keep the customer informed of progress and costs.
- ✓ Replace the faulty EGR valve.
- ✓ Correctly reassemble any dismantled components/systems.
- ✓ Clear any diagnostic trouble codes and adaptions.
- ✓ Thoroughly test the system to ensure correct function and operation.

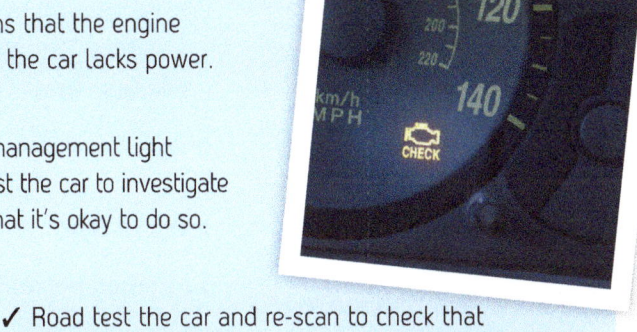

CHECK YOUR PROGRESS

1. What is a limited operating strategy?
2. List six engine management sensors.
3. What part of an ECU memory stores engine management adaptions?

157

Alternative fuel vehicles

Petrol and diesel are gradually becoming more expensive to produce. One of the main reasons for this is that the demand for these fuels is greater than the speed with which crude oil can be extracted from the ground and refined into petrol and diesel. This situation is known as 'peak oil'. Another factor affecting the use of petrol and diesel is the effect that these fuels have on the environment. Many manufacturers are therefore researching and designing cars that will run on alternative fuels.

> **NEW TECH**
>
> **Alternative fuel vehicles**
> An alternative fuel vehicle is a vehicle that runs on a fuel other than traditional petroleum-based fuels (i.e. petrol or diesel). Between 2008 and 2009, there were around 33 million alternative fuel and advanced propulsion technology vehicles on the world's roads. This represented around 5 per cent of the world's vehicles.

Liquefied petroleum gas (LPG)

Liquefied petroleum gas (LPG) is a low pressure liquefied gas mixture made up of mainly propane and butane. This fuel can be burned in a conventional internal combustion engine and produces less CO_2 than petrol. A standard petrol car can often be converted to run on LPG (this is known as **retrofitting**) by adding a second tank and fuel system for the LPG. Because the petrol tank and petrol fuel operating system stay, the car becomes 'dual fuel'.

Compressed natural gas (CNG)

High pressure compressed natural gas (CNG) is mainly composed of methane. It can be used to fuel normal internal combustion engines instead of petrol. The combustion of methane produces the least amount of CO_2 of all **fossil fuels**. Petrol cars can be retrofitted to CNG and become bi-fuel natural gas vehicles (NVG). Because the petrol tank and petrol fuel operating system stay, the car becomes 'dual fuel'.

Biogas

Biogas normally refers to a gas produced by the biological breakdown of **organic matter** in the absence of oxygen. Biogas normally comes from **biogenic** material, such as plant matter, and is a type of biofuel. After purification of the raw gas, compressed biogas can be used to power normal internal combustion engines, in a similar manner to compressed natural gas (CNG).

Key terms

Retrofitting – aftermarket vehicle conversion.

Fossil fuel – naturally occurring oil and fuels created from plant and animal life that has decayed over millions of years.

Organic matter – something that has come from a once living organism that is capable of decay.

Biogenic – produced or brought about by living organisms.

Biodiesel

Biodiesel (fatty acid methyl ester) is a way of making a form of diesel fuel from a very wide range of oils produced from plants, such as:

- Algae oil
- Artichoke oil
- Canola oil
- Castor oil
- Coconut oil
- Corn oil
- Cottonseed oil
- Flaxseed oil
- Hemp oil
- Jatropha oil
- Jojoba oil
- Karanj oil
- Kukui nut oil
- Milk bush oil
- Pencil bush oil
- Mustard oil
- Neem oil
- Olive oil
- Palm oil
- Peanut oil
- Radish oil
- Rapeseed oil
- Rice bran oil
- Safflower oil
- Sesame oil
- Soybean oil
- Sunflower oil
- Tung oil
- Waste vegetable oil (WVO).

Many of these oils can be commercially refined and sold as an alternative to standard diesel. Biodiesel has a lower energy density than fossil diesel fuel. This means that biodiesel vehicles are not quite able to keep up with the fuel economy of a normal fossil-fuelled diesel vehicle.

Bioalcohol/ethanol

Many internal combustion engines can be easily converted to run on alcohol or ethanol. While alcohol and ethanol can be made from petroleum or natural gas, they can also be easily obtained from sugar or starch in crops. This type of fuel is therefore considered to be a renewable energy source.

Ammonia

Ammonia Green (NH_3) is being used with success by some vehicle manufacturers because it can run in spark ignition or compression ignition engines with only minor modifications. Ammonia is considered to be very toxic, but if stored and handled correctly it is no more dangerous than petrol or LPG. It can be made from renewable electricity (solar, wind or hydro-electricity) and has half the density of petrol or diesel. This means that it can be easily carried in sufficient quantities in vehicles, unlike some other alternative fuels. On combustion it produces no emissions other than nitrogen and water vapour, making the exhaust non-polluting.

> **Did you know?**
> The first car that used ethanol as a fuel was the Model T Ford, produced from 1908 until 1927. It was fitted with a carburettor with adjustable jetting, allowing use of petrol or ethanol, or a combination of the two.

Electricity

Electric motors are a very effective method of providing a source of **propulsion** for cars and they produce no emissions. Unfortunately, many of the methods used to create the electricity needed to drive these motors are not very **efficient** or can be polluting to the environment. Currently, renewable sources of electricity such as wind, solar and hydro-electricity are unable to supply the demands needed to make electric powered vehicles truly non-polluting.

Methods used to create electricity for powering electric vehicles mainly come from:

- solar power
- mains supply
- hydrogen fuel cells
- hybrid drive.

Solar

A solar car is an electric vehicle powered by energy from the sun, which is obtained from solar panels on the car. A solar panel converts light energy into electricity that can be used as a source of power. Solar panels cannot currently be used to directly supply a car with enough power to provide drive to electric motors, but they can be used to extend the range of 'plug in' electric vehicles.

Mains supply

A number of manufacturers are producing a range of mains electricity charged electric cars. Instead of an internal combustion engine, these vehicles (often known as 'plug in') are powered from high capacity batteries that drive electric motors. Although these vehicles produce no emissions when they are driven, mains generated electricity is often created using fossil fuels (which creates pollution) or atomic energy (which is dangerous and radioactive). The main limitations of 'plug in' electric vehicles are the distance they can travel on a single charge (known as range) and the amount of time it takes to recharge the batteries, which can be many hours.

Hydrogen fuel cell

Some manufacturers are producing cars with a hydrogen fuel cell. The fuel cell creates electricity that can be used to power electric motors. A fuel cell is similar to a battery but it doesn't store its own internal electricity in the form of a charge. Instead, it uses hydrogen as a fuel source to create electricity. This means that if a fuel tank of hydrogen is kept topped up, the fuel cell, unlike a battery, will keep running and not go flat.

Fuel cell construction

The most common type of hydrogen fuel cell is created using a component called a **proton** exchange **membrane**. This is material that separates the two sides of the fuel cell. One side of the fuel cell is fed with oxygen from the surrounding air; the other side is fed with hydrogen from a fuel tank.

Key terms

Propulsion – the action of driving or pushing forward.

Efficient – how well something works.

Protons – the positively charged particles of an atom.

Membrane – a thin layer of material that is used to separate two connected areas.

Electrons – the negatively charged particles of an atom.

Electrolysis – chemical decomposition produced by passing an electric current through a liquid.

Molecules – the smallest component of a chemical element.

3 Diagnosis & rectification of light vehicle engine faults

Fuel cell operation

As hydrogen enters the fuel cell, a reaction takes place that strips the protons from the atom and moves them through the membrane towards the oxygen on the other side. This leaves the **electrons** from the hydrogen atoms, which travel through a different circuit and create electricity. After the energy has been created in the electric circuit, the electrons reattach themselves to the protons and the hydrogen atom combines with the oxygen to create H_2O.

This means that the only emission from the fuel cell is water, making it clean and non-polluting. The typical output from a single fuel cell is approximately 0.8V, so a number of fuel cells have to be combined (known as a fuel cell stack) to create a usable amount of voltage to drive electric motors.

Hybrid drive

A hybrid vehicle is one which combines an internal combustion engine with an electric motor to provide drive. This gives the flexibility of a petrol or diesel engine with the fuel economy and low pollution characteristics of electric motors. There are three main types of hybrid drive:

- **Series hybrid:** A small capacity internal combustion engine is used to act as a generator. This then charges batteries that are used to power the electric motors that drive the wheels. There is no direct connection between the engine and the wheels, meaning that a gearbox is not required. The advantage of this system is that no driving loads are placed on the engine and it can run at a constant speed. This reduces fuel consumption, emission output and engine wear.

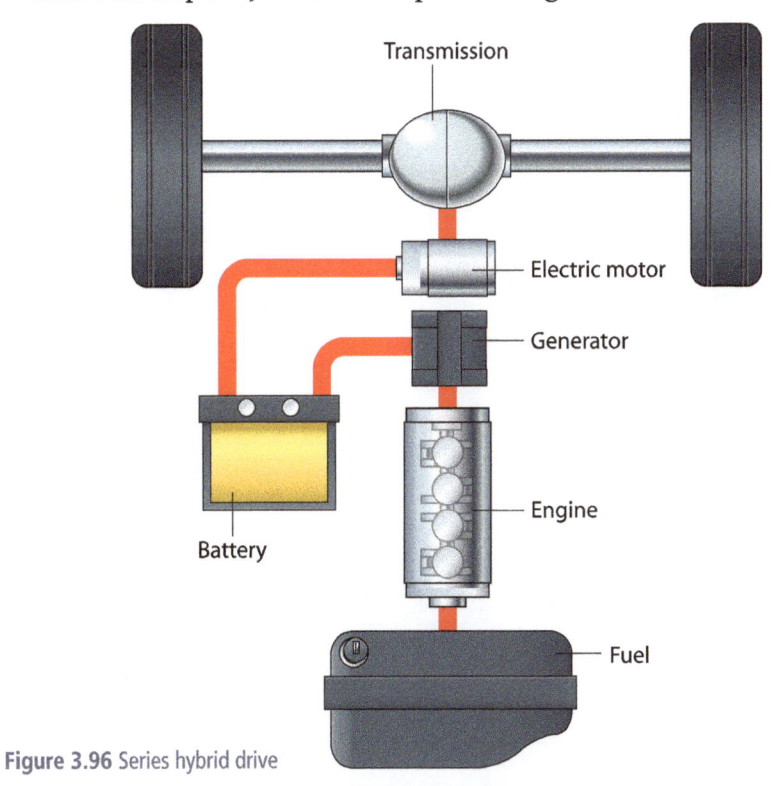

> **Did you know?**
>
> The process of making hydrogen is fairly straightforward. It normally involves separating the hydrogen and oxygen in water by **electrolysis**. Unfortunately, it takes about three times as much energy to make the hydrogen as can be obtained from the hydrogen when used in the fuel cell. This makes hydrogen an inefficient fuel. Also, hydrogen **molecules** are so small that they will leak through almost any container, which makes storage a problem. For example, if you filled up a standard fuel tank with hydrogen, even if you didn't use the vehicle, the tank would be empty in a few days.

> **Safe working**
>
> Many hybrid vehicles operate with high voltages of between 100V and 300V. Special care must be taken when working on hybrid vehicles as the high voltage systems can cause severe injury or death.

Figure 3.96 Series hybrid drive

161

Level 3 Light Vehicle Technology

- **Parallel hybrid:** An integrated electric motor is used to support or boost the performance of a small capacity internal combustion engine. When not required, the electric motor can be converted into a generator to recharge the high voltage electric batteries.

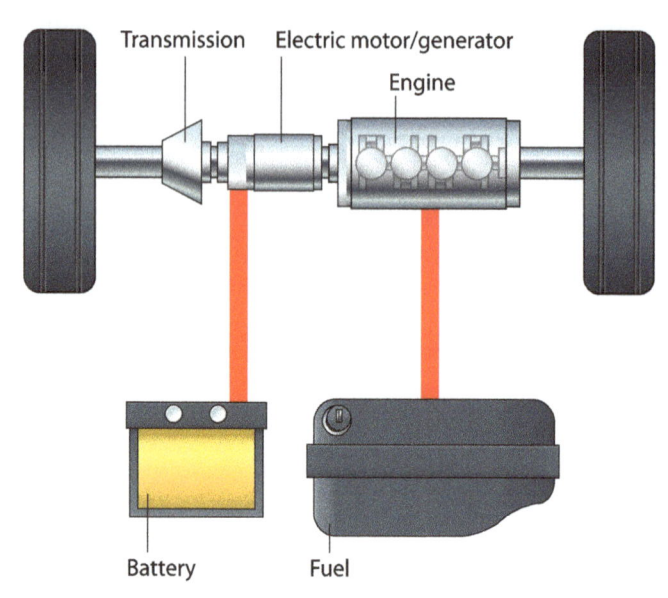

Figure 3.97 Parallel hybrid drive

- **Combination hybrid:** This type of hybrid uses the properties of both series and parallel hybrids. The car can operate on electric motors alone, internal combustion engines alone or a combination of both.

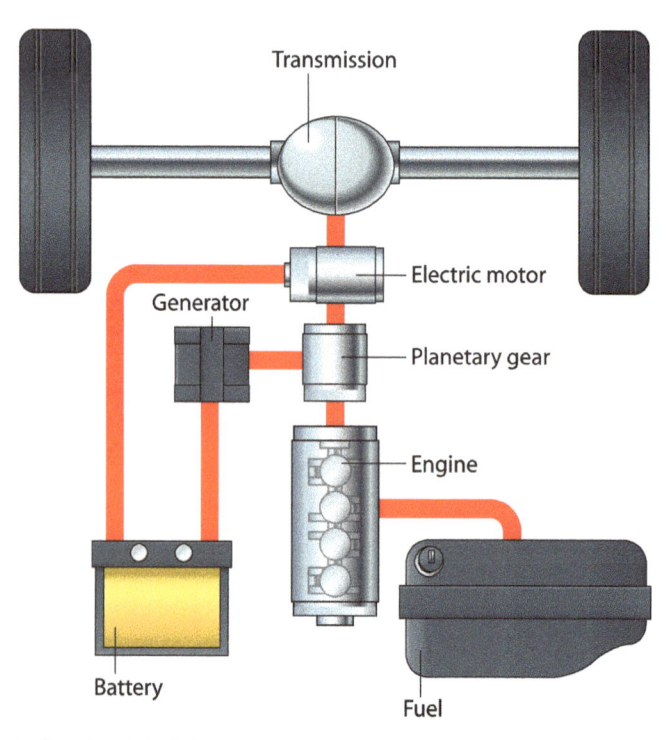

Figure 3.98 Combination hybrid drive

Did you know?

Vehicle designers are continuously creating unusual types of hybrid drive systems. This not only improves efficiency and operation, but also overcomes **copyright** issues.

Some manufacturers have been known to use the terms **series** and **parallel** in a slightly different context when describing the operation of their hybrid drive systems. Make sure that you have fully studied and understood the manufacturer's description of their system before you make any assumptions about how they operate. This will help reduce confusion and misunderstanding.

Key terms

Copyright – an exclusive legal right of design ownership and use.

Series – connected in a line (one after another).

Parallel – connected side by side.

Collaboration – working together in co-operation (this is also known as synergy).

3 Diagnosis & rectification of light vehicle engine faults

NEW TECH

Two common hybrid systems are described below.

Collaborative Motor Drive

A **collaborative** motor drive system has two separate motors that run parallel and are capable of producing drive: an internal combustion engine and an electric motor. When required, they work together to provide a smooth drive system.

- When pulling away, the petrol engine is not used and the electric motor provides the drive.
- When going up a hill, the engine and electric motor both power the vehicle to provide maximum performance.
- When decelerating or braking, the system recycles the kinetic energy to recharge the batteries.
- When overtaking, the engine and electric motor both power the vehicle to provide maximum performance.
- When the vehicle is stationary, both the engine and the electric motor automatically switch off to save power.

Integrated Motor System

An integrated motor system has a compact electric motor, sandwiched between an internal combustion engine and the gearbox, in a similar position to the flywheel. When needed, this electric motor is able to boost the performance from the engine.

- When pulling away, the electric motor provides maximum torque to assist the engine for strong acceleration and reduced fuel consumption.
- When the vehicle is cruising at low speed, the engine is stopped (valves remain closed) and the electric motor is used by itself for drive.
- When the vehicle is accelerating gently or cruising at high speed, the vehicle runs on engine power alone and the electric motor is switched off.
- When accelerating rapidly, the engine and electric motor both power the vehicle to provide maximum performance.
- When decelerating or braking, the engine is stopped (valves remain closed) and the electric motor acts as a generator to recharge the batteries.
- When the vehicle is stationary, the engine and electric motor automatically switch off to save power.

Heating, cooling and ventilation

Certain convenience systems can be constructed and powered by the engine in order to make the passenger compartment a comfortable and ambient environment. Heating, ventilation, air-conditioning and climate control bring added benefits; these include:

- warmth
- cooling
- dehumidifying
- air purification.

> **Safe working**
>
> Due to health and safety and environmental issues, legislation dictates who is able to conduct repairs and maintenance on air-conditioning or climate control refrigeration systems. In order to be able to work on a vehicle's refrigeration system, you must be appropriately qualified.

> **Action**
>
> Research the current requirements and qualifications needed in order to work on a vehicle's air-conditioning system.

Level 3 Light Vehicle Technology

> **Did you know?**
>
> It is possible for the rheostat to burn out over time. If this happens, a common symptom produced is that only the top speed of the ventilation fan is available. This is because the fastest speed requires 12V and does not normally pass through the rheostat itself.

Heating

A vehicle heater is created by including a second radiator called a heater matrix, which is connected to the cooling system inside the vehicle. Hot coolant from the engine is directed by a series of valves or controls so it passes through the heater matrix. An electric fan circulates air through the heater matrix – the air warms up and is used to heat the passenger compartment. This warm air can be directed through vents inside the vehicle to angle it towards the feet, face or windows.

Ventilation fan

An electrically powered fan can be used to regulate the amount of air passing into the passenger compartment. The speed of this fan can be controlled by the driver to suit comfort and driving conditions. Fan speed is normally controlled using a set of dropping resistors known as a rheostat, or may be controlled by the climate control ECU by using pulse width modulation (PWM) or duty cycle. A rheostat controls the voltage to the fan motor by converting the excess current to heat through resistor coils. The different resistor coils waste some of the electrical energy supplied and motor speed is reduced.

Principles associated with heating and cooling

The processes happening inside an air-conditioning or climate control system are listed in Table 3.14.

Figure 3.99 Ventilation fan

Table 3.14 The principles of heat transfer

Heat transfer is achieved through three main methods and will always move from hot to cold:	
Conduction	The transfer of heat energy through a solid
Convection	The transfer of heat energy through a liquid or gas. Convection normally creates a swirling motion in the liquid or gas as hot elements rise and cold elements fall. This motion is called convection currents.
Radiation	The movement of heat energy through a vacuum or space as electromagnetic radiation
Other heat transfer principles:	
States of matter	There are three main states of matter: solid, liquid and vapour.
Condensation	This is where a gas/vapour cools and turns into a liquid.
Evaporation	This is where a liquid heats up and turns into a vapour or gas.
Sensible heat	This is the temperature range within which a substance stays as a single state of matter, i.e. solid, liquid or vapour.
Latent heat	The heat required to produce a change of state from solid to liquid or liquid to gas (and back the other way). As gas condenses into a liquid, it gives up its heat energy to the atmosphere. And as liquid evaporates into a gas, it absorbs heat energy from the surrounding air.

Air-conditioning

Expansion valve type system

Air-conditioning uses an engine-driven pump called a compressor to raise the pressure of a refrigerant gas in a sealed system. The most common gas currently used is Tetrafluroethane, known as R134a. Other refrigerant gases include:

- dichlorodifluoromethane – R12 (now obsolete)
- carbon dioxide – R744.

The gas passes through a radiator, called a condenser, which is normally mounted just in front of the cooling system radiator. The high pressure gas is then cooled and condensed into a liquid. From here, it is transferred into a storage unit called a receiver drier until it is needed.

When the driver operates controls to lower the cabin temperature of the car, the refrigerant is released through a temperature-controlled expansion valve (TXV). As the pressure falls, the liquid refrigerant changes state in another small radiator inside the car, called an evaporator. The temperature in the evaporator falls, and as the cabin air is circulated through it, heat is removed. This helps cool the air inside the car. The refrigerant is then returned to the compressor, where the whole process starts once again.

> **Did you know?**
>
> The refrigerants used in air-conditioning systems are environmental pollutants. R12 creates ozone depletion and R134a contributes to the greenhouse effect. International agreements and protocols have led to legislation that restricts how refrigerants are used and controls their release to the atmosphere. The main legislation is often directed at the fluorinated gases contained in refrigerants, referred to as 'F-gas'.

> **Safe working**
>
> If a leak is suspected in an air-conditioning unit, you must not fill the system with refrigerant gas. Instead the system should be pressurised with oxygen-free nitrogen (OFN), which has no effect on the environment, and then tested for leaks before replacing it with refrigerant gas.

Figure 3.100 TXV air-conditioning circuit

Level 3 Light Vehicle Technology

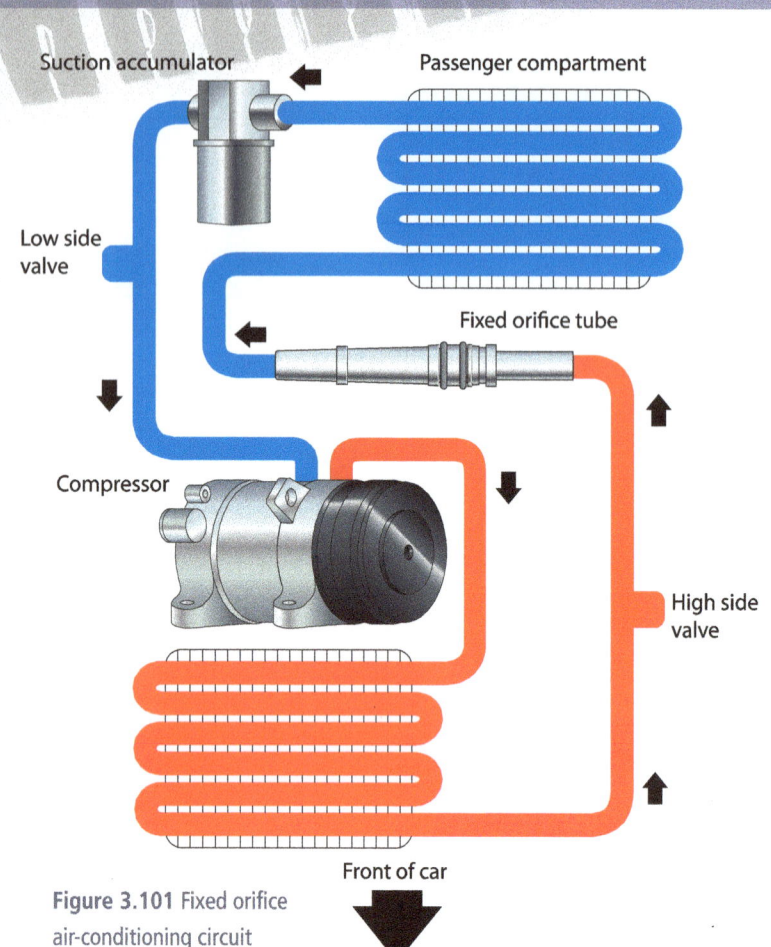

Figure 3.101 Fixed orifice air-conditioning circuit

Fixed orifice systems

An alternative type of air-conditioning uses a system known as a fixed orifice tube. An engine-driven pump, called a compressor, is used to raise the pressure of a refrigerant gas in a sealed system. The gas passes through a radiator, called a condenser, which is normally mounted just in front of the cooling system radiator. The high pressure gas is then cooled and condensed into a liquid. After the condenser, the gas is passed through an accurately sized restriction, called a 'fixed orifice', into the evaporator. In the evaporator, pressure falls and so does the temperature. As the cabin air is circulated through it, heat is removed. This helps cool the air inside the car.

When the cold gas leaves the evaporator, it enters a component called a suction accumulator. Here, any water moisture is removed and stored before returning to the compressor, where the whole process starts all over again.

The main components of an air-conditioning system are described in Table 3.15.

Table 3.15 Components of an air-conditioning system

Component	Description
Compressor	An engine-driven pump, used to raise the pressure of the refrigerant gas used in an air-conditioning system. Pump operation is usually controlled by an **electromagnetic** clutch attached to the drive pulley of the compressor. This means that the system pressure can be controlled within limits by **engaging** and **disengaging** the compressor as required.
Refrigerant gas	A gas that has a low boiling point, used as the main cooling medium in an air-conditioning system. R134a (tetrafluroethane) is the most common, but other gas types are also used.
Condenser	A radiator mounted at the front of the car, which is used to help remove some of the heat from the compressed refrigerant gas and turn it into a high pressure liquid.
Condenser cooling fan	An electric fan that is used to pass air through the cooling fins of the condenser radiator when air speed is low or the condenser becomes too hot.
Hoses and pipes	Pipes and rubber hoses that are used to transfer refrigerant gas around the system. They are chemically resistant and specially sealed to prevent any leakage of the refrigerant gas.
Receiver dryer	A storage area for the high pressure liquid refrigerant in an expansion valve type air-conditioning system. It also contains a **silicone desiccant**, which is designed to remove any water moisture from the system.
High pressure switch	A pressure switch that acts in conjunction with the compressor clutch to turn off the pump if system pressures rise too high. It is sometimes able to operate the condenser fan as well, because high pressures may be an indication that the system is running too hot.

CONTINUED ▶

3 Diagnosis & rectification of light vehicle engine faults

Component	Description
Low pressure switch	A pressure switch that acts in conjunction with the compressor clutch to turn off the pump if system pressures fall too low. This prevents the compressor being overworked if a leak occurs, reducing the amount of refrigerant in the air-conditioning system. It will also cut the compressor clutch out if a blockage appears in the system, leading to a vacuum on the low pressure side of the air-conditioning circuit.
Thermal expansion valve (TXV)	A temperature-controlled spray nozzle, designed to release specific quantities of refrigerant into the evaporator. A **thermocouple** connected to the evaporator regulates the expansion valve, opening it wider when evaporator temperature rises, and closing it when evaporator temperature falls.
Fixed orifice tube	Used in some air-conditioning systems to regulate the amount of refrigerant released into the evaporator. As the **orifice** tube is of a fixed size, system pressure must be controlled according to evaporator temperature. This is done by 'cycling' (turning) the compressor clutch on and off.
Evaporator	This is a small radiator mounted inside the car. As quantities of refrigerant are released into the evaporator, it changes state from a high pressure liquid to a gas. During this change of state, large amounts of heat energy are absorbed by a process called latent heat transfer. This makes the temperature of the refrigerant gas fall. As air from the passenger compartment is passed over the outside of the evaporator, heat is absorbed, lowering the temperature inside the car.
Thermistor	A temperature sensitive resistor (NTC or PTC) mounted on the evaporator. It can be connected to the compressor to engage and disengage the clutch when a certain temperature has been reached.
Suction accumulator	A storage area for refrigerant used in a fixed orifice tube type air-conditioning system. It is mounted after the evaporator and holds a quantity of refrigerant ready to be returned to the compressor for use. It also contains a silicone desiccant, which is designed to remove any water moisture from the system.
Ventilation fan	An electrically driven fan used to boost the amount of air entering the passenger compartment. The speed can be varied according to driver requirements.
Flaps and servo motors	Flaps that are used to direct the incoming air through the heater matrix or air-conditioning evaporator. They can also be used to direct air towards the windscreen, face or floor. If the system incorporates climate control, the operation of these flaps may be automatically controlled using small motors called **servos**.

Climate control

Air-conditioning and climate control use many of the same components and operate on the same refrigerant cycle. The main difference between the two systems is how they are controlled. With a standard air-conditioning system, the controls must be manually set by the driver to choose the amount of heating, cooling or ventilation. As the car is driven along the road and conditions change, it is down to the driver to make adjustments to try to maintain a comfortable environment within the passenger compartment.

With a climate control system, the driver sets the required temperature and sometimes ventilation speed, and a series of sensors monitor the passenger compartment environment. As the car is driven along the road, an ECU automatically makes adjustments to maintain a comfortable environment within the passenger compartment.

> **Key terms**
>
> **Electromagnetic** – describes magnetism that is produced by an electric current.
>
> **Engaging** – connecting.
>
> **Disengaging** – disconnecting.
>
> **Silicone desiccant** – a substance that is used to absorb water
>
> **Thermocouple** – a device used to measure and control temperature.
>
> **Orifice** – an opening or hole.
>
> **Servo** – a control system that converts mechanical motion into one requiring more power.

Did you know?

It is often possible to tell the difference between a receiver dryer and a suction accumulator by their temperature when the air-conditioning is operating. Because of their positions within the system, a receiver dryer will be hot and a suction accumulator will be cold.

Sensors used to monitor and adjust the climate control include:

- ambient air temperature sensors
- passenger compartment temperature sensors
- air quality sensors
- solar radiation sensors
- vehicle speed sensors
- condenser and evaporator temperature sensors.

Because of the complexity of many modern air-conditioning or climate control systems, many manufacturers are including a self-diagnosis feature to assist with repairs. If the fault occurs in an air-conditioning or climate control system, you should connect a scan tool to the data link connector and check for diagnostic trouble codes.

Common faults associated with air-conditioning or climate control systems include:

- refrigerant quantity too low
- refrigerant quantity too high
- ventilation fan not working correctly
- compressor clutch not actuating
- water contamination of the refrigeration system (this will freeze during the refrigeration cycle and block the system)
- climate control sensor failure.

When attempting to repair any of these faults, it is important to follow the manufacturer's repair instructions closely.

Faults and symptoms in air-conditioning and climate control systems

Table 3.16 gives an indication of some of the symptoms that you may experience when working on air-conditioning and climate control systems. This list is not exhaustive, and you should conduct a thorough diagnostic routine to ensure that you have correctly located the fault.

CHECK YOUR PROGRESS

1. Explain the terms latent heat and sensible heat.
2. What is the purpose of a receiver/dryer in an air-conditioning system?
3. What is the difference between air-conditioning and climate control?

Table 3.16 Air-conditioning symptoms and faults

Symptom	Possible fault
Leaks	Damage to condenser radiator caused by front end collision
Abnormal noise	Compressor drive belt worn and slipping
Ineffective operation	Ventilation flaps not moving due to failed climate control servo motors
Failure to operate	Compressor clutch not engaging due to low pressure/quantity of system refrigerant gas
Control faults	Fan speed un-adjustable due to burnt-out rheostat controller
Inadequate operation	Evaporator core frozen due to faulty temperature thermistor measurement

BEFORE YOU FINISH

Recording information and making suitable recommendations

At all stages of a diagnostic routine, maintenance or repair, you should record information and make suitable recommendations. The table below gives examples of how to do this.

Stage	Information	Recommendations
Before you start	Record customer/vehicle details on the job card. Make a note of the customer's repair request and any issues/symptoms. Locate any service or repair history.	Advise the customer how long you will require the car. Describe any legal, environmental or warranty requirements.
During diagnosis and repair	Carry out diagnostic checks and record the results on the job card or as a printout from specialist equipment. List the parts required to conduct a repair. Note down any other non-critical faults found during your diagnosis.	Inform your supervisor of the required repair procedures so that they can contact the customer and gain authorisation for the work to be conducted.
When the task is complete	Write a brief description of the work undertaken. Record your time spent and the parts used during the diagnosis and repair on the job card. (This information should be as comprehensive as possible because it will be used to produce the customer's invoice). Complete any service history as required.	Inform the customer if the vehicle will need to be returned for any further work. Advise the customer of any other issues you noticed during the repair.

Remember, any work you conduct on a customer's car should be assessed to ensure that it is conducted in the most cost-efficient manner. You should consider reconditioning, repair and replacement of components within units.

FINAL CHECK

1. What is the approximate volumetric efficiency of a naturally aspirated engine?
   a 50%
   b 80%
   c 100%
   d 120%

2. Which of the following valve control systems can vary how far the valve opens?
   a VVTi
   b VANOS
   c VTEC
   d VVC

3. A timing belt has broken after 30,000 miles. Which is the **least** likely to have caused the failure?
   a belt too loose
   b belt manufacturing fault
   c oil contamination
   d worn tensioner pulley bearing

4. When conducting a leakdown test, the most likely cause of air leaking from the dipstick tube is:
   a damaged piston rings
   b damaged exhaust valve
   c damaged inlet valve
   d damaged head gasket

5. Which is the most appropriate tool for measuring crankshaft bearing clearance?
   a internal micrometer
   b bore gauge
   c Vernier gauge
   d Plastigauge

6. Which is the most appropriate tool for checking the correct operation of an inductive engine speed sensor?
   a test lamp
   b noid light
   c laser thermometer
   d oscilloscope

7. Which of the following is **not** a type of wave form display for secondary ignition?
   a overview
   b parade
   c raster
   d superimposed

8. Which of the following terms only relates to the combustion of diesel fuel?
   a flame travel
   b detonation
   c flash point
   d cetane value

9. At approximately what temperature does a standard lambda sensor begin to operate correctly?
   a 300°C
   b 100°C
   c 20°C
   d 600°C

10. Which of the following is **not** a phase of injection on a high pressure common rail diesel engine?
    a post injection
    b common injection
    c pilot injection
    d main injection

3 Diagnosis & rectification of light vehicle engine faults

PREPARE FOR ASSESSMENT

The information contained in this chapter, as well as continued practical assignments, will help you to prepare for both the end-of-unit tests and the diploma multiple-choice tests. This chapter will also help you to develop diagnostic routines that enable you to work with light vehicle engine system faults. These advanced system faults will be complex and non-routine and may require that you work with and manage others in order to successfully complete repairs.

You will need to be familiar with:

- Diagnostic tooling
- Electrical and electronic principles
- Diagnostic planning and preparation
- SI fuel systems (petrol)
- CI fuel systems (diesel)
- Ignition systems
- Engine management
- Valve mechanisms
- Pressure-charged induction systems
- Exhaust emission reduction systems
- Alternative fuel vehicles
- Heating, ventilation and cooling

This chapter has given you an overview of advanced vehicle engine systems and has provided you with the principles that will help you with both theory and practical assessments. It is possible that some of the evidence you generate may contribute to more than one unit. You should ensure that you make best use of all your evidence to maximise the opportunities for cross-referencing between units.

You should choose the type of evidence that will be best suited to the type of assessment that you are undertaking (both theory and practical). These may include:

Assessment type	Evidence example
Workplace observation by a qualified assessor	Carrying out a two-stage diagnosis on an engine mechanical system
Witness testimony	A signed statement or job card from a suitably qualified/approved witness, stating that you have correctly tested and reset a turbocharger wastegate mechanism
Computer-based	A printout from a diagnostic scan tool showing the results from a system test to check the function of an electronic fuel injection system
Audio recording	A timed and dated audio recording of you describing the process involved when checking the pressures found in an air-conditioning system
Video recording	Short video clips showing you carrying out the various stages involved in an exhaust gas analysis to check for correct and legal engine operation
Photographic recording	Photographs showing you carrying out the stages of an engine mechanical system strip down and overhaul when the assessor is unable to be present for the entire observation (because this process may take several days to complete). The photos should be used as supporting evidence alongside a job card.
Professional discussion	A recorded discussion with your assessor about how you diagnosed and repaired an ignition system fault

CONTINUED ▶

Assessment type	Evidence example
Oral questioning	Recorded answers to questions asked by your assessor, in which you explain how you diagnosed an induction system air leak that was causing an engine to run badly
Personal statement	A written statement describing how you carried out the repair of a variable valve control system
Competence/skills tests	A practical task arranged by your training organisation, asking you to use an oscilloscope correctly to test a fuel injector
Written tests	A written answer to an end-of-unit test to check your knowledge and understanding of light vehicle engine systems
Multiple-choice tests	A multiple-choice test set by your awarding body to check your knowledge and understanding of light vehicle engine systems
Assignments/ projects	A written assignment arranged by your training organisation requiring you to show in-depth knowledge and understanding of a particular engine system (e.g. emission control)

Before you attempt a theory end-of-unit or multiple-choice test, make sure you have reviewed and revised any key terms that relate to the topics in that unit. Ensure that you read all the questions carefully. Take time to digest the information so that you are confident about what each question is asking you. With multiple-choice tests, it is very important that you read all of the answers carefully, as it is common for two of the answers to be very similar, which may lead to confusion.

For practical assessments, it is important that you have had enough practice and that you feel that you are capable of passing. It is best to have a plan of action and work method that will help you.

Make sure that you have the correct technical information, in the way of vehicle data, and appropriate tools and equipment. It is also wise to check your work at regular intervals. This will help you to be sure that you are working correctly and to avoid any problems developing as you work.

When you are undertaking any practical assessment, take care to work safely throughout the test. Light vehicle engine systems are dangerous and pose a potential risk of fire, and precautions should include making sure that you:

- always check that the engine is cold where possible and remove all sources of ignition
- observe all health and safety requirements
- use the recommended personal protective equipment (PPE) and vehicle protective equipment (VPE)
- use tools correctly and safely.

Good luck!

4 Diagnosis & rectification of light vehicle auxiliary electrical faults

This chapter will help you to gain an understanding of diagnosis and diagnostic routines that lead to the rectification of electrical and electronic auxiliary control system faults. It also explains and reinforces the need to test light vehicle electrical systems and evaluate their performance. It will support you with knowledge that will aid you when undertaking both theory and practical assessments. It will help you develop a systematic approach to complex diagnosis of light vehicle electrical auxiliary systems.

This chapter covers:

- Electrical principles related to light vehicle electrical circuits
- How to use electrical diagnostic tooling
- Multiplexing and network systems
- Electrical faults
- Batteries: starting and charging
- Lighting systems and technology
- Electrical comfort and convenience systems
- In-car entertainment (ICE) and satellite navigation systems
- Security systems
- Wiper, washer and central locking systems
- Airbags and supplementary restraint systems (SRS)

BEFORE YOU START

Safe working when carrying out light vehicle diagnostic and rectification activities

There are many hazards associated with the diagnosis and repair of advanced electrical systems. You should always assess the risks involved with any diagnostic or repair routine before you begin, and put safety measures in place. You need to give special consideration to the possibility of:

- **Contact with dangerous chemicals:** Many high intensity discharge (HID) light bulbs and SRS crash sensors contain mercury. Always take precautions when handling these components and treat old units as hazardous waste when disposing of them.
- **Electrical burns:** When you are connecting and disconnecting electrical system components, there is a risk of an unintended short circuit. As electrical discharge is created, the energy can be turned into heat. It is recommended that you use insulated tooling that is designed for use with electrical systems wherever possible.

You should always use appropriate personal protective equipment (PPE) when you work on electrical systems. Make sure that your selection of PPE will protect you from these hazards.

Electronic and electrical safety procedures

Working with any electrical system has its hazards, and you must take safety seriously. When you are working with light vehicle electrical and electronic systems, the main hazard is the possible risk of electric shock. For information on basic first aid for electrical injuries, see Table 1.3 in Chapter 1, page 18. Although most systems operate with low voltages of around 12V, an accidental electrical discharge caused by incorrect circuit connection can be enough to cause severe burns. Where possible, isolate electrical systems before repairing or replacing components.

If you are working on hybrid vehicles, take care not to disturb the high voltage system. You can normally identify the high voltage system by its reinforced insulation and shielding, which is often brightly coloured. These systems carry voltages that can cause severe injury or death. If you carry out repairs to hybrid vehicles, always follow the manufacturer's recommendations.

Always use the correct tools and equipment. Damage to components, tools or personal injury could occur if the wrong tool is used or a tool is misused. Check tools and equipment before each use.

If you are using measuring equipment, always check that it is accurate and calibrated before you take any readings.

If you need to replace any electrical or electronic components, always check that the quality meets the original equipment manufacturer (OEM) specifications. (If the vehicle is under warranty, inferior parts or deliberate modification might make the warranty invalid. Also, if parts of an inferior quality are fitted, this might affect vehicle performance and safety.) You should only carry out the replacement of electrical components if the parts comply with the legal requirements for road use.

Information sources

The complex nature of advanced light vehicle electrical systems requires you to have a comprehensive source of technical information and data. In order to conduct diagnostic routines and repair procedures, you will need to gather as much information as possible before you start. Sources of information include:

Information source	Example
Verbal information from the driver	A description of the symptoms that occur on the car when an electrical system is operated
Vehicle identification numbers	Restraint type taken from VIN plate
Service and repair history	A check of the service history that shows when the lighting system was last inspected
Warranty information	Is the car under warranty and is it valid? (Has the required service and maintenance been conducted?)
Vehicle handbook	To confirm how to correctly set the headlamp aims for different vehicle loading conditions
Technical data manuals	To find the recommended amperage for replacement circuit fuses
Workshop manuals	To find the recommended procedures for isolating the SRS system for airbag removal
Wiring diagrams	To trace the electrical circuit used for powering the brake light system
Safety recall sheets	To confirm which components need to be replaced for safe operation of an HID lighting system
Manufacturer-specific information	Vehicle-specific diagnostic trouble codes relating to the body electrical system
Information bulletins	Information on a common fault found on central locking systems
Technical help lines	Advice on the correct routine for setting up a built-in satellite navigation system
Advice from master technicians/colleagues	An explanation of how to use a multimeter to conduct a volt drop test
Internet	An Internet forum page where a number of people who had a similar problem with the failure of a vehicle security immobiliser system explain how it was resolved
Parts suppliers/catalogues	A cross-reference of light bulb part numbers, so that you can make sure that they comply with legal requirements
Job cards	A general description of the work to be conducted on a customer's in-car entertainment system
Diagnostic trouble codes	A fault code showing that the multiplex network system needs to be tested to ensure correct operation
Oscilloscope wave forms	A wave form from a CAN bus network system that shows communication is taking place

Remember that no matter which information or data source you use, it is important to evaluate how useful and reliable it will be to your diagnostic routine.

Operation of electrical and electronic systems and components

The operation of electrical and electronic systems and components related to light vehicle auxiliary systems:

Electrical/electronic system component	Purpose
ECU	The electronic control unit (ECU) is designed to monitor and control the operation of light vehicle electrical systems. It processes the information received and operates actuators that control auxiliary systems for comfort and convenience.
Sensors	The sensors monitor various vehicle auxiliary components against set parameters. As the driver makes demands on vehicle systems, dynamic operation creates signals in the form of resistance changes (voltage) which are relayed to the ECU for processing.
Actuators	The actuators are used to control auxiliary systems operation. Motors, solenoids, valves, transformers, etc., are operated by the ECU to help control the action of comfort and convenience systems, which also assist with dynamic safety.
Electrical inputs/voltages	The ECU needs reliable sensor information in order to correctly determine the action of the auxiliary systems. If battery voltage was used to power sensors, its unstable nature would create issues (battery voltage constantly rises and falls during normal vehicle operation). Because of this, sensors normally operate with a stabilised 5-volt supply.
Digital principles	Many vehicle sensors create analogue signals (a rising or falling voltage). The ECU is a computer and needs to have these signals converted into digital (on and off) before they can be processed. This can be done using a component called a pulse shaper or Schmitt trigger.
Duty cycle and pulse width modulation (PWM)	Lots of electrical equipment and electronic actuators can be controlled by duty cycle or pulse width modulation (PWM). This works by switching components on and off very quickly so that they only receive part of the current/voltage available. Depending on the reaction time of the component being switched and how long power is supplied, variable control is achieved. This is more efficient than using resistors to control the current/voltage in a circuit. Resistors waste electrical energy as heat, whereas duty cycle and PWM operate with almost no loss of power.
Fibre optic principles	The introduction of in-car entertainment (ICE) and information systems such as satellite navigation has increased the need for very fast transmission of data. Fibre optics use light signals transmitted along thin strands of glass to provide digital data transmission. (The light source is switched on and off.) In this way, information is transmitted essentially at the speed of light.

Electrical and electronic control is a key feature of all the systems discussed in this chapter.

Tooling

No matter what task you are performing on a car, you will need to use some form of tooling.

Always use the correct tools and equipment.

The following table shows a suggested list of diagnostic tooling that could be used when testing and evaluating light vehicle auxiliary, safety, comfort and convenience systems. Due to the nature of complex system faults, you will experience different requirements during your diagnostic routines and so you will need to adapt the list shown for your particular situation.

Tool	Possible use
Oscilloscope 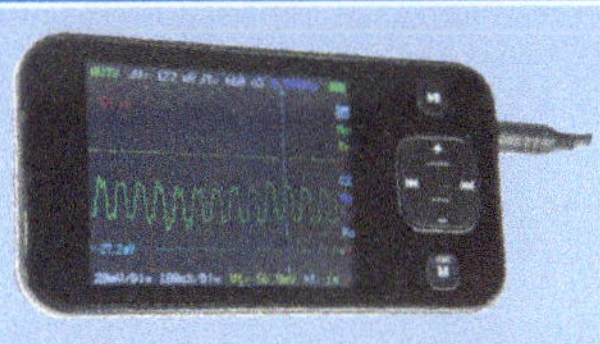	To test the signal produced by a multiplex network system
Multimeter 	To test the voltage output of an alternator
Test lamp/ logic probe 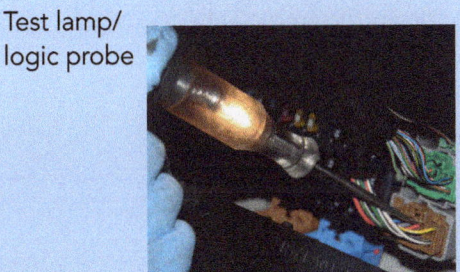	To test the existence of system voltage at a windscreen wiper motor (Always use test lamps with extreme caution on electronic systems, as the current draw created can severely damage components.)
Power probe	To power the motor of an electric window system and check its operation

CONTINUED ▶

177

Tool	Possible use
Vacuum gauge/pump 	To test the function and operation of a pneumatic central locking system
Code reader/ scan tool 	To retrieve diagnostic trouble codes (DTCs) related to the supplementary restraint system (SRS). To clear trouble codes, reset the malfunction indicator lamp, and evaluate the effectiveness of repairs.
Laser thermometer 	A non-contact thermometer, also known as a pyrometer, can help determine an electrical component that may be causing a parasitic drain. An electrical component which does not fully switch off may convert electrical energy into heat and become warm. With all other electrical components switched off and cold, you can use the thermometer to scan the car for the faulty part.
Inductive amps clamp 	Used to safely test for parasitic drain caused by faulty electrical auxiliary components.

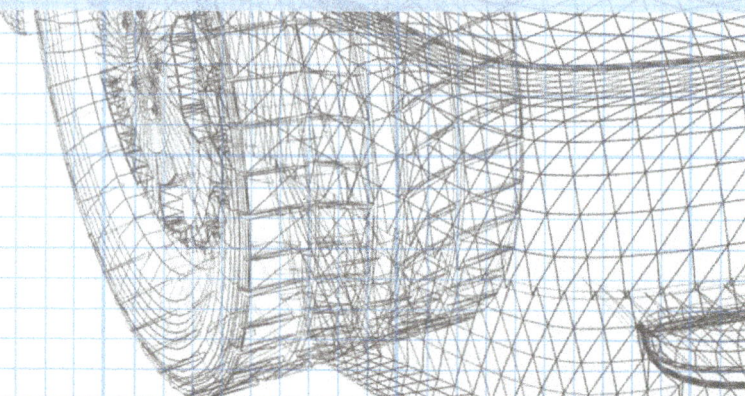

4 Diagnosis & rectification of light vehicle auxiliary electrical faults

Electrical principles related to light vehicle electrical circuits

You need to have a basic understanding of the principles of electricity before you can diagnose problems in an electrical circuit. This is crucial for anyone involved in the diagnosis and repair of automotive electrical systems. This knowledge will allow you to make the right judgements and arrive at a successful conclusion, leading to a first time fix.

In cars, electrical energy is created by a chemical reaction (in a battery, for example) or by the disruption of magnetic fields near electrical conductors (such as in a generator). You can measure how the electrical energy is created, moved and used using the electrical units shown in Table 4.1.

> **Did you know?**
>
> A difference in pressure (voltage) between two points in an electric circuit will create flow in the direction of the lower pressure (voltage). With conventional electrics, it is assumed that electricity flows from positive to negative. True electron flow is actually from negative to positive. You should remember this if the words 'electron flow' appear in a test or assignment.

Table 4.1 Electrical units

Unit	Description
Volts	Voltage is electrical pressure. It is the potential force in any part of an electrical circuit. Two main types of voltage occur in electrical circuits: • Electromotive force (EMF) is potential pressure, and is usually considered to be the open circuit voltage when all electrical consumers are switched off and no current is flowing. It should be higher than electrical system voltage when current is flowing. • Potential difference (Pd) is a circuit voltage measurement when components are switched on and current is able to flow. It is a measurement of voltage drop compared to the EMF at different positions within a circuit.
Amps	Amps are the units used to measure the amount of electricity in any part of an electrical circuit. Amps are measured when electricity is allowed to flow in an electrical circuit – this is known as current. There are two main types of electrical current: • Direct current (DC) is electricity that flows in one direction only. • Alternating current (AC) is electricity that moves backwards and forwards in an electric circuit. Amperage is the same wherever you measure it in the circuit (at the beginning, in the middle or at the end).
Ohms	Ohms are the units used to measure the resistance to electrical flow. Resistance has a direct effect on the operation of any electrical circuit. As resistance rises in a circuit, current and voltage fall, which can restrict the operation of electrical components. In some electrical circuits, resistance can be used as a method of control for electrical components, but in most circumstances a high resistance is undesirable.
Watts	Watts are the units used to measure electrical power made or consumed. Power is defined as the rate at which work is done. When referring to electrical components, the higher the wattage, the more powerful the component will be and the more electrical energy it will use.

Series and parallel circuits

Two main types of electrical circuit are used in the construction of motor vehicles:

- series
- parallel.

Series circuits

In a **series circuit** the consumers are connected in a line, one after another. Because they are all in the same circuit, they share the electricity provided depending on the amount of power that they use. If more than one consumer is fitted it will only get part of the voltage available.

> **Key term**
>
> **Series circuit** – a circuit with electrical consumers connected in a line, one after another.

Level 3 Light Vehicle Technology

If any one of the consumers fails, the circuit is broken and no electricity can flow. The rest of the consumers stop working. This makes series circuits unsuitable for many systems on cars. For example, if you wired a lighting circuit in series, not only would the bulbs glow dimly, but if one bulb broke all of the others would go out.

Figure 4.1 A simple series circuit

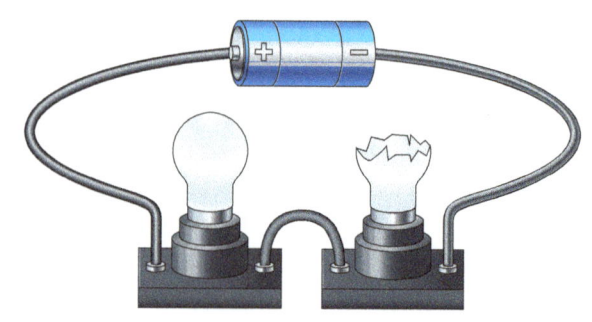

Figure 4.2 When a bulb in a series circuit is damaged the others will no longer work

Parallel circuits

Key term

Parallel circuit – a circuit where electrical consumers are connected side by side

In a **parallel circuit** the consumers are connected next to each other. Each has its own power supply and earth return back to the battery.

Because each consumer has its own power supply and earth, all the consumers receive the full voltage available and work at full power.

If one consumer in the circuit fails, the others keep working. For example, in a headlight circuit each bulb has its own 12-volt supply and earth return to the battery. If one bulb breaks, the other will keep working.

Figure 4.3 Bulbs in a parallel circuit

Figure 4.4 When one or more bulbs in a parallel circuit are damaged the others will still work

180

Ohm's law

There is a very strong relationship between the electrical units.

Ohm's law states that the current flowing in an electric circuit is in direct proportion to the voltage applied and inversely proportionate to the resistance. This means that:

- If you double the pressure (voltage), you double the quantity of current (amps) flowing.
- If you double the resistance (ohms), then you halve the amount of current flowing in the circuit.

This relationship was explained by Georg Ohm using the following mathematical formulas:

- amps = volts ÷ resistance
- resistance = volts ÷ amps
- volts = amps x resistance

If you know any two of the electrical measurements, you can use Ohm's law to calculate the third.

The Ohm's law triangle (shown in Figure 4.5) is a useful method for calculating the missing unit.

- V = volts (This is sometimes shown as the letter E to represent EMF, but still means volts.)
- I = amps (The letter I is used to represent instantaneous current flow.)
- R = ohms (The letter R is used for resistance because an 'O' could be confused for a zero.).

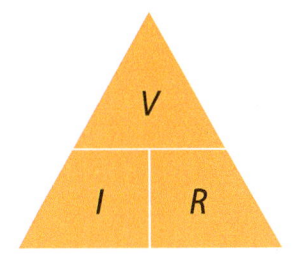

Figure 4.5 Ohm's law triangle

How to use Ohm's triangle

Cover up the unknown unit with your thumb and you are left with the calculation that you need to do.

For example, if amperage is unknown, cover the I and you are left with V/R (i.e. volts divided by resistance).

The power triangle

Watts or power can be calculated in a similar way:

    amps = watts ÷ volts
    volts = watts ÷ amps
    watts = amps × volts

A power triangle can be used in the same way as Ohm's law (shown in Figure 4.6) to calculate the missing unit.

- P = power (This is sometimes shown as the letter W to represent watts, but still means power.)
- V = volts (This is sometimes shown as the letter E to represent EMF, but still means volts.)
- I = amps (The letter I is used to represent instantaneous current flow.).

Figure 4.6 The power triangle

How to use the power triangle

Cover up the unknown unit with your thumb and you are left with the calculation required.

For example, amperage is unknown, so cover the *I* and you are left with *W/V* (i.e. watts divided by volts).

Using Ohm's law to help diagnose faults

The relationship between voltage, resistance and amperage can help you to diagnose faults within an electrical circuit. If you take measurements using the different electrical units and then compare them using the Ohm's law calculation, you will be able to work out if the fault is occurring because of:

- Pressure (volts)
  - If this is lower than expected, component performance is reduced.
  - If this is higher than expected, component damage can occur.
- Quantity (amps)
  - If this is lower than expected, component operation will normally be incorrect.
  - If this is higher than expected, component/system operation is being overworked.
- Resistance (ohms)
  - If this is lower than expected, current may be taking an alternative path to earth (short circuit).
  - If this is higher than expected, resistance will consume electrical energy and reduce system performance.

CHECK YOUR PROGRESS

1. Name the four main electrical units.
2. If an electric circuit supplied with 12 volts from a battery operates a bulb with a resistance of 2.5 ohms, how much current will the circuit draw?
3. What is the difference between AC and DC?

How to use electrical diagnostic tooling

Electricity is invisible so you will need to use specialist electrical diagnostic tooling to discover what is happening in a circuit. Some examples of electrical diagnostic tooling are described below.

Test lamps

One of the simplest diagnostic tools you can use is a test lamp. Whether this is a professionally built tool or a bulb and a couple of pieces of wire that you have put together yourself, this tool can be very effective. Its purpose is to check whether the circuit has power.

To use a test lamp:

- Connect one end of the test lamp to a good earth, such as the vehicle chassis or ground. (The negative terminal of the battery is better because this is the end of all electrical circuits on a car.)
- Connect the other end of the test lamp to the part of the circuit that needs to be checked.
- If power exists in the circuit, the test lamp will illuminate.

Figure 4.7 Test lamp

4 Diagnosis & rectification of light vehicle auxiliary electrical faults

> **Did you know?**
>
> Many modern test lamps have a point at the end of the probe to enable you to pierce wiring installation. Take care when using this probe as it is very easy to stab your finger. Also, if you pierce insulation, you will open up the wiring to the effects of **oxidisation** from the air. If the wiring is left open, this oxidisation can lead to a high resistance and electrical problems in the future. For this reason, it is best where possible to **back-probe** an electrical plug so as not to damage the wiring.
>
> If you have to pierce the insulation of the wiring, you should cover it with insulation tape or heat shrink. Wiping a small amount of silicon into the pierced insulation hole is not advisable as it can cause corrosion in the wiring and high resistance.

> **Key terms**
>
> **Oxidisation** – the effect of oxygen on metal, which can cause corrosion.
>
> **Back-probe** – a method of making a test connection at the rear of an electrical connection.
>
> **Polarity** – a term used to describe electrical connection to a circuit. It represents the positive and negative connections.

> **Safe working**
>
> You must not use a test lamp on electronic systems as it can cause severe damage to circuits and components.
>
> Every time you connect another consumer to an electrical supply wire, more electrical current will be drawn from that supply wire until eventually it can take no more. A test lamp contains a bulb and this is a consumer. A standard test lamp has a low resistance, usually around 6 ohms. This means that when testing an electrical circuit, 2 to 3 amps of electrical current may be drawn. If a test lamp is used on an electronic circuit, severe damage can be caused as this high amperage moves through the components.
>
> Always take care when using test lamps to diagnose electrical faults on vehicles. They should only be used when it is safe to do so. It is far safer to use an LED test light if you are likely to be testing near electronic circuits.

Power probe

A power probe is an advanced form of test light, with additional features and capabilities. Power probes are usually fitted with LEDs that are able to illuminate in different colours when connected to either a powered circuit (LED glows red) or an earth circuit (LED glows green).

Checking polarity

After a simple connection to the vehicle's battery, you will be able to see quite easily whether a circuit is positive or negative, without having to change **polarity** from one battery terminal to another. The power probe normally comes with two crocodile clips (red and black) – connect these to the appropriate positive and negative battery terminals.

To check that a correct connection has been made, quickly touch the tip of the power probe to each battery terminal in turn:

- The LED should illuminate red when touched to the positive terminal.
- The LED should illuminate green when touched to the negative terminal.

Figure 4.8 Power probe

183

Figure 4.9 Using a power probe – right illumination at positive terminal and green illumination at negative terminal

Checking continuity

In addition to checking for electrical feed and earth, you can also use a power probe to check for **continuity** (a continuous or unbroken conductor). You can check continuity on wires or components that have been disconnected from the vehicle's electrical system.

The power probe has an **auxiliary** ground wire – connect this to one end of the conductor, wire or component. Then connect the tip of the power probe to the other end. If continuity exists, the LED on the power probe will illuminate.

Conducting functional tests

You can also use the power probe to undertake functional tests of electrical components.

1. It is recommended that you disconnect the component from the vehicle's electrical system when conducting this test.
2. Connect the auxiliary ground to one terminal of the component and connect the tip of the power probe to the other terminal.
3. Check that the LED illuminates to show that the component has continuity.
4. Keeping an eye on the LED, quickly rock the power switch and immediately release.
5. If the LED indicator changed momentarily from green to red, you may proceed with the test.
6. By rocking the power switch forwards and holding it down, electrical potential will be supplied to the component and you can check its operation.
7. If during the initial rocking of the power switch the LED turned off, this normally indicates that the current being drawn by the

> **Key terms**
>
> **Continuity** – when an electrical circuit conducts current easily and is unbroken (i.e. continuous).
>
> **Auxiliary** – something that functions in a supporting capacity.

> **Safe working**
>
> Always remember to turn off power first before disconnecting a wire component on the vehicle's electrical circuit.

component is too high for the power probe and the internal circuit breaker has tripped. This may require a manual reset and you will need to check the manufacturer's instructions.

Multimeters

The multimeter is a piece of electrical test equipment designed to measure a number of different units within an electrical circuit. There are two types of multimeter: analogue and digital.

Analogue multimeters

Analogue multimeters use a needle that moves across a graduated scale to record electrical readings within a circuit. The old-fashioned name for this type of unit was an 'AVO meter', which stood for amps, volts and ohms.

The problem with analogue meters is that they are only as good as the operator. The graduated scale can be difficult to read and this may result in inaccurate readings. Depending on the range of the scale provided by the manufacturer, a needle that lies somewhere between two units could be reading any fraction available. Analogue multimeters also have an upper range limit. If the needle flicks all the way to the end of this scale, this is known as full-scale deflection (FSD).

Digital multimeters

Digital multimeters display digits (numbers) on a liquid crystal display (LCD) screen. These numbers are clearly displayed and are easy to read accurately.

Two types of digital multimeter are common: manually operated and autoranging.

- With a manual multimeter, the operator selects the unit and the scale to be measured, normally by turning a dial on the front of the unit.
- With an autoranging multimeter, the operator selects the unit but the scale of that unit is automatically selected by the multimeter.

Figure 4.11 Manual multimeter

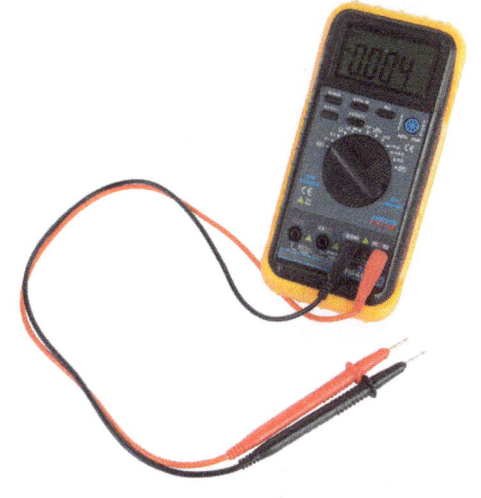

Figure 4.12 Autoranging multimeter

> **Safe working**
>
> Power probes are only designed to test components which draw relatively small amounts of current. Never use them to test starter motors, power winches, etc.

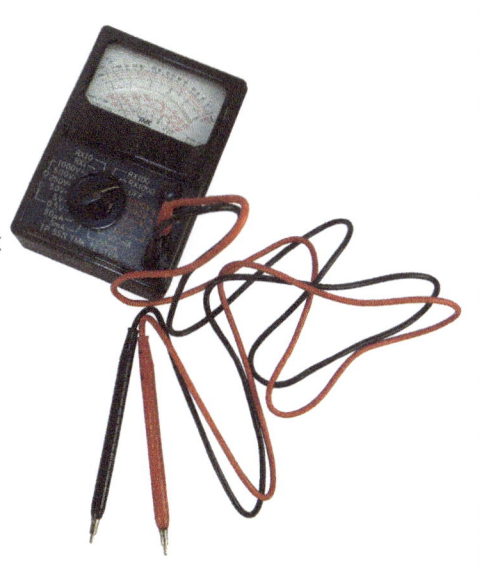

Figure 4.10 Analogue multimeter

> **Did you know?**
>
> It is quite normal for the last digit on the far right of the screen to continuously change. This is a feature common to most digital multimeters. In a lot of cases, as really high accuracy is not required, this figure can be ignored.

> **Action**
>
> Examine a manual multimeter and an autoranging multimeter. Make a list of all of the settings/tests available on each type. Use your list to compare the two multimeters.
>
> - Which do you think has the most functions?
> - Which do you think will be the easiest to use and why?

185

When using an autoranging multimeter, you must take care that your reading is accurate by taking note of the scale of the unit being used. For example, if voltage is measured, the scale might be in:

- millivolts
- volts
- kilovolts
- megavolts.

Connecting the multimeter

When connecting the test leads to a digital multimeter make sure that they are plugged into the correct sockets for the type of measurement you will be taking. There are normally three sockets:

- **Socket 1, marked 'common' or 'ground'** – used with the black test probe
- **Socket 2, marked 'volts, ohms and milliamps'** – used with the red probe
- **Socket 3, marked '10 amps'** – also used with the red test probe, but only when measuring amperage. It is separate from the others to help protect the multimeter from damage if you connected it wrongly to a circuit.

When connecting the multimeter to an electrical circuit:

- For checking volts – it should be connected in parallel (across the circuit)
- For checking amps – it should be connected in series (circuit opened and the ammeter connected in line)
- For checking resistance/ohms – the power should be switched off, the component disconnected and connected in parallel (across the component).

Using a manual multimeter

When you use a manual multimeter, if you do not know the scale to be used, always follow this procedure:

- For testing volts and amps, first select the highest scale on the dial, then rotate the dial slowly down through the scales until you obtain an accurate reading.
- For testing ohms, first select the lowest scale on the dial and then rotate the dial slowly up until you obtain an accurate reading.

Using a digital multimeter

You can measure a number of electrical units on a digital multimeter, including volts, amps and ohms, but other measurements may also be taken. Extra facilities on a digital multimeter may include: temperature, frequency, diode testing, transistor tests and audio continuity testing.

The electrical units of volts and amps are often broken down into two further areas: DC and AC.

- The DC scale is normally shown on the meter as a straight line with a number of dots underneath it: ═══ This symbol is designed to prevent confusion. If just a single line was used, it might be mistaken for a minus sign; if two lines were used; it might be mistaken for an equals sign.

- The AC scale is normally shown on the meter as a wavy line: ∿
- The ohms scale on a multimeter is normally represented by the Greek letter omega (Ω) (if the letter 'O' was used, it might be confused with zero).

Using a multimeter to check voltage

You can use a multimeter as a voltmeter to measure the pressure difference in an electric circuit between where you place the black probe and where you place the red probe.

Checklist			
PPE	VPE	Tools and equipment	Source information
• Steel toe-capped boots • Overalls • Latex gloves	• Wing covers • Steering wheel covers • Seat covers • Floor mat covers	• Multimeter capable of reading volts	• Manufacturer's technical data • Wiring diagrams

1. Connect the probes to the correct sockets on the front of the multimeter.
   - Connect the black lead and test probe to the common socket.
   - Connect the red probe and test lead to the voltage socket.

2. Most voltage that you will measure on a light vehicle is DC, so select the scale with the straight and the dotted lines (= = =).

3. Connect the voltmeter in parallel.

4. Connect the tip of the black lead to a good source of earth, such as the battery terminal, metal bodywork or engine.

5. Use the tip of the red lead to probe the electrical circuit being tested.

Using a multimeter to check for electrical resistance

You can use a multimeter as an ohmmeter to measure resistance. When checking for electrical resistance, always make sure that the power is switched off first and disconnect the component to be tested from the circuit.

Checklist			
PPE	**VPE**	**Tools and equipment**	**Source information**
• Steel toe-capped boots • Overalls • Latex gloves	• Wing covers • Steering wheel covers • Seat covers • Floor mat covers	• Multimeter capable of reading ohms	• Manufacturer's technical data

1. Connect the probes to the correct sockets on the front of the multimeter.
   - Connect the black lead and test probe to the common socket.
   - Connect the red probe and test lead to the socket marked with the omega symbol (Ω).

2. Before you take any measurements, you need to calibrate the ohmmeter to check that it is accurate.

3. Turn the selector dial to the lowest ohms setting and join the tips of the two probes together.

4. When the leads are connected, the readout should show zero or very nearly zero. (If any figures are shown on the screen, you will need to add or subtract them from your final results.)

5. When the leads are disconnected, you should see OL (meaning off limits) or the number 1, which is used to represent the letter 'I' (meaning infinity).

6. Now connect the ohmmeter in parallel across the components so that you can measure the resistance.

- You can also use the ohmmeter to check for continuity. To check a piece of wire for continuity, place the red and black probes at each end of the wire. The screen should display a very low resistance reading.
  - To check a switch for correct operation, connect the red and black probes across the terminals and operate the switch.
  - In the off position, the display should read OL (off limits). In the on position, the reading on the display should be very close to zero.

4 Diagnosis & rectification of light vehicle auxiliary electrical faults

Using a voltmeter to measure electrical current

When measuring the electrical current in a circuit, use the amps setting on the multimeter so that it is used as an ammeter. Take care when using an ammeter because it can be damaged if it is connected incorrectly.

Checklist			
PPE	**VPE**	**Tools and equipment**	**Source information**
• Steel toe-capped boots • Overalls • Latex gloves	• Wing covers • Steering wheel covers • Seat covers • Floor mat covers	• Multimeter capable of reading amps	• Manufacturer's technical data

1. Connect the probes to the correct sockets on the front of the multimeter.
   - Connect the black lead and test probe to the common socket.
   - Connect the red probe and test lead to the socket used for measuring amps. (This socket is normally separate from the one used to measure volts or ohms.)

2. Turn the selector dial to amps measurement.

3. You need to break into the circuit being tested, being careful to avoid short circuits.

4. Connect the ammeter in series, turn on the circuit and measure the current.

A good place to connect an ammeter is at the fuse box – remove the fuse completely and replace it with the ammeter.

> **Safe working**
>
> Never connect an ammeter in parallel (across a circuit). A good ammeter has a very low internal resistance, so if the ammeter is connected in parallel, excessive current will flow and the ammeter will be damaged. Also remember that, depending on the quality of your ammeter, the amount of current that you can measure may be restricted to around 10 amps.

Other functions of a multimeter

Many multimeters are capable of other functional tests in addition to checking voltage, amperage and resistance. Some examples of extra functions are described below.

Audible continuity testing

Some multimeters include an audible continuity tester. This allows you to test the continuity of an electrical component without having to look at the screen.

189

Level 3 Light Vehicle Technology

> ⚠️ **Safe working**
>
> If diodes need to be unsoldered from a circuit for testing, care should be taken to ensure that the electrical connections are not damaged.

- Connect the test probes to the multimeter: black to the common or ground socket and red to the ohms socket.
- Turn the dial to the audible continuity test setting.
- To calibrate the meter and check correct operation, touch the probes together. You should hear an audible tone.
- As with ohms testing, you must switch off circuit power and remove the component being checked from the circuit.
- Now connect the red and black probes to the terminals of the conductor. If continuity exists, you will hear the audible tone.

Diode testing

Most multimeters include a diode test facility. A diode is a one-way valve for electricity. Conduct the test in a similar manner to the continuity test.

- Connect the test probes: black to the common or ground socket and red to the ohms socket.
- Turn the dial to the diode testing setting.
- To calibrate the meter and check correct operation, touch the probes together. The display should show an ohms reading of zero.
- As with ohms testing, you must switch off circuit power and remove the diode from the circuit. You may need to unsolder the diode to remove it.
- With the diode removed, connect the probes to the terminals. If the diode is operating correctly, the display should show a low ohms reading.
- When the polarity of the probes is swapped over, the display should show an off limits or infinity reading.
  - If it shows zero in both directions, the diode has become short circuited.
  - If it shows off limits or infinity in both directions, the diode has become open circuited.

Figure 4.13 Electrical diode symbol

Frequency testing

Some multimeters have a **frequency** test facility. Frequency is a measurement of how quickly a circuit switches. The reading is normally measured in **hertz** (Hz). 1Hz is equal to one complete cycle of operation (for example on and off) occurring in one second.

- Connect the test probe leads to the appropriate sockets on the multimeter.
- Turn the dial to the frequency setting.
- Test the component while the circuit is operating.

Temperature measurement

Some multimeters have a temperature measurement facility. This normally requires an additional probe to be connected. The temperature probe usually has its own socket for connection. Once you have turned the dial to the appropriate setting, you can measure temperature by placing the end of the probe where the measurement is to be taken. Temperature measurement can be useful for diagnosing cooling system faults, for example.

> **Key terms**
>
> **Frequency** – how often something happens. It also describes the distance/time between the peaks of a wave.
>
> **Hertz** – a measurement of frequency.
>
> **Transistor** – an electronic component with no moving parts that can operate as a switch or an amplifier.
>
> **Electromagnetic interference (EMI)** – a disturbance that affects an electrical circuit due to either electromagnetic conduction or electromagnetic radiation emitted from an external source. It is also called radio frequency interference (RFI).

4 Diagnosis & rectification of light vehicle auxiliary electrical faults

Transistor testing

Some multimeters have a **transistor** testing facility. This facility is rarely used by automotive technicians.

Transistors are small electronic switches with no moving parts. They are normally soldered to an electrical circuit board and have three connections: collector, emitter, and base.

There are two types of transistor in common use: positive negative positive (PNP) and negative positive negative (NPN). If the multimeter has a transistor test facility, a six connector socket will be available, marked PNP or NPN. The transistor must be unsoldered from its circuit and connected to one of these diagnostic sockets. Once this is done, the transistor can be tested following the multimeter manufacturer's instructions.

Inductive amps measurement

Using an ammeter to check electric current is intrusive and the circuit must be broken. Also, incorrect connection may cause damage to your ammeter. For these reasons, an alternative method of testing for amperage has been developed: some multimeters come with an inductive amps clamp, which can also be purchased separately as an additional unit.

The amps clamp uses **electromagnetic interference (EMI)** to measure current flow within a circuit. It does not require connection in series but is simply clamped around the wire to be tested. When the circuit is switched on and current flows, you can read the amperage measurement from the display.

The amps clamp is not always as accurate as connecting an ammeter in series, but it is quicker and should not cause damage if connected incorrectly. It is also capable of taking much higher amperage readings than a standard multimeter.

Inductive amps clamps are quick and easy to use, and very good for testing electrical faults such as the ones shown in Table 4.2.

Figure 4.14 Inductive ammeter

> **Did you know?**
>
> There is normally a plus or minus sign on the amps clamp to show which way round it should be connected to an electric circuit.

> **Safe working**
>
> Due to the nature of high voltage systems used in the operation of hybrid vehicles, you should always conduct current measurement with extreme caution. If you have to take a current measurement, the inductive clamp testing method is recommended as this should not involve disconnecting any high voltage circuits.

Table 4.2 Electrical faults that can be tested using an inductive amps clamp

Symptom	Amps clamp use/test
Slow cranking	When a customer presents their car because the engine is cranking over slowly, how do you know whether it is the battery, the wiring or the starter motor that is at fault?
	Connect the inductive amps clamp around the positive or negative battery lead and crank the engine. Initial peak amperage will be shown on the display, indicating the strain that has been put on the electrical system.
	• If the current draw is low, then the starter motor is not struggling and it is the battery that is unable to supply the correct amount of amperage required to turn the engine over at speed. This can indicate that the battery needs replacing.
	• If the current draw is high, then you should suspect the wiring or starter motor and test them. (An example of a high current draw on a four-cylinder petrol engine is a peak amperage of around 140 amps; this should not be exceeded.)
	If you do not know the amount of current that should be drawn on a starting system, connect your inductive amps clamp to another vehicle of similar size and type and compare the two.

CONTINUED ▶

Symptom	Amps clamp use/test
Glow plugs	When a customer presents their diesel engine car and says that it is suffering from poor cold starting, how do you know that the glow plugs are at fault? The normal way of testing the glow plugs would be to leave the car to go cold, sometimes overnight, before confirming the symptoms. If the glow plugs are suspected, they are normally removed and tested across a battery. Not only is this time-consuming, it can also be very dangerous. (If the tip of a glow plug is contaminated, it may explode when tested. Excessive heat caused by testing a glow plug across a battery may also ignite hydrogen sulphide gas produced within the battery, leading to an explosion.) The inductive amps clamp allows you to test the glow plugs more quickly and safely. Allow the engine to cool for around an hour, connect the inductive amps clamp around the feed wire to the glow plugs and turn on the ignition. As a general rule of thumb, each glow plug will draw approximately 20 amps of current. With a four-cylinder engine this means that you are looking for around 80 amps of current in total if the glow plugs are working correctly. If the inductive amps clamp records a reading significantly lower than this, you can suspect that one or more glow plugs have failed. You will then need to remove and replace all of the glow plugs. The advantages of this type of test are: the glow plugs do not have to be removed, saving time and cost; there is less danger of personal injury from incorrect test methods; the vehicle does not have to be left overnight to confirm the symptoms.
Alternator output	When testing an alternator for correct operation, amperage output is often overlooked. Many vehicle technicians only conduct voltage tests when checking alternator operation. A voltmeter is connected across the battery, the engine is started and a load is placed on the system by switching on the headlights. The engine idle is normally raised to around 2000 rpm and, if a voltage of approximately 14.2V is obtained, it is assumed that the alternator is charging correctly. However, the alternator may still be at fault. Voltage is electrical pressure, and a pressure higher than that which is coming from the battery is required to push electrons back into the battery for charging. Electrical pressure is not the same as electrical quantity. The pressure or voltage may be high enough to charge the battery, but if the quantity is not available, the electrical components may be using it up quicker than it is going in, resulting in the battery running down over time. Many modern alternators use a three-phase system, meaning that three coils of wire are connected internally to the charging circuit. This has the advantage of providing three times as much electricity as a normal generator. If one or more of these phase windings fails, correct electrical voltage may still be produced but the correct quantity might not. To ensure that an alternator is operating correctly, conduct the following amps test: With the engine turned off, switch on a number of electrical consumers, such as headlights, heated rear screen, etc. Leave the car for around 10 minutes so that the battery partially discharges. Once the battery has partially discharged, connect the inductive amps clamp to one of the battery cables. Switch off all the consumers and start the engine. Because the battery is now partially empty, the alternator should put out a high quantity of electric current (amps). Check this peak current output to see if it is similar to the one marked on the side of the alternator. If the amperage is well below the amperage indicated on the alternator, the unit should be replaced. As the battery recharges with current from the alternator, the output from the alternator will fall.

CONTINUED ▶

4 Diagnosis & rectification of light vehicle auxiliary electrical faults

Symptom	Amps clamp use/test
Parasitic drain (see pages 202–203)	A common fault with a car's electrical system is parasitic drain. The symptoms normally include the battery going flat over a period of time. An inductive amps clamp can be used to assist with diagnosis. It should be connected around one of the battery leads. With all electrical consumers switched off, current draw should be almost zero. If a current draw exists, then something is using the electrical energy; this component is considered to be a parasite on the battery. Gain access to the fuse box and systematically remove fuses until the current draw disappears. Once the current draw disappears, the circuit causing the problem can be identified based on the fuse that has been removed. With the fuse reconnected, and the current draw reinstated, you can now isolate or disconnect consumers within that circuit until the fault has been found.
Lazy fuel pump	If the customer presents a car with fuelling issues, and you suspect that the fuel pump may be at fault, the normal course of action would be to conduct a fuel pressure test. The current draw on a fuel pump can give an indication of its operation. A high current draw indicates that the pump is struggling, and a low current draw indicates that the pump is finding it too easy. In many cases, you can measure current draw with an inductive amps clamp without removing the fuel pump. With the rear seat lifted, gain access to the wiring running to the fuel pump in the tank. Place the inductive amps clamp around the feed wire. If there are a number of wires, try each one in turn until you obtain an appropriate reading. Start the engine and record the readings. If the fuel pump is fused individually, you will not need to gain access at the fuel tank. You can remove the fuel pump fuse and use a small loop of wire in its place. It is recommended that you use an inline fuse in this loop of wire to protect the circuit. Now connect the inductive amps clamp around the loop of wire, start the engine and take readings. You should expect a current draw of between 3 and 5 amps for a standard petrol pump. If you do not know what to expect, try another vehicle of a similar type and compare the readings.

Dedicated battery testing equipment

Tooling manufacturers are now producing specialist battery testers. These run fully automated checks that help show the condition of lead-acid batteries and charging circuits. The testers can include an internal charger which brings the battery up to the required state of charge before any checks on the battery are made.

Tests can include:

- capacity
- voltage
- heavy discharge (drop testing)
- charge system voltage/amperage
- cold cranking amps (CCA).

The results will often be displayed on a screen as pass or fail. Because the testers use fully automated procedures, they can be operated by people with limited technical knowledge.

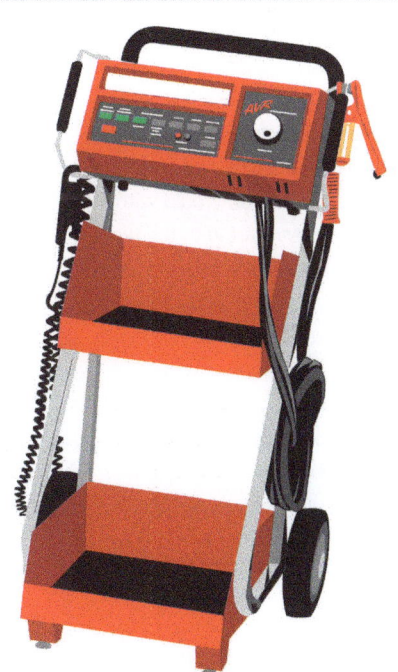

Figure 4.15 Battery tester

Level 3 Light Vehicle Technology

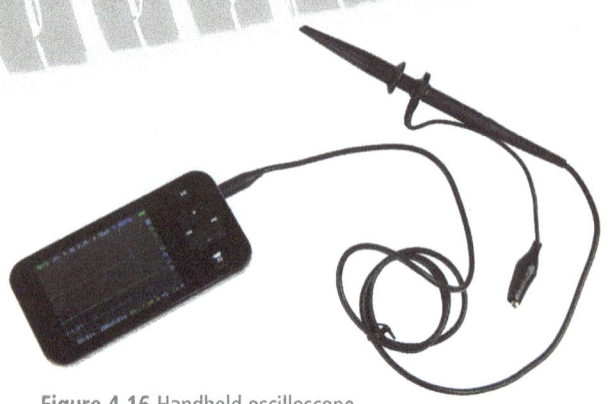

Figure 4.16 Handheld oscilloscope

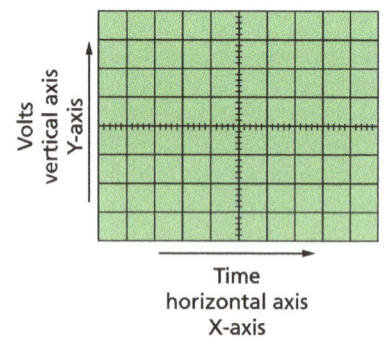

Figure 4.17 An oscilloscope screen

> **Did you know?**
> An easy way to remember which axis is which on a graph is to say 'X is across' (a cross).

> **Did you know?**
> If you don't know what voltage or timescale to use on an oscilloscope, find out in the same way as you would with a multimeter. Start with the highest setting available and work downwards until you can see an image on the screen.

> **Key term**
> **Amplitude** – the height of a wave form, measured in volts or amps.

Oscilloscopes

An oscilloscope is a piece of electrical test equipment designed to act like a voltmeter or an ammeter. A multimeter's measurement readout can't change fast enough to deal with modern electronic systems on motor vehicles – the numbers on the screen can't keep up. The answer to this is to use an oscilloscope.

Unlike a voltmeter, oscilloscopes not only show volts or amps, they also show time. Instead of a digital readout, the results are shown as a graph of volts or amps against time on a screen (as shown in Figure 4.16).

- The graph normally shows voltage or amperage at the side of the screen (on the *y*-axis) – this axis is often called **amplitude**. Use the scale setting switch in a similar way to the dial on a manual multimeter to choose the amount of volts or amps that are shown on the screen.
- The graph normally shows time across the bottom of the screen (on the *x*-axis). This axis is often called frequency. Use the timescale switch in a similar way to the dial that is used to choose the amount of volts on a multimeter.

Although many motor vehicle workshops own an oscilloscope, it is often not used to its full potential, if at all. This can be due to lack of time, knowledge or ability.

Lots of people are put off using oscilloscopes by the large box containing many wires and connectors. They feel that it will be complicated and time-consuming to set up, so they don't bother. However, to use an oscilloscope for simple electrical testing, you only need two probes – a common and voltage wire – just like with a multimeter. To measure amperage, you may need an inductive clamp.

Most of the diagnostic sockets found on oscilloscopes are colour-coded, so after a quick check of the manufacturer's instructions it should be fairly easy to know where to plug these probes in.

Using an oscilloscope for electrical testing

Checklist			
PPE	**VPE**	**Tools and equipment**	**Source information**
• Steel toe-capped boots • Overalls • Latex gloves	• Wing covers • Steering wheel covers • Seat covers • Floor mat covers	• Oscilloscope	• Manufacturer's technical data

Note: The oscilloscope probes may come in different colours, but for the sake of simplicity we will call them red and black here.

4 Diagnosis & rectification of light vehicle auxiliary electrical faults

1. Connect the tip of the black lead to a good source of earth, such as the battery terminal, metal bodywork or engine. This will then only leave you with the red wire to worry about.

2. Now connect the red probe to the circuit to be tested.

3. Adjust the scales until you see an image on the screen.

4. After some practice, you will become familiar with the patterns and wave forms created by different vehicle systems.

Case study

Mr Amor's car is recovered to your garage on the back of a breakdown lorry. The car will not start and the recovery driver thinks that the crankshaft position sensor has failed. Your boss asks you to check it out.

Here's what you do:

- ✓ Listen to the customer's description of the fault.
- ✓ Question the customer carefully to find out the symptoms.
- ✓ Gather information from technical manuals (including vehicle fault codes) before you start and take it to the vehicle.
- ✓ Devise a diagnostic strategy.
- ✓ Use a scan tool to see if any diagnostic trouble codes have been stored. (P0337 Crankshaft position sensor A circuit low input)
- ✓ Check the quick things first (10-minute rule) and conduct a visual inspection of electrical connections, etc.
- ✓ Use an oscilloscope to test the operation of the crankshaft position sensor. (There is no output.)
- ✓ Return to reception and explain the results of the diagnostic tests.
- ✓ Give an estimate for the cost of repair and gain authorisation to carry out the work.
- ✓ Strip out and replace the faulty crankshaft position sensor.
- ✓ Correctly reassemble any dismantled components/systems.
- ✓ Thoroughly test the system to ensure correct function and operation.

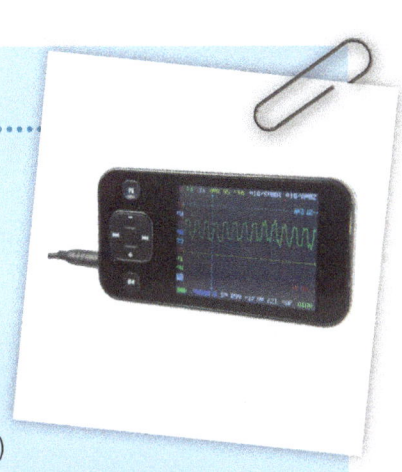

Level 3 Light Vehicle Technology

Scan tools and fault code readers

Faults with many modern vehicle systems would be difficult to diagnose without the aid of a scan tool. The electronic processes that take place within electrical and electronic circuits mean that these systems are being controlled many thousands of times a second, and faults can occur so quickly that you could miss them.

Since the 1980s, manufacturers have been including on-board diagnostic (OBD) systems as part of their vehicle design. The computers that control the vehicle's electrical systems have a self-diagnosis feature. This allows them to detect certain faults and store a code number. Because these electronic control units (ECUs) are monitoring functions, they are able to record intermittent faults and store them in a keep alive memory (KAM) for retrieval by a diagnostic trouble code (DTC) reader.

It is a common misunderstanding to think that plugging a fault code reader into the vehicle's OBD system will tell you what the fault is. It actually only points you in the direction of the fault. You must test the system and components to find the fault.

Figure 4.18 Scan tool

To use a scan tool you will need to locate the diagnostic socket. Since the year 2000 the type and position of the diagnostic socket, also known as the data link connector (DLC), has been standardised. A 16-pin socket should be located inside the vehicle, within reach of the driver, somewhere between the centre line of the car and the driver's seat.

The scan tool should be connected to the diagnostic socket and the ignition switched on. The scan tool will then attempt to communicate with the vehicle's on-board computer systems.

Once communication has been established you should follow the on-screen instructions to retrieve information and operate the on-board diagnostic system as required.

Figure 4.19 Diagnostic scan tool and standardised 16-pin connector

There are two main types of diagnostic information available on many systems:

- OEM
- E-OBD.

OEM

Did you know?

Not all data link connectors are found on the driver's side of the car. This is often a design issue, particularly if the car was originally intended as a left-hand drive.

Manufacturer's data is often available to help you locate the diagnostic socket.

OEM is information from the original equipment manufacturer. To gain access to this information, you will need to enter vehicle specific information such as make, model, engine type and vehicle identification number (VIN). Once this has been done a large amount of manufacturer/vehicle specific data is often available. Many of the diagnostic trouble codes (DTC's) have been standardised so that:

- codes beginning with P relate to powertrain faults
- codes beginning with C relate to chassis system faults
- codes beginning with B relate to body system faults
- codes beginning with U relate to network communication system faults.

This standard is not mandatory and as a result some manufacturers use their own coding system. Also the information from OEM is not generic, meaning that each manufacturer and vehicle type will have its own set of codes. When using diagnostic trouble codes it is good practice to follow the steps shown in Figure 4.20.

E-OBD

European legislation states that any faults with an engine management system that might lead to excessive exhaust pollutants being released to atmosphere must be stored as a diagnostic trouble code. A generic standardised list of codes and a diagnostic connector were produced to be used by manufacturers selling cars in Europe. In this way, information was made available to all service and maintenance repair facilities. This system has become known as E-OBD. The diagnostic trouble codes for E-OBD start with the coding P0 and should be the same for every manufacturer and vehicle type. (For a list of diagnostic trouble codes and their meanings go to hotlinks and click on this chapter.)

Vehicle types which are E-OBD compliant include:

- petrol engine vehicles:
  - vehicles **type approved** from January 2000
  - all new car registrations from January 2001
- diesel engine vehicles:
  - all new car registrations from January 2003
- liquefied petroleum gas (LPG) vehicles:
  - all new car registrations from January 2005.

Features of scan tools

Typical features of scan tools include:

- retrieval of electronic control unit (ECU) fault codes
- erasing of system ECU fault codes
- displaying serial data/live data
- displaying readiness monitors
- resetting of ECU adaptions
- displaying freeze frame data
- coding of new components, such as fuel injectors
- access to information on various vehicle electronic systems
- resetting of service reminder lights
- vehicle key coding.

Did you know?

E-OBD is designed to detect emission-related faults only and should not be confused with original equipment manufacturer (OEM) serial data or diagnostic trouble codes (DTCs).

Key term

Type approved – refers to a product which has met the legal requirements to be sold in a particular country. Type approval is granted to a product that meets a minimum set of regulatory, technical and safety requirements.

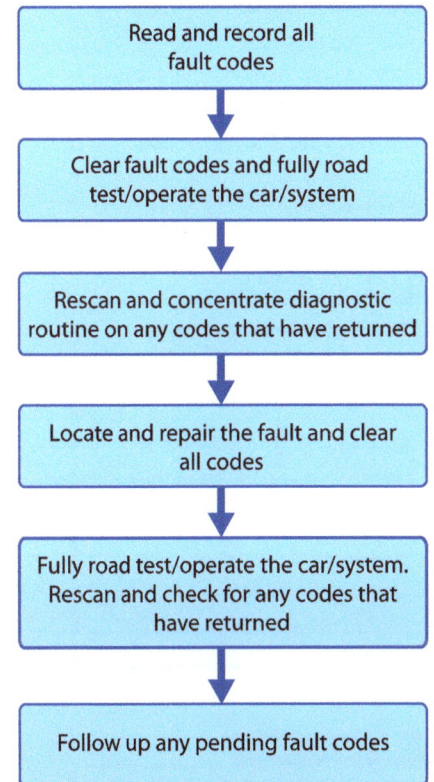

Read and record all fault codes
↓
Clear fault codes and fully road test/operate the car/system
↓
Rescan and concentrate diagnostic routine on any codes that have returned
↓
Locate and repair the fault and clear all codes
↓
Fully road test/operate the car/system. Rescan and check for any codes that have returned
↓
Follow up any pending fault codes

Figure 4.20 Good practice flow chart for using diagnostic trouble codes

CHECK YOUR PROGRESS

1. What is the difference between an analogue and a digital multimeter?
2. List three tests that can be conducted using a specialist battery tester.
3. List four functions of a diagnostic scan tool.

Level 3 Light Vehicle Technology

Figure 4.21 Electronic control unit (ECU)

Multiplexing and network systems

As the amount of technology on cars has increased, demand for faster computer operation and processing has also risen. Advances in vehicle management include:

- engine management
- body control
- chassis systems
- safety systems.
- transmission
- infotainment
- traction control

ECUs had to become bigger to cope with system requirements, and large amounts of wiring were needed to distribute electrical power around the car. These demands also generated a rise in the number of sensors required, leading to complication, extra weight and increased cost of manufacture.

To reduce the amount of sensors and wiring needed for system operation, **multiplexing** was introduced. Multiplexing simply means carrying out more than one operation at a time (for example, a multiplex cinema has more than one screen, and is able to show more than one film at once).

Instead of a single large ECU in a vehicle, smaller ECUs were developed that managed individual systems. These single ECUs became known as **nodes**. The nodes are connected to each other by a communication wire, which allows information to be shared in a **network**. When one of the ECUs receives information from a sensor, it processes the signal and acts if required. It then passes on this information to the communication network wire linking the other ECUs, which then use that information if required, and once again pass it on. This means that signals from a single sensor can be shared across a number of different vehicle systems.

CAN bus

Controller area network (CAN) was introduced by Robert Bosch in the 1980s. There are a number of different network types and manufacturers available, but the name CAN bus has been adopted by many technicians to describe nearly all network systems.

The nodes are connected by a single communication line, which allows the exchange of multiple pieces of data. The communication line can link these ECUs in the following layouts:

- a large loop, known as a **daisy chain** (see Figure 4.22), often used for body systems
- a **star** pattern, known as a server system (see Figure 4.23), often used for infotainment systems
- connected in parallel to a single **bus line** (see Figure 4.24), often used for powertrain control systems.

Did you know?

ECUs were becoming larger and larger because of the need for more connections and pins where they joined the wiring. There was a limit to how small these connections could be made, and how closely the wires in the loom could be bundled. Multiplexing has allowed a reduction in sensors and wiring, as it allows the ECUs to share information on a network.

Key terms

Multiplexing – a method of carrying out more than one operation simultaneously.

Nodes – ECUs connected to a computer network (from the Latin word *nodus*, which means knot).

Network – several computers connected so they can communicate with each other.

Did you know?

The word 'bus' is used in various situations. One meaning of bus is a vehicle that collects you from one place and delivers you to another. This is very similar to its meaning within a communication network. Information is picked up at one point on the communication line; it then takes a route around the system and stops at various ECUs (like bus stops).

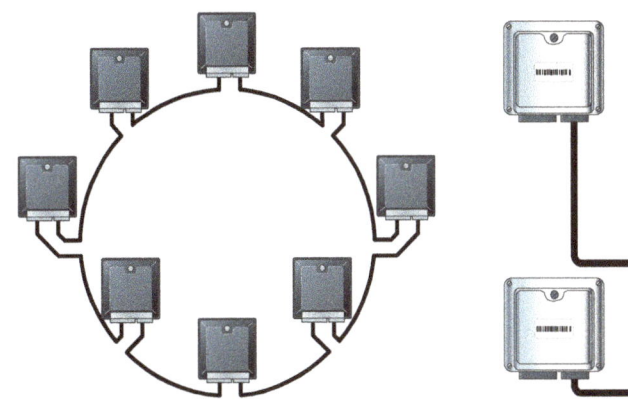

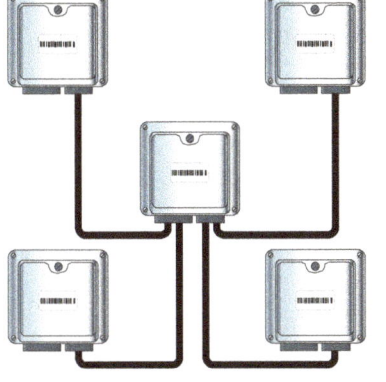

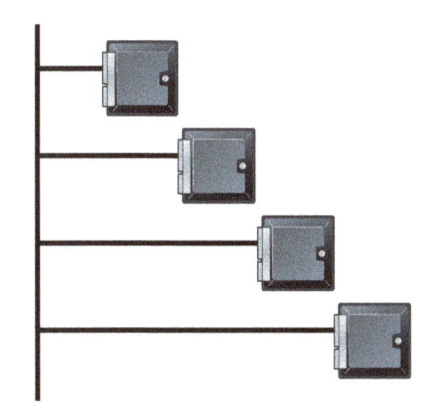

4 Diagnosis & rectification of light vehicle auxiliary electrical faults

Figure 4.22 Daisy chain network

Figure 4.23 Star network

Figure 4.24 Bus line network

When a daisy chain layout is used, the data sent travels in both directions at once, which gives much greater reliability. If one wire is damaged or broken within the loop, the information can still arrive at the appropriate ECU as it comes from the other direction. In addition to better data reliability, this system also improves malfunction diagnosis.

Communication data

When an ECU receives a signal from a vehicle sensor, it processes this and places the information on the network bus as a data packet. The data packet is usually made up of the following:

- A header: the equivalent of 'hello, I am transmitting a message'.
- The priority: how important this message is, e.g. vehicle safety information will be more important than a bodywork communication such as a command to open an electric window.
- Data length: this is so the receiver knows it has not lost or 'misheard' any of the information.
- Address for the receiver ECU: the final destination for the message.
- Data type: what type of information is contained, e.g. speed, temperature, etc.
- Data: the actual sensor information itself.
- An error detection code: this says 'has all the information been received?' and is known as a cyclic redundancy check (CRC).
- End of message: 'goodbye'.
- Finally, a request for a response from the receiving ECU (the equivalent of 'thank you, I got your message').

Figure 4.25 Communication data packet

199

Level 3 Light Vehicle Technology

> **Key terms**
>
> **Corruption** – a breakdown of integrity or communication.
>
> **Potential difference** – the difference between two voltage values.

> **Did you know?**
>
> The data is sent as an on and off signal. It would be seen on the screen of an oscilloscope as a square wave form.

To help reduce the possibility of data **corruption** caused by misinterpretation or external electromagnetic interference, a CAN bus system uses two communication wires instead of one, twisted over and over each other in a spiral. The same data is sent on both of these communication wires as an on and off voltage signal. One signal is sent as a positive switch and one is sent as a negative switch. These provide a mirror image on each network wire, which are known as CAN high and CAN low. The **potential difference** between the voltages on the two lines produces a digital signal that can be processed into information. The communication wires are exactly the same length, and as the data travels at the same speed, both versions of the data should arrive at the receiver at the same time. These messages can now be compared to help identify data corruption. The opposing voltage of the signals transmitted down the communication wires will also help cancel out electromagnetic interference from other systems.

Bus speeds

There are three main bus speeds:

- Low speed: used for instrumentation, body control and comfort, etc. It operates at a rate of 33,000 bits of information per second (33 kbps).
- High speed: used for powertrain control and safety critical information, etc. It operates at a rate of 500,000 bits of information per second (500 kbps).
- Very high speed: used for high volumes of data transmission in infotainment systems (such as streaming video and music, etc). It will operate at a rate of 25,000,000 bits of information per second (25 Mbps).

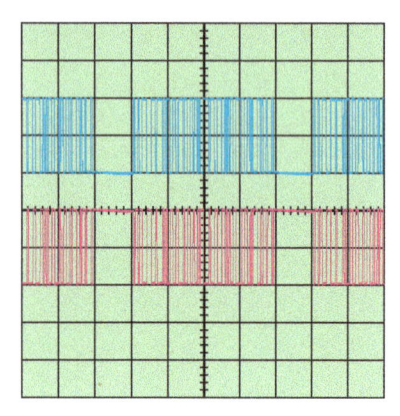

Figure 4.26 CAN bus wave form

System reliability

The CAN bus system is more reliable than a standard wiring system due to the fact that a single open bus wire would not stop communication. Two open bus wires can stop communication, but as more ECUs are used for control, only part of the system may fail.

Short circuits can have a catastrophic effect on network communication. A short to either positive or earth will disrupt the communication on the bus wire, as an on and off signal can no longer be transmitted. If viewed on an oscilloscope screen, this would be a flat line at either 0V or 5V.

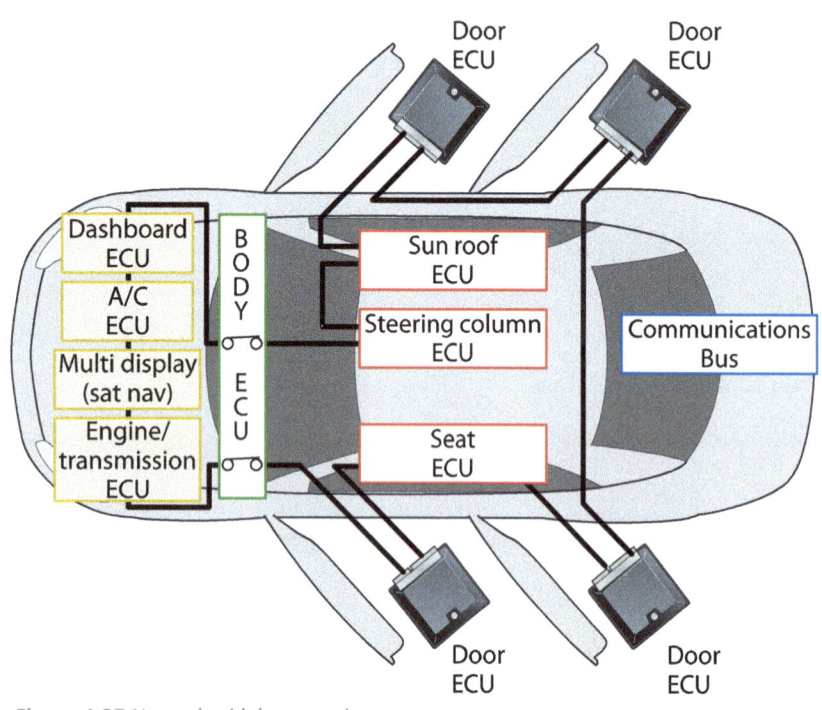

Figure 4.27 Network with bus cut relays

To avoid total failure of the system, bus-cut relays can be used. These are a type of circuit breaker that isolates part of the network, allowing the rest of the system to continue communicating.

Multiplex and networked diagnosis

If a critical network failure occurs, such as short to positive or earth, the vehicle may suffer a complete communication loss.

With a networked system, if communication is lost within a certain area, numerous items will not work and several fault codes may be generated. Having connected a scan tool and retrieved the diagnostic trouble codes, you should look for the code that is the root cause.

Communication failures are normally an effect of the original fault. You should ask yourself, 'Is this the cause or an effect created by the fault?' CAN bus systems report communication faults as live data. This means that once you have identified the cause fault code, you should be able to conduct a diagnosis by disconnecting and isolating components or sections of **wiring loom** until communication is re-established.

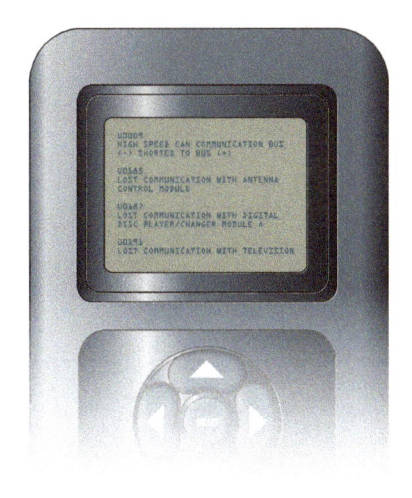

Figure 4.28 Network communication error

NEW TECH

Faster and faster network systems...

Although CAN bus systems are very fast, and can receive, process and **disseminate** one million bits of information per second, they are not fast enough to keep up with modern technology demands. Manufacturers are producing faster and faster network systems that are able to process over 10 million bits of information per second. These systems work in exactly the same way as a normal CAN bus, just faster.

Key terms

Wiring loom – a number of electrical wires bundled together to assist with their routing around the car.

Disseminate – spread out or disperse widely.

Electrical faults

Many technicians get very confused when trying to diagnose an electrical fault, as the symptoms can be wide and varied. Diagnostic routines can be considerably shortened if you fully understand the nature of the original fault.

There are four main electrical faults:

- open circuit
- high resistance
- short circuit
- parasitic drain.

If you have a good understanding of these four faults, then you may be able to remove three-quarters of your diagnostic routine just from a description of the symptoms.

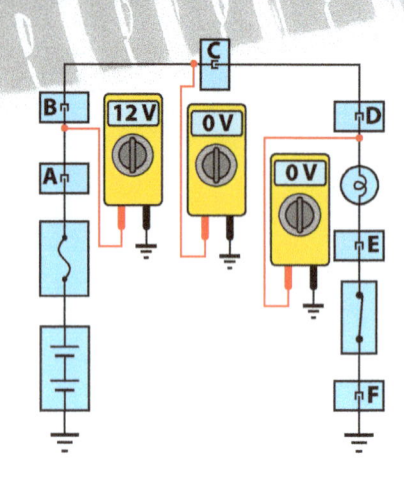

Figure 4.29 Using a voltmeter to check an open electric circuit

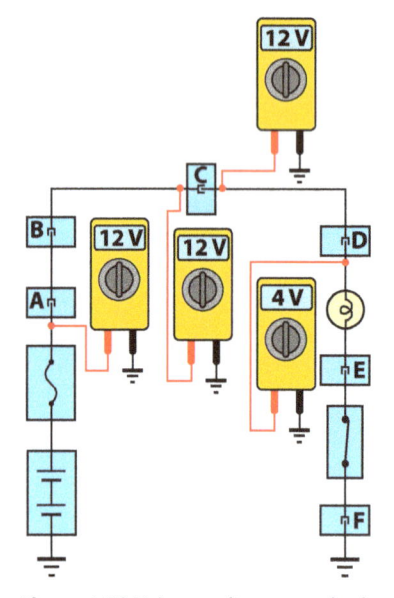

Figure 4.30 Using a voltmeter to check a high resistance electric circuit

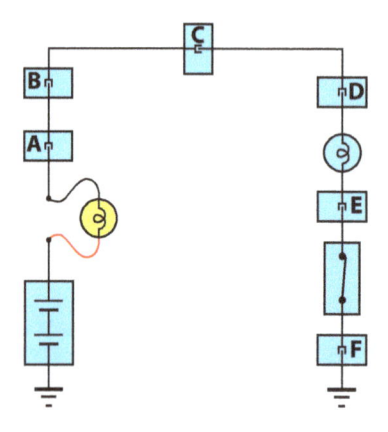

Figure 4.31 An electric circuit with a test lamp bulb connected in the place of the fuse, to check for an electrical short circuit

Open circuit

In an open circuit, electricity cannot flow. This is usually because there is a physical break in the system.

An example of a symptom caused by an open circuit is a light on the car not working. To diagnose this fault, you can use a voltmeter.

If the circuit is working correctly, voltage will exist up to the consumer (in this case the bulb), which will use up the electrical energy. If you connect a voltmeter to the bulb feed and there is no voltage, the open circuit must be somewhere before this point in the system. Return to the battery – if a normal 12 volts is recorded, the fault must lie after this point in the system. Slowly move the voltmeter along the circuit, taking readings at various points (for example, see Figure 4.29). The point where the voltage disappears somewhere before the consumer is the location of the open circuit.

This test is sometimes known as a volt drop test.

High resistance

In a high resistance circuit, the electricity slows down. This is usually because of a partial restriction in the system.

An example of a symptom caused by high resistance is a dim light on the car. To diagnose this fault, you can use a voltmeter.

Test for high resistance in the same way as for an open circuit. The point where the voltage falls below 12 volts somewhere before the consumer is the location of the high resistance. (See, for example, Figure 4.30)

Short circuit

Electricity is lazy, and will always take the path of least resistance. (Why travel the full length of the circuit when a shortcut can be taken?)

In a short circuit, the electricity does not make it all the way to the end. As the electrical energy is not used up by a consumer, it is normally converted into heat, which will cause circuit damage.

An example of a symptom caused by a short circuit is wiring or a fuse that has burnt out. To diagnose this fault, you can use a test lamp.

If a **dead short** exists, remove the blown fuse completely and replace it with a test lamp (as shown in Figure 4.31). The bulb on the test lamp will light up. Because the test lamp bulb is consuming electrical energy, the circuit should not overheat and burn out. Starting at the far end of the circuit, you should systematically disconnect or move components and wiring until the test lamp goes out. This will help you find the position of the short circuit.

> **Key term**
>
> **Dead short** – an electrical short circuit that goes straight to earth without passing through a consumer.

Parasitic drain

A parasitic drain is similar to a short circuit: electricity will continue to flow even when the system is switched off.

An example of a symptom caused by a parasitic drain is a battery that goes flat if left for a period. To diagnose this fault, you can use an ammeter.

Insert the ammeter into the circuit (connected in series) to measure if any current is flowing when everything is switched off. If any current (measured in amps) is shown, there is a parasitic drain. To isolate the parasitic drain, disconnect the suspected circuits (isolate/unplug components) until flow stops and amps draw falls to zero. You can then replace the faulty component.

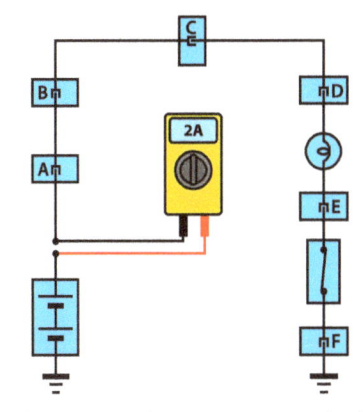

Figure 4.32 Using an ammeter to check for a parasitic drain

CHECK YOUR PROGRESS

1 What does the acronym EMF stand for?
2 Explain the term multiplex.
3 What is the data transmission speed of a high speed powertrain network?

Batteries: starting and charging

The electrical systems found on most modern cars are supplied from a 12-volt battery which stores the electrical energy in chemical form.

Lead-acid battery

Many car batteries are of the lead-acid type. For a description of their construction and operation, see *Level 2 Diploma in Light Vehicle Maintenance & Repair Candidate Handbook*, Chapter 6 page 373.

Table 4.3 explains some battery terms and ratings.

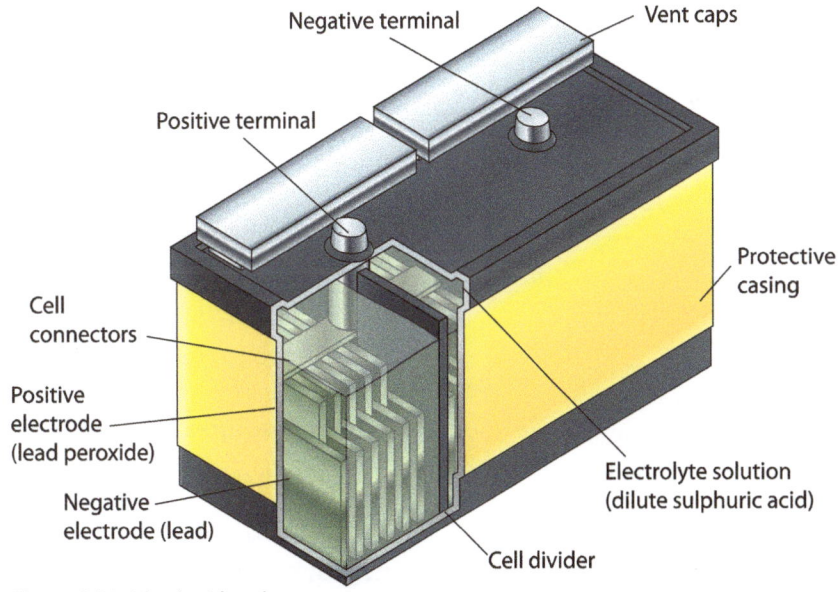

Figure 4.33 A lead-acid car battery

Did you know?

Electronic car battery testers are now available. When they are connected to a battery and programmed with details found on the battery casing, they run through a series of checks and display the results of the analysis on a screen.

Table 4.3 Battery terms and ratings

Term	Description
Amp hours (Ah)	These are a measurement of the electrical current that a battery can deliver. This quantity is one indicator of the total amount of charge that a battery is able to store and deliver at its rated voltage. The amp hours value is the total of the discharge-current (in amps) multiplied by the duration (in hours) for which this discharge-current can be sustained by the battery. For example, a car battery could be rated as 100 Ah, which should be enough electricity to provide: • 100 amps for one hour • 1 amp for 100 hours • 10 amps for 10 hours • any other combination that multiply together to make 100 (e.g. 25 amps for 4 hours). The amp hours rating is required by law in Europe.
Cranking amps (CA)	A number that represents the amount of current a lead-acid battery can provide at 0°C (32°F) for 30 seconds while maintaining at least 1.2 volts per cell (7.2 volts for a 12-volt battery).
Cold cranking amps (CCA)	A number that represents the amount of current a lead-acid battery can provide at −18°C (0°F) for 30 seconds while maintaining at least 1.2 volts per cell (7.2 volts for a 12-volt battery). This test is more demanding than those conducted at higher temperatures.
Hot cranking amps (HCA)	A number that represents the amount of current a lead-acid battery can provide at 27°C (80°F) for 30 seconds while maintaining at least 1.2 volts per cell (7.2 volts for a 12-volt battery).
Reserve capacity minutes (RCM), also known as reserve capacity (RC)	A lead-acid battery's ability to sustain a minimum stated electrical load. It is defined as the time (in minutes) that the battery at 27°C (80°F) will continuously deliver 25 amperes before its voltage drops below 10.5 volts.

Did you know?

Silver calcium batteries are commonly used with smart charging systems. The addition of silver to a standard lead acid battery helps reduce internal corrosion and resistance. The addition of calcium helps improve the ability of the battery to cycle from charge to discharge.

Nickel cadmium battery

An alternative to the lead-acid battery is the nickel cadmium (nicad) type, which is found in some cars. It works in a similar way to a standard battery but requires less maintenance and cannot be overcharged.

This type of battery is made of the following materials:

- positive plates: nickel hydrate
- negative plates: cadmium
- electrolyte: potassium hydroxide and water.

Nickel cadmium batteries tend to be larger and more expensive than normal lead-acid batteries. However, they are better at coping with the extreme loads placed on them by modern electrical systems, especially in hybrid or battery electric vehicles.

4 Diagnosis & rectification of light vehicle auxiliary electrical faults

NEW TECH

Metal hydride battery

Metal hydride batteries are becoming the most popular for use with hybrid drive vehicles. They are very similar in construction to a nickel cadmium battery, but use a metal hydride (hydrogen atoms stored in metal) as the negative plate. A metal hydride battery has the following advantages over other battery types:

- They have a high electrolyte conductivity. This allows them to be used in high power applications (such as hybrid drive vehicles).
- The battery system can be sealed, which minimises maintenance and leakage issues.
- They operate over a very wide temperature range.
- They have very long life characteristics when compared with other battery types – this offsets their higher initial cost.
- They have a higher energy density and lower cost per watt than other battery types.

Current hybrid drive systems (depending on manufacturer) are using batteries with a high voltage potential somewhere between 100V and 300V.

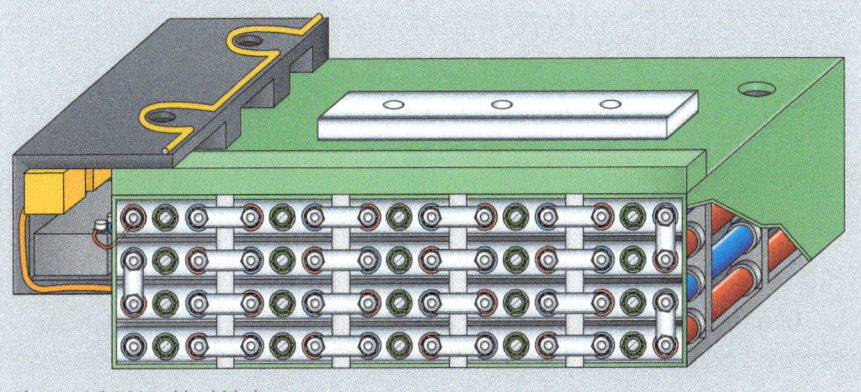

Figure 4.34 Metal hydride battery

Charging

Batteries do not hold limitless power, and so they need a system for recharging. Most modern cars use an **alternator** for this purpose. An alternator works on the principle of rotating an electromagnet inside a number of copper windings. As the North and South poles of the electromagnet pass the windings (known as **phases**), an electric current is induced in one direction and then the other.

This alternating current (AC) must be converted into direct current (DC) before it can charge the battery – this is done by the **rectifier**. As engine speed increases, the amount of voltage produced in the alternator also increases. To prevent the voltage becoming too high and damaging the battery, a **regulator** is used to stop charging when voltage reaches a preset limit. For a description of the construction and operation of an alternator, see *Level 2 Diploma in Light Vehicle Maintenance & Repair Candidate Handbook*, Chapter 6, pages 398–400.

Key terms

Alternator – an engine-driven electrical generator used to charge batteries and supply electricity for vehicle systems.

Phases – the separate electrical windings used in an alternator to generate electricity. (Most systems operate three phases.)

Rectifier – the unit inside an alternator that converts alternating current (AC) to direct current (DC).

Regulator – the unit inside an alternator that controls the output voltage.

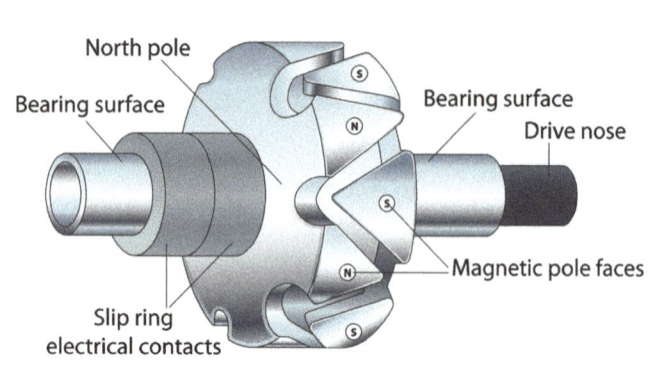

Figure 4.35 An alternator electromagnetic rotor

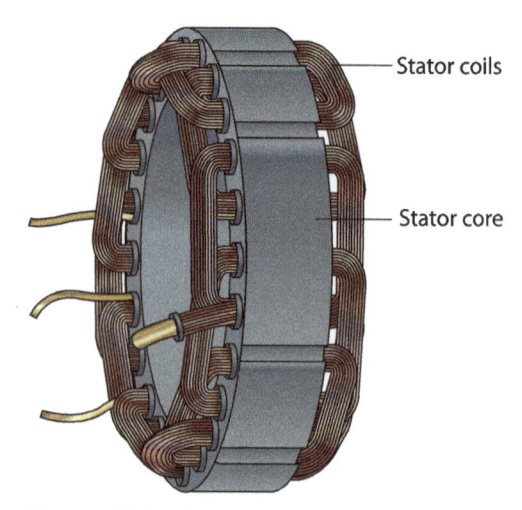

Figure 4.36 An alternator stator unit

NEW TECH

Liquid-cooled alternators

When an alternator is generating electricity, it produces a great deal of heat. Some manufacturers are now producing liquid-cooled alternators that are connected to the engine's cooling system. This helps prevent overheating and improves efficiency.

Smart charging

The demands placed on a modern charging and battery system vary considerably depending on how the car is used. Some manufacturers are now controlling the regulated output voltage of alternators for different driving and operating situations. On a standard system, the regulated alternator output voltage is kept below 14.2V to help prevent overheating and gassing of the battery. In a smart charging system, the battery temperature is estimated by the regulator by analysing its own temperature. The regulated voltage can be increased and decreased as demands are placed on the electrical system without overheating the battery; strain on the battery is considerably reduced. When extra performance is required the alternator can be prevented from charging, which reduces the loads placed on the mechanical drive system. Engine running can be smoother and more fuel efficient. The charging system can operate using a 'closed loop' method: it communicates with an ECU using pulse width modulation and output to the regulator is controlled by duty cycle to allow for different operating conditions. This controlled alternator output has benefits, including:

- reduced battery charge times
- increase in battery service life
- battery temperature regulation
- improved idle stability
- increased vehicle performance
- better fuel economy
- a self-diagnosis ability is often available.

Hybrid drive battery charging

Not all hybrid vehicles use an alternator to charge the 12V battery system. In many models, the electric motors used to drive the road wheels can be converted into generators. They operate when power is being provided by the internal-combustion engine or regenerative braking is occurring (see Chapter 2, page 50). These motors are able to generate the electricity required to recharge the high voltage system on a hybrid car (somewhere between 100V and 300V). The high voltage batteries are then able to charge the 12V battery system through a transformer called a DC to DC converter. This system reduces the amount of strain on the standard 12V circuits, which means that smaller batteries with a longer service life can be used for the low voltage system.

Starting

To start the engine, the crankshaft must be rotated at speeds higher than 180 rpm to initiate combustion. This is the job of the starter motor. For a description of starter motor construction and operation, see *Level 2 Diploma in Light Vehicle Maintenance & Repair Candidate Handbook*, Chapter 6, pages 400–405.

> **Safe working**
>
> Remember that, when diagnosing a starter motor system fault, the immobiliser system should be disabled, as this could prevent the starter from operating.

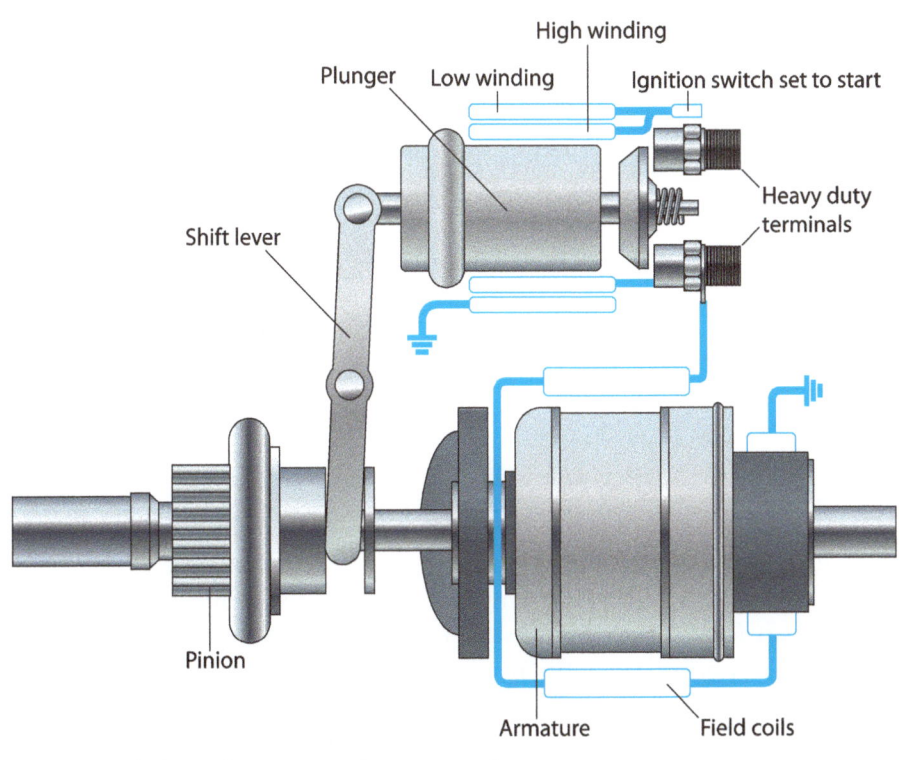

Figure 4.37 The internal components of a pre-engaged starter motor

Hybrid drive starting

Although some hybrid vehicles are fitted with a starter motor, they are often only there for emergency purposes. If the high voltage battery system becomes discharged, the 12-volt starter motor cuts in to get the engine running. Now electricity can be generated to operate the charging system.

Level 3 Light Vehicle Technology

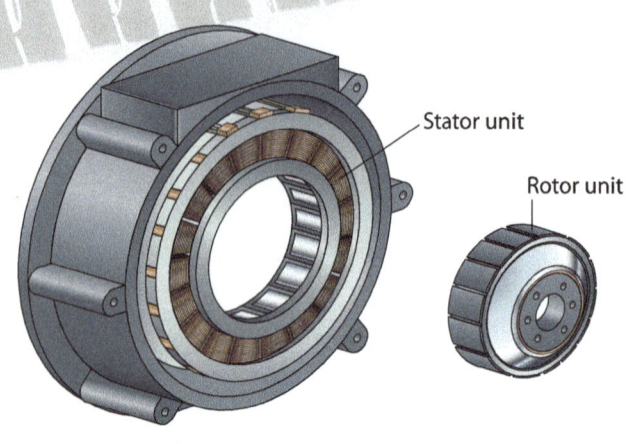

Figure 4.38 A hybrid vehicle motor/generator

CHECK YOUR PROGRESS

1. What do the acronyms CCA, HCA and RCM stand for?
2. What component converts AC to DC in an alternator?
3. Why does a hybrid vehicle often have a starter motor?

Key terms

Obligatory – compulsory by law.

Tungsten – a metal with a very high melting point (3410°C) that is used to make electric light filaments.

Filament – a thin, finely spun wire.

Incandescently – emitting visible light as a result of being heated.

Inert gas – a gas that does not react chemically with other substances.

Diffuse – to spread out over a wide area.

Converging – coming together from different directions to meet at one place.

During normal operation, the starter motor is no longer needed. This is because hybrid vehicles often use a combination electric motor/generator as part of their drive operation.

When the internal combustion engine needs to be started to provide the appropriate power delivery, the hybrid electric motor can be used instead of the starter motor. The instant and powerful rotation of the crankshaft created by this motor ensures that the 'stop–start' operation of the internal combustion engine is smooth and precise. As the hybrid electric motor doesn't use a starter pinion and ring gear, the characteristic whine created by a normal starter motor is no longer apparent.

For more information on the operation of hybrid drive vehicles, see Chapter 3, pages 161–163.

Lighting systems and technology

In order to be used on a public highway, **legislation** states that the vehicle must have an external lighting system that shows presence, position and direction of travel. For a description of **obligatory** lights and their operation, see *Level 2 Diploma in Light Vehicle Maintenance & Repair Candidate Handbook*, Chapter 6, page 417, Table 6.6.

External vehicle lights include:

- sidelights, including number plate lights and marker lights
- dipped beam headlamps
- main beam headlamps
- dim/dip or running lights
- indicators and hazard lights
- high intensity driving lamps and fog lights.

Many external lighting systems use standard light bulbs, where electricity is forced through a thin **tungsten filament**. The resistance created causes the tungsten filament to get hot and glow **incandescently**. When it reaches a temperature of 2300°C, the filament gives off white light. To ensure that the filament does not burn out or vaporise, it is covered in a glass or quartz bulb and all of the oxygen is removed. Some light bulbs are pressurised with **inert gases**, so that the temperature of the filament can be raised still further.

- Exterior lights that are designed to show vehicle position usually have a lens that is used to **diffuse** the light produced.
- Exterior lights that are designed to illuminate the road ahead usually have a reflector and a **converging** lens to produce a light beam that can be directed to precise areas for night-time vision.

Headlamps

Lighting that is designed to help the driver see where they are going uses a method to focus a beam of light to assist with night-time vision. Focused lighting is used for headlamps, and the way that they create the beams depends on the headlamp design.

To ensure that headlights operate efficiently without dazzling oncoming drivers, the headlamp aim/alignment must be correct. You need to adjust the headlamp aim if:

- the vehicle has been involved in a collision to the front
- a headlamp unit is replaced
- a suspension component has been replaced
- new wheels/tyres have been fitted
- the headlamp bulb has been replaced.

You can adjust the headlamp beam accurately by using a headlamp beam setter, or aligner, as shown in Figure 4.39.

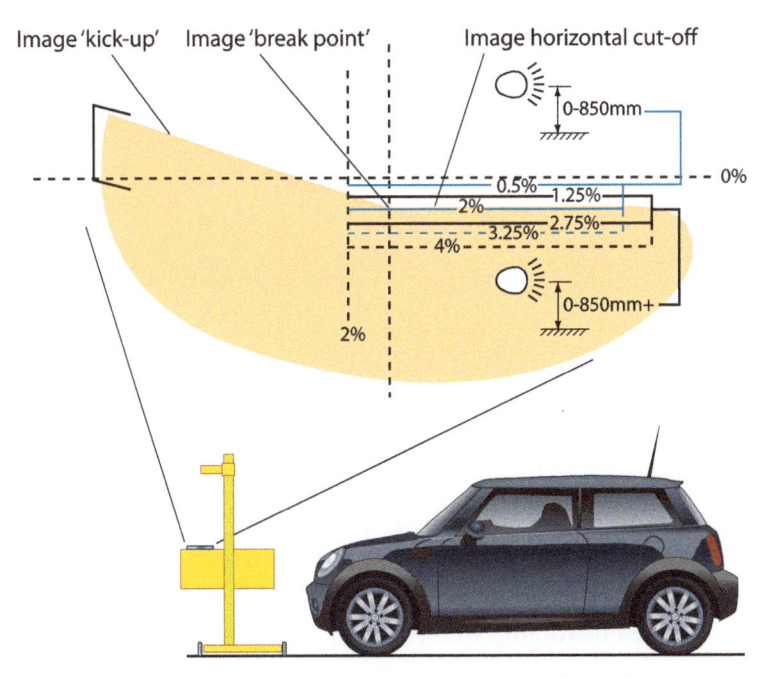

Figure 4.39 Checking aim of the headlamp beam using a headlamp aligner

> **Did you know?**
> The beam patterns produced by headlamp units are required by law to conform to certain shapes and tolerances. These headlamp beam patterns should be regularly checked to ensure that the lighting systems are functioning correctly and will not dazzle other road users.

> **Did you know?**
> Tyre pressures and vehicle load can have an effect on headlamp aim. If tyre pressures are low at the rear, it is possible that headlamp aims will be too high; if excessive load is placed in the rear of a car, it is possible that headlamp aims will be too high. Many cars now come equipped with a headlamp levelling device. This can be manually set by the driver or automatically controlled by the lighting management system.

Some examples of headlamp design are described below.

Freeform headlamps

Complex reflectors have been developed to accurately focus and scatter light beams to improve night-time vision. The reflectors are divided into segments to illuminate sections of the road. All of the surface area is used to provide a beam pattern.

Level 3 Light Vehicle Technology

> ⚠️ **Safe working**
>
> Be careful when working with high intensity discharge lighting (HID) systems. The bulbs and associated components can become very hot during use, leading to the possibility of burns. The electrical ballast system will use a controlled alternating current with a voltage high enough to cause electric shock.

> **Key terms**
>
> **Prism** – a transparent solid body, often with a triangular base, used for bending light into a headlamp beam pattern.
>
> **Ellipsoidal** – in the curved shape produced by an ellipse.
>
> **Convex** – having a shape that curves outwards like the exterior of a sphere.
>
> **Arc** – the visible spark seen between two electrodes.
>
> **Electrode** – a conductor through which electricity enters or leaves an object.
>
> **Xenon** – a specialised gas used in some headlamp bulbs.

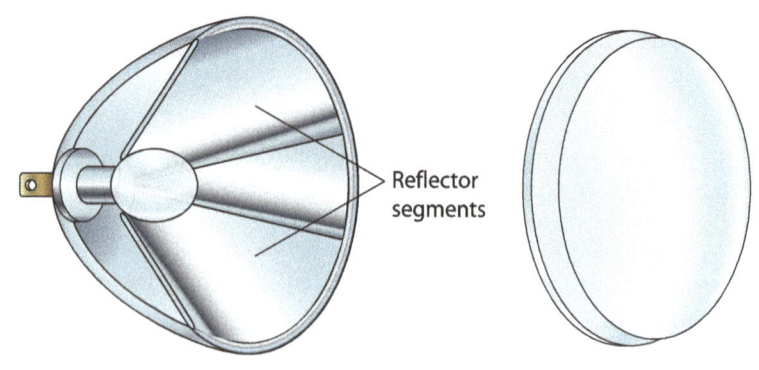

Figure 4.40 A freeform headlamp reflector

These headlamps no longer use **prisms** etched into the headlamp lens to bend the light output. They are known as freeform headlamps.

Projection headlamps

Projection headlamps are small in diameter, but produce high output. They use an **ellipsoidal** reflector that reflects light onto a lens. The **convex** lens acts like a magnifying glass and projects light in the required pattern. Shields are used to cut off any unwanted light.

Light and temperature

As the temperature of the light source increases, its colour begins to change from white to blue and then to purple and violet. As the lighting colour approaches the blue part of the spectrum, it begins to represent natural daylight more closely. This means that your eyes are able to function better at night when a headlight system with a blue/white light is used. Although a standard bulb can be pressurised with certain gases that allow the light emitted to approach the blue/white spectrum, other lighting systems are now being developed which produce far greater intensity.

High intensity discharge (HID)

High intensity discharge (HID) is a system that uses an electric **arc** to produce light rather than a filament. They produce more light for the same level of power consumption than a standard tungsten filament or tungsten halogen headlamp bulb.

Enclosed inside a glass bulb are **electrodes** capable of producing a spark across an air gap. The high intensity arc comes from metallic salts that are vaporised inside the bulb, which also usually contains **xenon** gas. When the initial spark has been produced, the arc is maintained at a constant strength which causes the HID system to give off light.

In order for this system to operate, a component called an igniter is required to create a spark across the electrodes within the bulb. The igniter is a step-up transformer that usually works in a similar manner to an ignition coil. Once the initial spark has been created, a component called a ballast takes over and keeps the arc burning.

Figure 4.41 High intensity discharge bulb HID

Ballast and igniter operation

HID lights operate in three main stages, as described in Table 4.4.

Table 4.4 The three main stages of HID lights operation

Stage	Description
Ignition	A high voltage pulse from the igniter is used to produce a spark. This is similar to spark plug operation. The spark **ionises** the xenon gas, creating a conducting tunnel between the tungsten electrodes.
Warm-up	The bulb is supplied with a controlled overload, whereby extra current is delivered for a short period. Because the arc is operated at high power, the temperature in the arc chamber rises quickly. The metallic salts vaporise, and the arc is intensified. The resistance between the electrodes now falls. The electronic ballast control unit senses this and automatically switches to continuous operation.
Continuous operation	All the metal salts are now in the vapour phase. The arc will have reached its steady shape. The ballast now supplies stable electrical power so the arc will not flicker and the high intensity light is given off.

Advantages of HID

- HID-xenon lights provide around three times more light than a standard halogen headlight bulb.
- They last for around 3000 hours compared to around 350 hours for a halogen bulb. This is because there is no filament to be heated.
- They use less power, which contributes to greater fuel economy and reduced emissions from the engine.

Disadvantages of HID

- If not kept clean and correctly adjusted, HID-xenon systems can produce glare that will dazzle oncoming drivers.
- HID bulbs may contain mercury, which is extremely toxic, so must be treated as hazardous waste.
- HID headlamps are significantly more costly to produce, install, purchase and repair.
- The failure of an HID component (bulb, igniter or ballast) often results in damage to at least one of the other components. For this reason, it is common for ballasts, bulbs and igniters to be replaced at the same time at great expense.
- If an HID system fails, it cannot easily be converted back to a standard halogen lighting system.

> **Key term**
>
> **Ionise** – to change a chemical substance by adding or removing charged particles.

> **Did you know?**
>
> Light can be measured in lumens (lm), which is a measurement of candlelight power. HID-xenon lights produce 100 lm per watt. Halogen lights produced around 25 lm per watt.

> **Safe working**
>
> Vehicles equipped with HID headlamps are required by ECE Regulation 48 to be equipped with headlamp lens cleaning systems and automatic beam levelling control.

> **Did you know?**
>
> Some manufacturers are now including ambient light sensors in their vehicle design. This gives the driver the option of controlling the operation of the lighting system manually, or when switched to automatic mode the lights can be turned on and off by the body control system depending on the amount of available light.

Level 3 Light Vehicle Technology

Skills for work

A customer has asked you to fit a set of aftermarket HID headlamps to his car. Because the car didn't originally come with HID lighting, it has no self-levelling system or headlamp washer. This means that if you carry out the work requested, the car will no longer meet the legal requirements for use on a UK road. The customer is willing to overlook this problem and says that he is happy to pay for the work.

1. You will need to refuse to conduct the work for the customer, as it is illegal. Describe how you could explain this to the customer without creating offence.

2. This situation requires that you use particular personal skills. Some examples of these skills are shown in Table 6.1 at the start of Chapter 6, on pages 304–305. Using the examples given in Table 6.1, choose one skill from each of the following categories that you think you need to demonstrate in this situation.
   - General employment skills
   - Self-reliance skills
   - People skills
   - Customer service skills
   - Specialist skills

3. Now rank these skills in order of importance, starting with the one that it is most important for you to have in this situation.

4. Which of the skills chosen do you think you are good at?

5. Which of the skills chosen do you think you need to develop?

6. How can you develop these skills and what help might you need?

LED

A light emitting diode (LED) is a form of bulb with no filament. It is made of electronic semi-conductor material that gives off light when current moves electrons across it. As there is no filament to create resistance and get hot, a standard LED draws very little current. This makes LEDs efficient and low power-consuming devices.

Because these components are diodes, they only accept current in one direction. This means that they must be connected to an electric circuit with the correct polarity in order to work. The movement of electrons is instantaneous, so an LED illuminates very quickly. This makes them ideal for use in high level brake light systems, as they can be seen by others as soon as the brake light switch is operated.

Depending on how they are manufactured, LEDs can be designed to give off light in different colours. This means that when they are grouped within a light cluster, they can be easily used as different light sources (indicators, tail lights and brake lights, for example).

Although LEDs can be damaged by high voltage, they are generally cheaper and more reliable than standard light bulbs.

Figure 4.42 Light emitting diode (LED) headlamp bulb arrangement

White light emitting diodes (LED)

Using special manufacturing processes, some diodes are produced that give off a bright white light. Their output is somewhere between that of halogen and HID bulbs. Unlike standard LEDs, white light diodes can consume a large quantity of power and produce lots of heat. Because of their small size, to make these LEDs suitable for use within headlamp systems, they must be grouped together to produce a suitable light source. Currently cost, size and packaging are a disadvantage, which means white light diodes are only used on some expensive vehicles.

Intelligent front lighting

A recent development in lighting systems is the ability to move the headlight beam in response not only to vehicle steering and suspension dynamics but also to weather and visibility conditions, vehicle speed, and road curvature and contour. This development is known as advanced front-lighting systems (AFS).

This system of moving headlamps has been around for many years, but early operation relied on mechanical rods and linkages. In contrast, AFS uses dynamic vehicle information from various chassis, speed and acceleration sensors that is available on the vehicle communication network to determine the most effective use of headlamp aim and angle. The headlamps can then react to vehicle movement and operation using actuators and motors built into the headlamp units.

Some manufacturers have been producing vehicles equipped with AFS since 2002. These auxiliary systems may be switched on and off as the vehicle and operating conditions call for light or darkness at the angles covered by the beam. Developments now mean that AFS systems can use GPS signals to anticipate changes in road curvature, making them truly intelligent.

> **Did you know?**
> Many modern lighting system components contain their own electronic control unit (ECU). This means that the individual lights are able to control their own operation when they receive a signal from the network.

> **Action**
> Investigate and name three cars (makes and model) that use intelligent front-lighting systems.

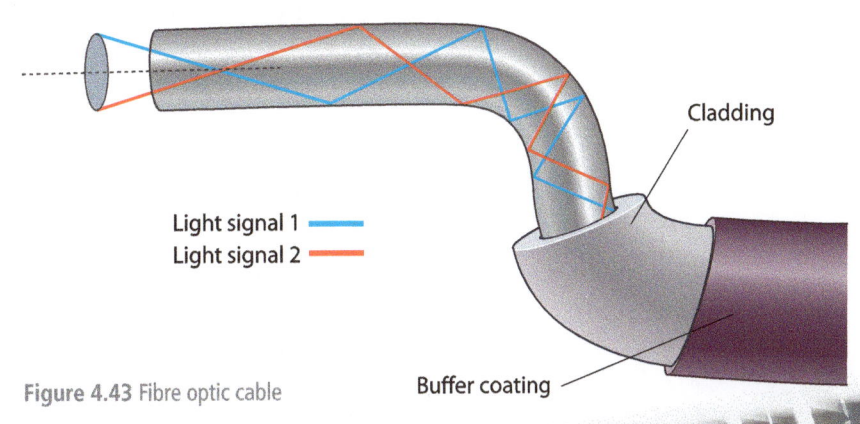

Fibre optics

Fibre optics is a system that allows a light source produced in one area of a vehicle to be directed to other points on the car by thin glass fibres. When glass is drawn into long thin strands, it becomes flexible and can be routed around the car like wiring. Each strand is coated on the outside with a mirrored cladding that reflects inwards. This cladding is then covered with an outer insulation called a buffer coating (see Figure 4.43).

When a light is applied to one end of the fibre strand, it is transferred along its entire length and emitted from the opposite end. Because the surface of the strand has a mirrored coating, it is able to reflect light through the fibre even if it goes around a bend.

Figure 4.43 Fibre optic cable

The use of fibre optics means that a single light source can be used to feed a number of different lighting outputs. This system is currently very limited as a useful source of light for headlamps, but is more commonly used for interior and dashboard lighting or as a method of creating network communication for infotainment systems.

Faults associated with lighting systems

The main faults associated with lighting systems can be separated into four areas. Examples of these faults, symptoms produced and possible causes are shown in Table 4.5.

Table 4.5 Faults associated with lighting systems

Fault	Symptom	Possible cause
Open circuit	Sidelight not working	Filament burnt out in the bulb, breaking the circuit and stopping current flow
High resistance	Nearside dip beam headlamp is dim	Corroded earth connection on nearside dip beam circuit that shares the voltage with the bulb and reduces its performance
Short circuit	Brake light circuit fuse blows when brakes are pressed	Brake light wire chafed on bodywork before the bulb, creating a short cut to earth
Parasitic drain	Car battery goes flat overnight	Boot light switch incorrectly adjusted so that the bulb stays illuminated even when the boot is closed

CHECK YOUR PROGRESS

1 List two advantages of an HID lighting system.
2 What colour light allows the human eye to see best at night?
3 Why does an HID headlamp system need to have a self-levelling and washer system?

Electrical comfort and convenience systems

The use of electricity has improved the control of vehicle systems provided for driver comfort and convenience. The following section describes the operation of some of these systems.

Electric windows and mirror systems

The operation of nearly all opening windows and wing mirrors is now controlled by electric motors and regulator mechanisms. These are sometimes referred to as 'power windows' and 'power mirrors'.

Electric window operation

At the press of a switch, windows can be opened and closed for ventilation. A main control panel is often available to allow the driver to operate all of the windows. In addition, rocker switches are mounted on each door for the passengers to open and close individual windows as required.

The side window glass is attached to a mechanical gearing mechanism known as a window regulator. This regulator mechanism is used to turn

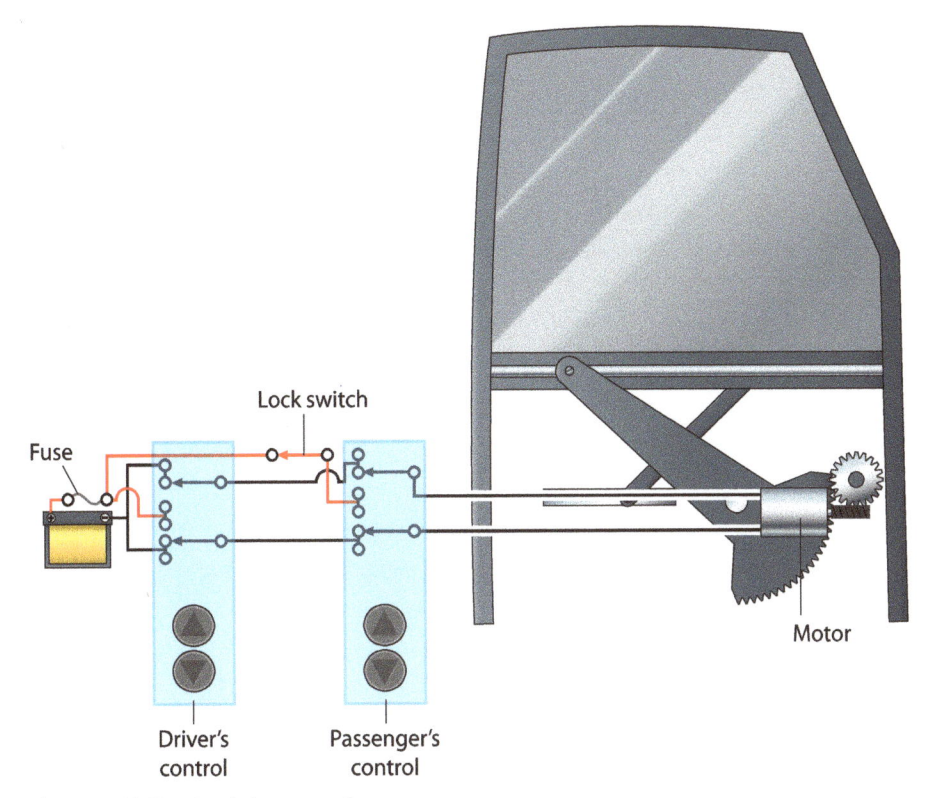

4 Diagnosis & rectification of light vehicle auxiliary electrical faults

Figure 4.44 Electric window operation

the rotational movement of a motor into an up and down movement of the window glass while multiplying the effort produced by the motor.

A direct current motor is connected to the window regulator gear mechanism in each door and, depending on the direction the motor is turned, the window is moved up or down. A 12-volt feed is supplied to both of the wires attached to the motor and, when the rocker switch controlling the window is operated, one wire is switched to earth. This completes the electric circuit and operates the motor. When the rocker switch is operated in the opposite direction, polarity is switched (power and earth are swapped over) and the motor turns the other way. (See Figure 4.44 above.)

One touch and inch back

Many electric window systems now incorporate **one touch** operation: if the rocker switch is pressed firmly once, the window will fully open or fully close. To ensure the window knows that it has reached the open or closed position, a method is needed to sense how far it has moved. The rotations of the motor can be counted between fully open and fully closed using a Hall effect sensor or by measuring output from the commutator.

To reduce the possibility of accidental entrapment in the window mechanism if the one touch operation has been activated, safety cut-outs are used. If the window control system senses that excessive current is being drawn by the electric motor, it assumes that something has become trapped and cuts the power. Some systems are able to reverse the window motor slightly so that pressure on the regulator is reduced and anything trapped is released. This is known as **inch back**.

Key terms

One touch – an electric window operating system that allows the window to fully open or fully close with a single press of the switch and without keeping the switch depressed.

Inch back – an electric window safety system that opens the window slightly if it senses something has become trapped in the mechanism.

Did you know?

If power is removed from the electric window control unit, for example by disconnecting the battery, the fully open and fully closed positions may need to be reprogrammed before one touch operation is available. This can sometimes be achieved by holding the rocker switch for around three seconds once the window has reached the full extent of its travel (both fully open and fully closed).

215

Level 3 Light Vehicle Technology

Case study

Mrs Ryan brings her car into your garage on her way home from shopping. She had been parked in a car park. After she wound her electric window down to pay on exit, the window would not close. She is concerned that her car is not secure with the window in the down position, and the weather forecast is for rain. She has an important job interview tomorrow and needs to use her car, so asks if it can be fixed while she waits? Your boss asks you to check it out.

Here's what you do:

- ✓ Listen to the customer's description of the fault.
- ✓ Question the customer carefully to find out the symptoms.
- ✓ Carry out a visual inspection (check to see if the other windows operate correctly).
- ✓ Gather information from technical manuals (including wiring diagrams) before you start and take it to the vehicle.
- ✓ Devise a diagnostic strategy. (The window is not working so you suspect an open circuit.)
- ✓ Use a scan tool to see if any diagnostic trouble codes have been stored. (There are no codes present.)
- ✓ Check the quick things first (10-minute rule) and conduct a visual inspection of connections, fuses, etc.
- ✓ Conduct as much diagnosis as possible without stripping down.
- ✓ Unclip the electric window switch from the panel and check its operation using a volt drop test.
- ✓ Power is not being switched, so you unclip the switch from the circuit and check for continuity. (The switch has failed.)
- ✓ Return to reception and explain the results of the diagnostic tests.
- ✓ Give an estimate for the cost of repair and gain authorisation to conduct the work. (Your boss orders a switch but it won't arrive for three days.)
- ✓ Using a power probe as a switch, you close the window so that the car is secure and can be used by Mrs Ryan for her job interview tomorrow.
- ✓ Book the car in for the end of the week to complete the repairs.
- ✓ When the car is returned later that week, replace the faulty switch.
- ✓ Correctly reassemble any dismantled components/systems.
- ✓ Thoroughly test the system to ensure correct function and operation.

> **Did you know?**
>
> An electric sunroof mechanism operates in a similar manner to power windows, but a **micro-switch** is often used to sense when the sunroof has reached the fully open or fully closed positions.

> **Key term**
>
> **Micro-switch** – a very small on/off switch used in controlling mechanisms.

Mirror operation

The position and angle of the rear view mirrors mounted on the driver's and passenger's doors can often be adjusted electronically from the driver's seat. Each mirror has two motors mounted in the mirror housing behind the glass. Each motor is attached to a small rack and pinion gearing system: one controlling the mirror glass up and down, and one controlling the mirror glass left and right. When operated by the driver via a switch mechanism the electric motors rotate the pinion gears along the rack mechanisms, pulling and pushing the mirror glass backwards and forwards. A combination of up, down, left and right allows the driver to set the desired angle of the mirror glass.

Faults associated with electric window and mirror systems

The main electrical faults associated with window and mirror systems can be separated into four areas. Examples of the faults, symptoms produced and possible causes are shown in Table 4.6.

> **Did you know?**
> Some electric door mirror glasses contain a demister element that operates on a timer cycle in a similar manner to front and rear screen heaters.

Table 4.6 Faults associated with electric window and mirror systems

Fault	Symptom	Possible cause
Open circuit	Electric mirror adjustment not working	Electric feed wire to the drive motor broken at the door hinge as it is flexed backwards and forwards as the door is opened and closed
High resistance	Electric window slow to operate	Poor connection at the window motor creates a resistance which shares the voltage with the window motor and reduces its performance
Short circuit	Wiring to offside electric mirror damaged (burnt/melted)	Wire trapped against the metal bodywork where the mirror is bolted to the frame, creating a short cut to earth
Parasitic drain	Car battery goes flat overnight	Faulty electric window switch, causing power to be transmitted to the motor in the up direction even when it is not pressed

If faults occur in an electric window system, you should assess mechanical issues such as tight glass runners and guides or partially seized regulators before suspecting electrical faults. When checking the power supply to window motors, it is common to have a 12-volt supply on both wires. The voltage on one of the wires will switch to earth when actuated.

Heating, cooling and ventilation

Sensors and actuators used to control the operation of light vehicle heating cooling and ventilation systems should be regularly checked. The diagnosis of faults on many modern systems can be assisted by the use of diagnostic trouble codes, which can be retrieved using a dedicated scan tool. If you find any electrical problems, you should consider the four main electrical faults (open circuit, high resistance, short circuit and parasitic drain).

For a description of the operation of heating, cooling and ventilation systems, see Chapter 3, pages 163–168.

> **Safe working**
> Remember that in order to work on a system containing fluorinated refrigerant gas, you must be appropriately qualified.

Screen heating systems

During the manufacture of front and rear windscreens, a thin electrical element can be embedded or etched onto the surface of the glass. During cold or damp weather, electrical current can be supplied to these elements, which convert the energy into heat, thereby helping to demist the windscreens. The operation of these screen demisters is controlled by the driver via a switch on the dashboard, but as large amounts of electric current are required by the screen elements, a **relay** is normally fitted to the circuit. Many systems also incorporate a timer to cut power to the demister after a preset period, to reduce the possibility of overheating and excessive current draw.

> **Did you know?**
> One difference between front and rear screen demisters is that the element used on the front screen is much thinner than the one on the rear screen. This is so that it doesn't affect forward visibility.

> **Key term**
> **Relay** – an electromagnetic switch that is used to control current in some electric circuits.

Level 3 Light Vehicle Technology

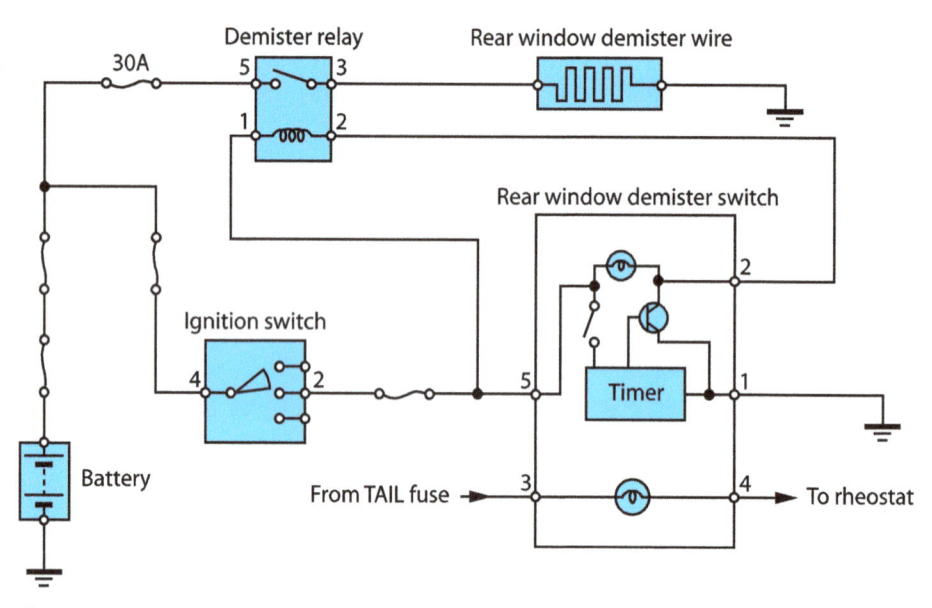

Figure 4.45 Heated rear window wiring diagram with timer circuit

Figure 4.46 Heated windscreen element

Faults associated with heated windscreen systems

The main electrical faults associated with heated windscreen systems can be separated into four areas. Examples of these faults, symptoms produced and possible causes are shown in Table 4.7.

Table 4.7 Faults associated with heated windscreen systems

Fault	Symptom	Possible cause
Open circuit	Heated rear windscreen not working	Window element broken/damaged at the mid-point of the windscreen, caused by load/luggage rubbing against it
High resistance	Demister slow to operate	Burnt contact in the circuit relay, creating a resistance which shares the voltage with the heater element and reduces its performance
Short circuit	Heated rear screen fuse blown/burnt out	Wire rubbed against the bodywork before the front screen element, creating a short cut to earth
Parasitic drain	Car battery goes flat overnight	Circuit timer failure, causing the screen not to switch off after a predetermined timescale

4 Diagnosis & rectification of light vehicle auxiliary electrical faults

Other comfort, convenience and safety systems

Some of the other comfort, convenience and safety systems you will come across are shown in Table 4.8.

Table 4.8 Other comfort, convenience and safety systems

Rear view camera systems	A camera can be mounted at the rear of the car and connected to a visual display unit such as a TV or DVD player. It can be linked to the reversing light system so that when backing up, additional vision is available for safety.
Lane change control	A series of sensors can be used that identify road lane markings. The information supplied to an ECU can inform the driver if they begin to drift across lanes by mistake. A vibration or buzzer can be used to alert the driver, while some systems may make small corrections to the steering via the electronic power-assisted steering (EPS).
Self-parking	The addition of a Doppler parking radar means that manufacturers are able to combine engine, transmission and electronic power steering control to allow a car to park itself in a parallel parking place. The driver just needs to find a parking space of suitable size, press a button and the car will automatically reverse into the space.
Anti-collision	Many manufacturers are now beginning to include active safety systems that use sensor information to apply the brakes and avoid a collision. How the system operates normally depends on a number of parameters, including speed and direction.

CHECK YOUR PROGRESS

1. With reference to electric windows, what does the term one touch mean?
2. With reference to electric windows, what does the term inch back mean?
3. Why is a timer circuit often used in heated rear window construction?

In-car entertainment (ICE) and satellite navigation systems

In-car entertainment and information systems are often grouped under the title 'infotainment'. They include:

- speakers
- radio and aerial systems
- amplifiers
- cassette tape players
- CD players and multi disc systems
- DVD players
- MP3
- television
- satellite navigation and GPS.

Figure 4.47 In-car entertainment (ICE)

219

Speakers

Sound normally travels as an air pressure wave which is directed into your ear, where it vibrates the ear drum. This vibration is transmitted through three small bones to the cochlea and auditory nerve. Your brain then interprets the electrical signals from the auditory nerve as sound.

In order for infotainment systems to work, a method of transmitting sound waves is required, and this is the job of the speakers. A speaker consists of a diaphragm which can be vibrated to create air pressure waves that represent sound. An electromagnet is attached to the centre of the diaphragm. When it is switched on and off, the electromagnet is attracted to or moved away from a permanent magnet attached to the frame of the speaker, producing a sound wave. The electromagnet can be controlled by connecting it to an electrical output from a radio, for example. The size of the diaphragm used determines the tonal range of the speaker:

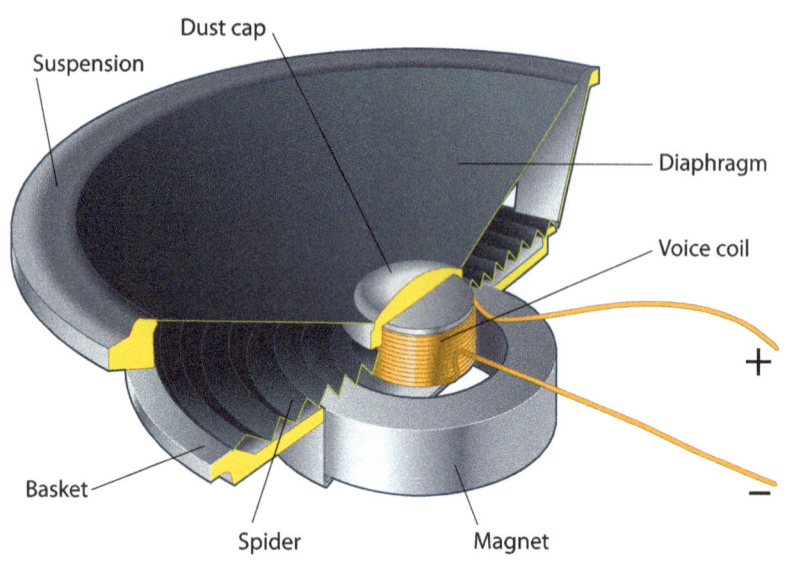

Figure 4.48 Speaker unit

- Very small speakers (sometimes called 'tweeters') produce sound in the high-pitched treble range.
- Medium-sized speakers (sometimes called 'squawkers') produce mid-range tones.
- Large speakers (sometimes called 'woofers') produce sound in the deep bass tonal range.

A combination of speaker sizes is often used with in-car entertainment systems to produce high-quality sound output.

Radio and aerial systems

Radio signals created at the broadcasting centre are imprinted on an electromagnetic carrier wave with a set frequency. Early carrier waves varied the amplitude (height) of the wave and this was known as amplitude modulation (AM). Later carrier waves varied the frequency (width) of the wave and this is known as frequency modulation (FM).

> **Did you know?**
>
> Radio data system (RDS) is a method of embedding small amounts of digital information in conventional FM broadcast signals. The RDS system is a standard that covers several types of information, including: time, station identification and programme information. If you have in-car entertainment equipment which is capable of interpreting this information, details may be displayed on the radio head unit screen.

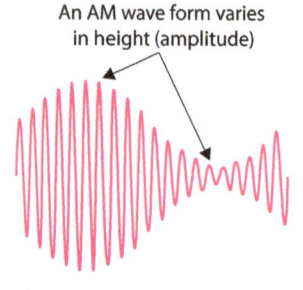

Figure 4.49 AM wave form

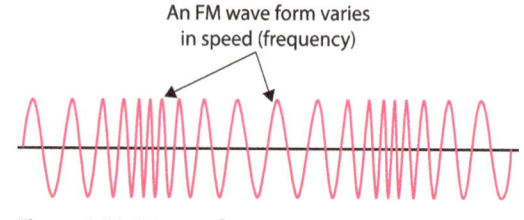

Figure 4.50 FM wave form

The broadcast signal leaves a transmitter aerial in all directions at once at the speed of light. A conducting antenna on the car is able to pick up the broadcast signal and transmit it to the radio unit. Inside the radio, the sound signal is separated from the carrier wave and converted into electrical pulses that can be used to drive the electromagnets in the speakers to reproduce sound.

NEW TECH

Digital radios

Analogue radio signals are subject to external interference and as a result many broadcasting organisations are slowly switching to digital signals.

This means that to continue to receive radio in cars manufacturers will have to fit radio equipment that is able to operate and interpret the digital signals broadcast. Analogue adapters are being developed so that digital signals can be used with existing radio equipment, but vehicle manufacturers will be fitting digital radios as standard to new cars from 2013.

Amplifiers

Amplifiers are often used to help increase the volume of sound created by in-car entertainment systems. The small electric pulses that are sent from a radio unit are stepped up in stages inside the amplifier unit using transistors. The transistor is a small electronic semi-conductor component which is able to act as a switch with no moving parts. It has three connections: the collector, the emitter and the base.

- The collector terminal is connected to a large stabilised source of power.
- The emitter terminal is connected to the speaker circuit.
- The base terminal is connected to the output of the radio.

As the radio operates, the small electrical output pulses are transmitted to the base of the transistor inside the amplifier unit. The large stabilised power source is unable to pass to the emitter until it receives a small electrical current at the base. As the base is switched on and off by the radio, the large stabilised power source is also switched on and off; this controls the electromagnet in the speaker. The amplifier usually contains a number of transistors which step the power out in small stages, increasing the speakers' output.

Legislation

Because radio signals do not stop at borders, international laws have been created to ensure that interference is not caused by two radio signals being broadcast on the same frequency. Some radio frequencies are restricted, for example those designed to be used by the emergency services. In order to broadcast on certain frequencies, a licence may be required. The Wireless Telegraphy Act 2006 is the main legislation on the regulation of radio spectrum in the UK.

Did you know?

Radio signals are broadcast at the speed of light and, depending on their wavelength, are able to travel considerable distances. The wavelength of a radio signal allows the car radio to be tuned in to a particular station.

- Short waves have a very close frequency and can only travel a short distance. They will easily pass through the Earth's ionosphere and out into space.

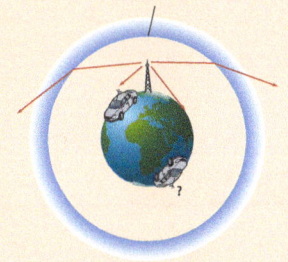

Short waves

- Medium waves have an average frequency and can travel further than a short wave. Some medium waves are reflected off the Earth's ionosphere back towards the ground, while small amounts escape into space.

- Long waves have a frequency which is spread out. Long waves are reflected off the Earth's ionosphere, trapping them and sending them round the world. A long wave is able to travel all around the world but as distance increases, signal quality is lost.

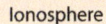

Medium and long waves

Level 3 Light Vehicle Technology

> **Did you know?**
>
> A CD has a single spiral track of data, circling from the inside of the disc to the outside. The fact that the spiral track starts at the centre means that the CD can be made smaller if needed. The incredibly small dimensions of the bumps make the spiral track on a CD extremely long. If you could lift the data track off a CD and stretch it out into a straight line, it would be 5 kilometres long!

> **Did you know?**
>
> Manufacturers of in-car entertainment systems produce equipment to standard sizes so that they have a universal fitment between different makes and models of cars. The 1 DIN standard is normally the width of the unit which makes it able to accept a full size CD. The 2 DIN standard refers to systems that are twice the height of a normal car radio/CD unit.

Cassette tapes, MP3, CDs and DVDs

Cassette tapes

Some older in-car entertainment systems use cassette tapes. These consist of a thin plastic tape covered with a magnetisable coating, which can be used to record music. When the tape is moved against a playing head, the magnetised recording can be converted back into sound by the cassette system and played over the car's speaker system.

MP3

A popular form of music storage for playback through an in-car entertainment system is MP3. This is a patented digital audio encoding format which uses a form of data compression. The computerised recording of music as digital information takes up an extremely large amount of storage space. The use of MP3 compression is designed to greatly reduce the amount of data required to represent the audio recording, while still sounding like a faithful reproduction of the original uncompressed audio for most listeners. The compression works by reducing the accuracy of certain parts of sound that are considered to be beyond the hearing ability of most people.

CD players

Compact discs (CD) are a method of storing digital media that can be played back as music through the in-car entertainment system. To produce CDs, a master glass disc is first created – the disc is covered in photo-resistant material and is etched with a laser to make pits on the surface. A disc made of nickel is then produced from the master glass disc to form a template. The nickel disc is used to stamp the reproduction copies that are made from transparent plastic. Each plastic disc is then coated with a thin reflective metallic layer and a protective coating of plastic. The microscopic pits and bumps are arranged as a single, continuous, extremely long spiral track of data.

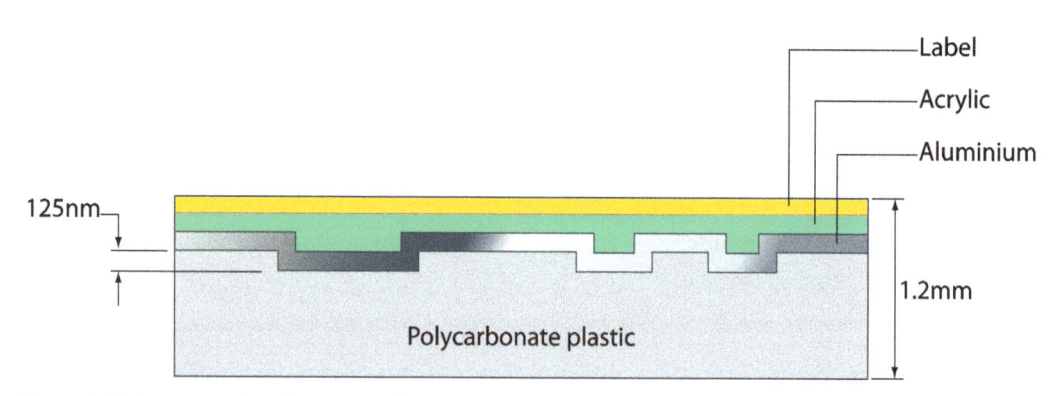

Figure 4.51 A cross-section of a compact disc

A drive motor in the CD player spins the disc. It is precisely controlled to rotate between 200 and 500 rpm, depending on which track is being read. A laser and a lens system focus in on and read the bumps/pits. The laser beam passes through the clear plastic layer, reflects off the

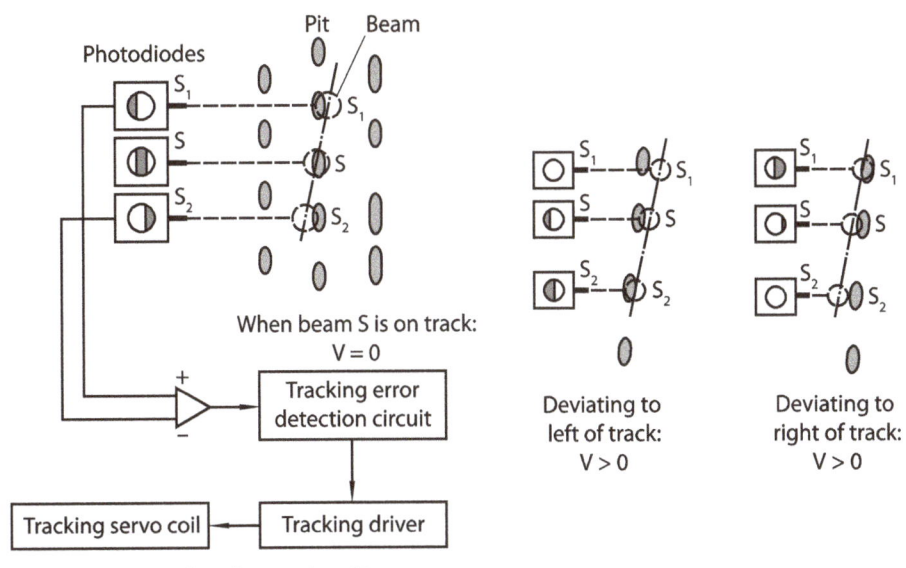

4 Diagnosis & rectification of light vehicle auxiliary electrical faults

Figure 4.52 A CD player laser and tracking system

metallic layer and hits a lens that detects changes in light. The bumps reflect light differently than the pits and the light sensor detects that change in reflectivity. The electronics in the drive interpret the changes in reflectivity in order to read the data. A processor inside the CD unit then converts this data into music and plays it through the speakers. A tracking mechanism moves the laser assembly so that the laser's beam can follow the spiral track.

DVD

Digitally versatile discs (DVDs) work in fundamentally the same way as compact discs. They are normally dual layer and require dedicated equipment for data reading and playback. Because of their increased capacity, DVDs are often used to store video data, which can be played back through a visual display screen inside the car.

Navigation

Driver navigation aids can be manufactured as part of the vehicle integrated design or fitted by the owner as auxiliary aftermarket equipment. Auxiliary navigation equipment relies on Global Positioning System (GPS) satellites, whereas integrated systems can use dead reckoning, GPS or a combination of the two.

Dead reckoning

Dead reckoning is a system that makes use of vehicle sensor data such as vehicle speed, steering angle and wheel speed, and matches this to maps that are pre-programmed into the navigation system. If dead reckoning is used without GPS combination, it has to be calibrated by setting a start position. As the car is driven along the road, its location is then calculated from the information supplied by the vehicle sensors.

> **Safe working**
>
> Regulation 109 of the Road Vehicles Construction and Use Regulations 1986 states that you must not drive a motor vehicle if you are able to see, either directly or by reflection, a television or similar apparatus. Exceptions to this regulation allow the driver to view a television or similar apparatus if it aids:
>
> - knowledge of the state of the vehicle or its location or the road location (head-up display)
> - vision of the road adjacent to the vehicle (external cameras for blind spots)
> - them reach their destination (satellite navigation).

> **Did you know?**
>
> Distance should not be confused with length. True distance is calculated by multiplying speed by time. The problem is that speed may be increased or decreased. So, a car travelling at 60 mph will travel:
>
> - 60 miles in 1 hour
> - 30 miles in 30 minutes
> - 10 miles in 10 minutes
> - 1 mile per minute.

223

Level 3 Light Vehicle Technology

Figure 4.53 Dead reckoning navigation

Did you know?

The coriolis force can be seen in the generation of swirling wind and weather patterns. The wind and weather rotates clockwise in the northern hemisphere and anticlockwise in the southern hemisphere.

Action

Research the names and countries of origin of two other satellite systems that are used for GPS navigation.

Direction of travel can be determined by a gyroscope mounted inside the navigation unit. The gyroscope is sensitive enough to be aware of the coriolis force created by the rotation of the Earth. As long as the navigation system knows whether it is being used in the northern or southern hemisphere, the gyroscope is able to sense its position in relation to the Earth's rotation and establish north, south, east and west.

Global Positioning System (GPS)

Navigation using GPS relies on radio signals sent from satellites in space. The system is based on a series of American military satellites, but other countries are now beginning to launch their own satellites for navigation purposes. At any one time, 24 satellites are used for GPS, with a number of spares also in orbit in case one goes wrong. Each of these solar-powered satellites circles the Earth at a distance of about 12,000 miles (19,300 km), making two complete rotations every day. The orbits are arranged so that at any time, anywhere on Earth, there are at least four satellites 'visible' in the sky.

The job of the GPS receiver, mounted in the car, is to locate four satellites and work out its distance from them. It can then use this information to establish its own position and map this against information in its memory. This process is called trilateration. Once the position of the GPS receiver has been determined, software programmed into the unit is able to calculate the route to a desired destination. Many systems are equipped with voice commands that give timely instructions to the driver, allowing them to reach their destination without having to concentrate on a map.

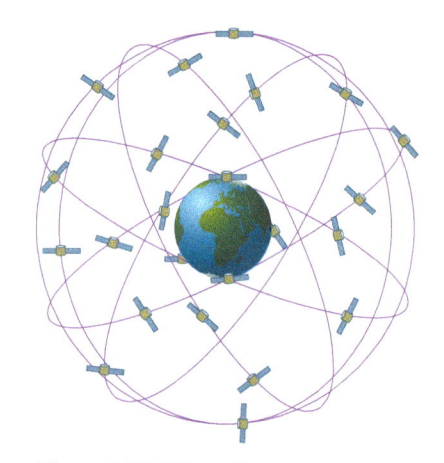

Figure 4.54 GPS satellites

Trilateration

Each satellite sends out radio signals containing time, position and climate information. When the GPS unit in the car receives a signal from a satellite, it can work out its distance from the satellite by the length of time it has taken the time signal to travel. (Radio signals sent from the satellites travel at the speed of light: 299,792,458 metres per second, 1,079,251,985 kilometres per hour or 186,282 miles per second). When the distance from four satellites is combined,

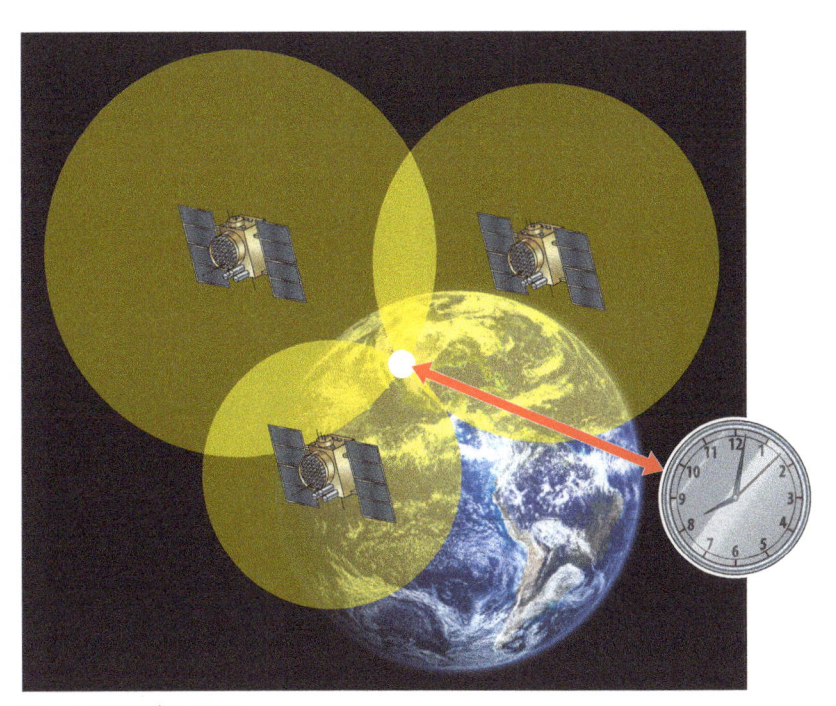

Figure 4.55 Trilateration

your exact position can be accurately calculated and matched to a map stored in the receiver's memory.

Advantages and disadvantages of satellite navigation

The advantages and disadvantages of satellite navigation are shown in Table 4.9.

Safe working

1. You must not mount the sat nav receiver where it will obscure the view of the road ahead.
2. You must not update the sat nav system while the vehicle is moving.

Table 4.9 Advantages and disadvantages of satellite navigation

Advantages	Disadvantages
Guides the user to unfamiliar destinations	Limited use for the majority of journeys
Avoids the use of maps when driving	May take driver down unsuitable roads
Gives accurate and timely instructions	Some systems may not be aware of current traffic problems

Communication

The use of mobile phones when driving is considered a distraction and so legislation has been introduced to restrict the use of handheld phones while driving. A number of methods now facilitate the use of mobile telephone systems while driving, including:

- **Built-in infotainment:** With built-in systems, car manufacturers design a telephone system that is directly integrated with the entertainment and information system. No external telephone is required as the microphone and speaker are mounted inside the passenger compartment. Many systems have a method for placing or receiving a call by voice control or buttons mounted on the steering wheel.

Level 3 Light Vehicle Technology

- **Auxiliary phone connections:** Usually a cradle system is mounted inside the car, which the telephone is attached to while in use. The cradle may be connected to the in-car entertainment system to make and receive calls, or can have its own speaker and microphone system.
- **Bluetooth:** Many mobile phones have a Bluetooth capability. Bluetooth uses standardised short range radio communication to connect two devices. With the Bluetooth switched on, the phone is able to act as a remote transmitter and receiver for telecommunication. An appropriate Bluetooth speaker and microphone device is required to use this system, which can be stand-alone or built into the in-car entertainment system.

Figure 4.56 Car telephone system

Did you know?

If the in-car entertainment and information system is part of the original vehicle design, it can sometimes store diagnostic trouble codes which can be retrieved with a dedicated scan tool. Diagnostic routines should be based on any codes stored that are related to the infotainment systems.

Faults associated with infotainment systems

As with all other electrical devices, the supply and use of electrical power is vital to the way the in-car entertainment and information system works. The four main electrical faults (open circuit, high resistance, short circuit and parasitic drain) cause different symptoms with equipment operation.

If the infotainment system uses fibre optics for the transmission of audio and video media between different units, the cabling should be checked for damage, routing and correct connection. If signal faults are suspected, you will need to use a scan tool to retrieve live data and fault codes.

Another issue you should consider when dealing with faulty DVD, CD and cassette tape players is mechanical failure of the playing mechanisms.

Interference caused by external radio waves can disrupt the playback operation of all in-car entertainment and information systems. These radio waves may be created by the car's own electrical system, as electrical and electronic components switch on and off, or may be caused by external broadcast signals. The use of mobile phones and telephone transmitter masts can sometimes prevent the operation of entire light vehicle electronic systems, causing complete failure. Many systems are carefully shielded to try to prevent this.

CHECK YOUR PROGRESS

1. List four types of in-car entertainment.
2. What is the name of the navigation system that uses vehicle sensors to determine the car's location?
3. Give one advantage and one disadvantage of satellite navigation.

4 Diagnosis & rectification of light vehicle auxiliary electrical faults

Security systems

Theft of and theft from cars are always very high in the crime statistics. Manufacturers are therefore continually improving the anti-theft and security systems on their vehicles. Systems that are integrated with the original design of the car prove to be the most effective, although professionally installed aftermarket systems may also provide acceptable vehicle protection.

Aftermarket systems

When fitting an aftermarket security/warning device, you must make sure that it will not affect the operation of the original vehicle systems or electrics.

The security components should be fused to prevent electrical short circuits, which could damage the car's original electrics/electronics.

Ensure that the security devices are suitable for the type of car to which they will be fitted. A warning device that senses voltage drop to activate an alarm may not be suitable for a vehicle where electrical systems operate independently after the car has been switched off. (An electric cooling fan cutting in once the vehicle has been left can cause an alarm to sound.)

If you fit an aftermarket security system, you should attach the immobiliser to at least two different electrical systems. This may not stop a determined thief, but it might slow them down enough that they give up.

Figure 4.57 Aftermarket car alarm system

Types of security device

Security devices can be broken down into two types: physical and electronic. Examples of these two systems are shown in Table 4.10.

Table 4.10 Types of security device

Physical security	Electronic security
Chassis/VIN numbers	Alarms/sirens
Security marking	Immobilisers
Steering locks	Volumetric sensing
Dead locks	Doppler radar
Wheel clamps	Lazy locking
Steering wheel/handbrake clamps	Transponder key codes
Complex key/lock design	ECU coding
Tamper-proof nuts and bolts	Keyless entry/starting

Did you know?

Some manufacturers are now fitting their vehicles with tracking systems to help recover stolen cars. The tracker uses Global Positioning System (GPS) (see page 224) to trace the location of the stolen car and relay this information to security organisations.

Security devices can be split into four further categories:

- **Active:** includes systems such as alarms which try to prevent the car from being broken into in the first place.
- **Passive:** includes immobilisers, which don't stop the car from being broken into, but try to prevent it being taken away.
- **Hardware:** includes the physical components such as sirens or locks that form part of a vehicle security system.
- **Software:** includes the electronics and programs which form part of a vehicle security system.

Types of integrated security/warning systems and components

If an integrated security/warning system is used, it can be embedded in the car's main control units and electronics. If a dedicated ECU is used for security, it is normally mounted in a position that is hard to access. Some manufacturers use wiring of a single colour at the control unit to ensure that it cannot easily be traced and disconnected by a potential thief.

The methods used for sensing intrusion or theft are shown in Table 4.11.

> **Did you know?**
>
> Accelerometers (see Table 4.11) are used inside many smart phones. They are sensors which are able to detect movement and rotate the orientation of the screen as the phone is tilted.

Table 4.11 Methods of sensing theft/intrusion

Shock sensing	This alarm system contains a device which is sensitive to vibration, and when doors are opened and closed the movement is enough to set off the alarm. This type of sensor can be prone to false triggers, which means that the alarm can sound when not required.
Entry point switch operated	Doors, tailgates, boot lids and bonnets normally have a switch mechanism fitted to allow a courtesy light to illuminate when opened. These switches can be wired to the alarm system to activate a siren when an entry point is opened.
Voltage sensing	When an electric component is switched on, a potential difference (Pd) is created in the electrical system voltage. This voltage drop is sensed by the security system and an alarm can be activated.
Volumetric sensing	Volumetric sensing is normally designed to detect intrusion into the passenger compartment. This can be achieved using a small microphone that measures the variations in air pressure or a dedicated pressure sensor which detects the opening and closing of doors, for example, and triggers the alarm.
Movement and tilt sensing	To help prevent the car just being towed away, some manufacturers use movement and tilt sensors in their alarm systems. A switch can be fitted which is turned on when a predetermined angle of tilt is reached. When the switch has been triggered, it allows mercury to flow and connect two electrical contacts, which sets off the alarm. Some manufacturers use accelerometers in their designs to sense movement. These can detect movement using electrostatic forces and trigger an alarm.
Doppler radar	A Doppler system is able to detect proximity and is sometimes able to give an audible warning when someone is too close. It works by sending out an electromagnetic radar wave and analysing the signal that is bounced back to the receivers. If the frequency of the signal bounced back increases, someone is moving towards the car. If the frequency of the signal bounced back reduces, someone is moving away from the car.

Alarms

Alarm systems may have a dedicated siren or speakers that can give audible warnings, or they may be directly connected to the vehicle horn. Some sirens contain their own battery backup so that if the wiring to the alarm is cut, the alarm continues to sound. When the alarm sounds, many systems also flash external lights as an added deterrent. Lights often also flash to show that the system has been armed, and LEDs can blink to show that the security system is operational.

Immobilisers

Immobilisers are a passive anti-theft system. They are designed to prevent the vehicle from being started and driven away. There are a number of ways to disable the vehicle, including:

- ignition circuit cut off
- fuel system cut off
- starter circuit cut off
- ignition, fuel, starter ECU
- engine ECU code lock. (Some engine management ECUs require programming with dedicated equipment before they are compatible with the vehicle.)

To set the immobiliser/alarm, a separate switch or transmitter can be used. This switch mechanism is often mounted in the vehicle's key fob. When a button is pressed on the key fob the system is armed and by pressing another it can be disarmed. The key fob will send out an infrared beam or more likely a short range radio signal that can be picked up by the car's security system. This coded signal is then used to enable or disable the system. To diagnose faults with a key fob specialist tools are available that will flash an LED if they pick up an infrared or radio signal when the button is pressed.

Many modern alarms and immobilisers are set automatically after a short delay once the ignition is switched off or when the doors are locked.

Legislation

It is a requirement under the Road Vehicles (Construction and Use) Regulations 1986 to fit a five-minute cut out device to all vehicle alarms. It is also an actionable nuisance for the registered keeper to allow an alarm to sound frequently or for a prolonged period so as to cause a nuisance to local residents and people working in the immediate vicinity.

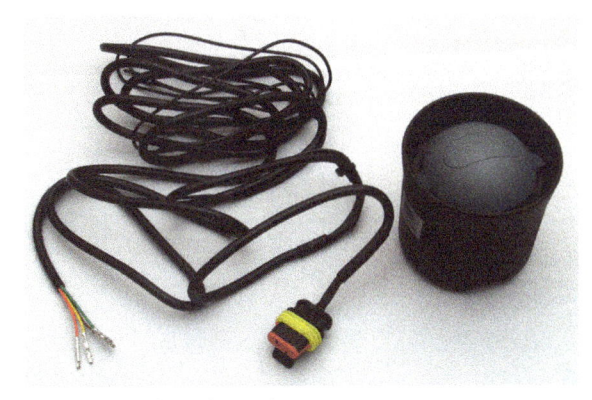

Figure 4.58 Alarm siren unit

Figure 4.59 A key fob alarm remote

> **Did you know?**
>
> The Noise and Statutory Nuisance Act 1993 allows councils to turn off vehicle alarms if they are causing a nuisance. An alarm which has been sounding for more than five minutes and is affecting residents is considered to be a nuisance. Once a nuisance has been established, the council can serve a Noise Abatement Notice on the owner or, if the owner cannot be found, they can place the notice on the car.
>
> This notice requires the owner to deactivate the alarm within one hour. If after one hour an alarm is still sounding and the owner has not been found, the law allows the council to deactivate the alarm by calling out a specialist engineer to do so. They may try to open a door but, if necessary, they can break a window. Some vehicles have sophisticated alarms that cannot be deactivated by alarm specialists, and on some vehicles the methods needed to silence the alarm may leave the vehicle unsecure after deactivation. If this is the case, the council will tow the vehicle away. The council does not have to prove that the alarm is faulty to take this action and it can recover any costs incurred from the owner.

Level 3 Light Vehicle Technology

> **Did you know?**
>
> In case of a security system failure, for example due to the loss of a key, many manufacturers are able to bypass immobilisers and alarms using dedicated equipment or pass codes. This allows alarms and immobilisers to be switched off in certain circumstances.

> **Action**
>
> Examine the cars in your workshop and investigate their security. Do they use alarms or immobilisers?
>
> Find out if the security systems are:
> - active
> - hardware
> - passive
> - software.

> **Key terms**
>
> **Reciprocating** – moving backwards and forwards or up and down.
>
> **Crank** – a shaft that is bent at a right angle.
>
> **Torque** – turning effort.

Faults associated with vehicle security systems

Alarms and immobilisers should be regularly tested to ensure their correct function and operation. As with all other electrical devices, the supply and use of electrical power is vital to the way vehicle security systems work.

The four main electrical faults (open circuit, high resistance, short circuit and parasitic drain) cause different symptoms with equipment operation (see pages 201–203).

Diagnostic scan tools are sometimes able to retrieve fault codes, which can help locate faulty system components. Due to the nature of vehicle alarms and immobilisers, you will need to have manufacturer's technical information available in order to undertake any system repairs.

Wiper, washer and central locking systems

Other electrical auxiliary comfort and convenience systems include:

- windscreen wipers and washers to ensure that the windscreens stay clean and clear during wet conditions
- central locking to assist with security.

Windscreen wiper system components and operation

Most windscreen wiper motors, front and rear, consist of a standard direct current permanent magnet motor. When supplied with electricity, they rotate in one direction only. To provide the **reciprocating** motion required to move the windscreen wiper arms backwards and forwards across the windscreen, the motor is attached to a **crank** arm which operates a linkage to control the wipers. A large amount of effort is required to move the windscreen wiper blades, so the motor is normally connected to a gear mechanism which increases **torque** but reduces speed.

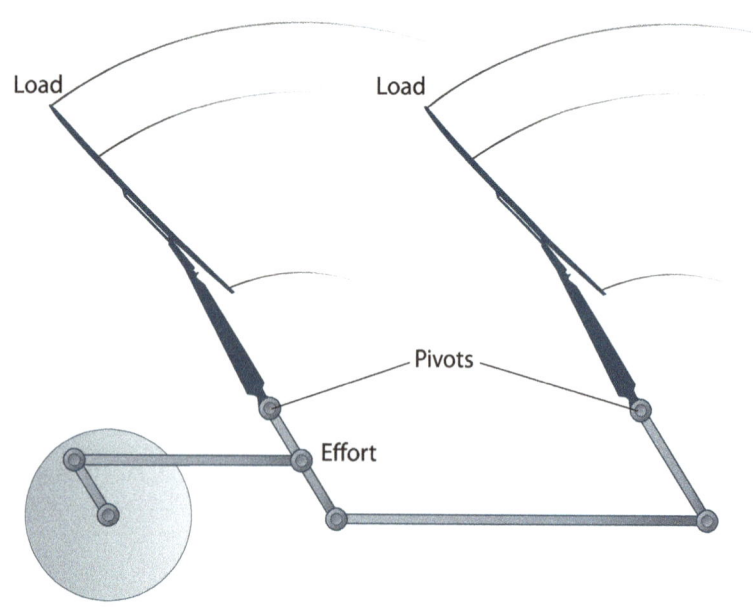

Figure 4.60 Windscreen wiper crank mechanism

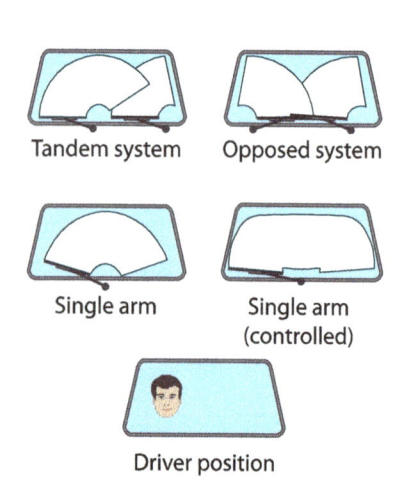

Figure 4.61 Windscreen wiper layouts

To achieve two-speed wiper operation, a three-brush connection can be applied to the wiper motor's commutator (two feed and one earth). The feed and earth brushes are placed opposite each other and, when powered, they drive the motor at standard speed with high torque output. When switched to high speed operation, power is supplied to a third brush which is mounted closer to the earth brush. This means the electric current takes a shorter path through the windings of the electric motor. The reduced resistance created by taking a short cut increases the current flowing in the motor, which makes it run faster but with a reduced torque output. (See Figure 4.62).

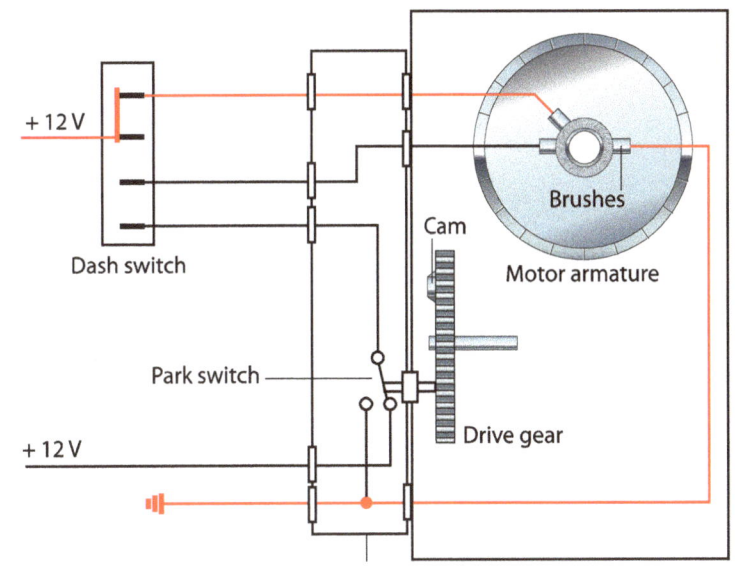

Figure 4.62 Three-brush, two-speed windscreen wiper motor

Switching off

When the windscreen wipers are switched off, a limit switch is often used to stop the arms at the bottom of the screen. Power is cut to the motor and any momentum created by the wiper arms turns the motor into a basic electric generator. The motor now generates electricity moving in the opposite direction to the supplied current flow. This places resistance on the motor and immediately stops the wipers moving. Some advanced wiper systems are then able to reverse the direction of the motor and move the wiper blades completely out of the driver's field of vision; this is known as **parking**.

Did you know?
The electricity created in the wiper motor when it is switched off is similar to the regenerative braking used in hybrid electric vehicles (see Chapter 2, page 50).

Intermittent wipe

If incorporated with a basic timer circuit, a system of intermittent wipe can be achieved. Some manufacturers are now providing electronic control of windscreen wiper systems, which allows the time delay of the wipers to be increased or decreased depending on how wet the screen is. Rain sensors are mounted in or near the windscreen and this information is relayed to the ECU, which alters the wiper speed accordingly.

Key term
Parking – a system which moves the windscreen wipers out of the driver's field of vision when switched off.

Self-activated and rain-sensing wipers

Some cars are now being produced with rain-sensing windscreen wiper operation. A small area of the front windscreen glass, usually located on the outside of the vehicle opposite the rear-view mirror, is monitored by an optical sensor. The sensor is designed to project infrared light at the windshield at an angle and then read the amount of light that is reflected back. A clean windscreen will reflect nearly all of the infrared light back, while a wet or dirty windscreen will cause the light to scatter. The optical sensor can determine the necessary frequency and speed of the windscreen wipers by monitoring the amount of light reflected back into the sensor.

Did you know?
As part of the headlamp cleaning system, some manufacturers supply small wiper blades that operate across the lens of the headlamps. They work on the same principle as normal windscreen wipers.

Level 3 Light Vehicle Technology

As a safety precaution, and to prevent damage to the wiper mechanism, nearly all rain-sensing wipers must be activated each time they are used. The activation process prevents the system from automatically wiping a frozen windscreen or triggering while the vehicle is in a car wash as both instances could damage the blades or electric motor powering the wipers.

Windscreen washers

To help clean the windscreen when it becomes dirty, a washer system is used. A small permanent magnet DC motor drives a pump submerged in a reservoir of washer fluid. The washer fluid is pumped through pipes to jets, which spray out across the windscreen. A one-way valve is often incorporated in the windscreen washer pipes so that fluid doesn't drain back to the reservoir after each use. In this way, the pipes are already full, making washer fluid instantly available next time it is required.

Figure 4.63 Windscreen washers

Faults associated with windscreen wiper systems

It is important to check that the mechanical linkages and motor are moving freely when you are diagnosing windscreen wiper faults. Anything that can prevent the wipers from moving and cause the system to stall can increase current draw, which may blow the system fuse.

As with all other electrical devices, the supply and use of electrical power is vital to the way the windscreen wiper systems work. The four main electrical faults (open circuit, high resistance, short circuit and parasitic drain) cause different symptoms with equipment operation (see pages 201–203).

You should check windscreen wipers on all speeds, including intermittent wipe, and you should ensure they stop or park in the correct positions.

Door locking

Central locking is a method for locking and unlocking all doors in one operation, and it is common on most cars. It can be done manually by turning a key in the door lock or remotely using a key fob with infrared or radio signal capabilities. When operated from the key fob, a coded signal is broadcast that is picked up by the central locking ECU, which then operates solenoids or actuator motors in each door to lock or unlock the car. As added security, many systems incorporate a 'rolling code' system so that the broadcast signal is different every time, making it hard for potential thieves to copy key remotes. Most systems also immobilise the door locking system if the incorrect code is received more than three times. This helps prevent thieves from using equipment that produces scanning codes to unlock cars.

> **Safe working**
>
> If you have to remove the windscreen wipers, arms or linkages for maintenance and repair, it is important that they are in their normal rest/park position before you begin work. You should mark the positions of the arms and linkages to ensure that you put them back in the same place when refitted. If you don't do this, the arms may collide and become entangled the first time they are used.

> **Did you know?**
>
> Some central locking remote systems are able to close windows, sunroofs, cabriolet roofs, etc., when the button on the key fob is pressed. This system is known as 'lazy locking'.

Advanced key (keyless entry)

Some cars have a locking system that is triggered if a transducer type advanced key is within a certain distance of the car. Sometimes called hands-free, this system allows the driver to unlock their vehicle without having to physically push a button on the key fob. The driver is also able to start or stop the engine without physically having to insert the key and turn the ignition. As the driver approaches the car, the vehicle senses that the key is approaching (located in a pocket, for example). When inside the car's required distance, there are two methods typically used by manufacturers to unlock the doors:

1. Once the driver is within the required distance for the key to be recognised, the car automatically unlocks the driver's door.
2. Once the driver is within the required distance for the key to be recognised, the car doesn't unlock the door until the key holder touches a sensor located behind one of the door handles. If passengers attempt to gain entry, the system senses that the driver is within the required proximity and, as they touch the sensors behind their door handles, the car unlocks their door.

Pneumatic systems

A number of manufacturers use air pressure to actuate central locking systems. A small pump is located inside the car and, when the driver operates the key in a lock or presses the button on a key fob remote, the air pump runs and drives the door locks open or closed.

Faults associated with door locking systems

As most central locking system components are hidden behind panel work and can be shielded for security reasons, when you are carrying out diagnosis you should check the items that are easy to access first, such as fuses. It is important to check that the mechanical linkages and motors are moving freely when diagnosing door locking systems, and remember to check the function of the pump and pipework for pneumatic systems.

As with all other electrical devices, the supply and use of electrical power is vital to the way central locking systems work. The four main electrical faults (open circuit, high resistance, short circuit and parasitic drain) cause different symptoms with equipment operation (see pages 201–203).

When checking the power supply to solenoids or actuators, it is common for both wires to have a 12-volt supply on both wires. The voltage on one of the wires will switch to earth when actuated.

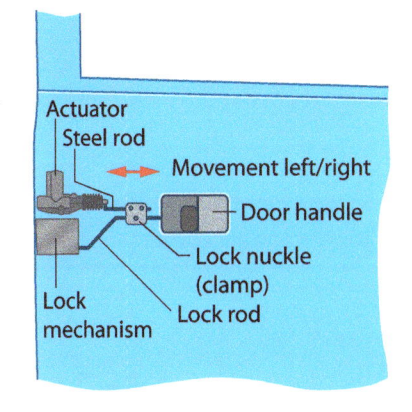

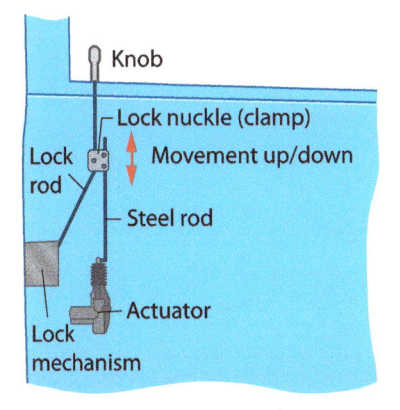

Figure 4.64 Central door locking mechanisms

CHECK YOUR PROGRESS

1. What does the term 'lazy lock' refer to?
2. With reference to windscreen wipers, what does the term parking mean?
3. How is reciprocating motion achieved on most windscreen wiper systems?

Airbags and supplementary restraint systems (SRS)

Modern cars come equipped with advanced systems to try to protect the occupants in the event of an accident. There are two types of vehicle safety: active and passive.

Active safety attempts to prevent an accident occurring in the first place and includes systems such as:

- anti-lock brakes
- traction control
- electronic stability programs
- emergency brake assist.

Passive safety attempts to protect the occupants of a car in the event of an accident or impact and includes such systems as:

- safety cells
- side impact protection
- crumple zones
- head rests.
- airbags
- seatbelts
- pre-tensioner systems

The term supplementary restraint systems (SRS) is applied to one or more individual systems that provide additional protection in the event of an accident. The most important of these systems are airbags and seatbelt pre-tensioners.

Figure 4.65 Car airbag system

Mechanical systems

Early systems for activating airbags and **pre-tensioners** were mechanical and used a triggering mechanism similar to a gun. In the event of an impact, metal weights were used to activate the airbag or pre-tensioner by:

- igniting a small explosive which started a chemical reaction in the airbag, creating a gas to inflate it
- operating a piston in the pre-tensioner unit of the seat belt.

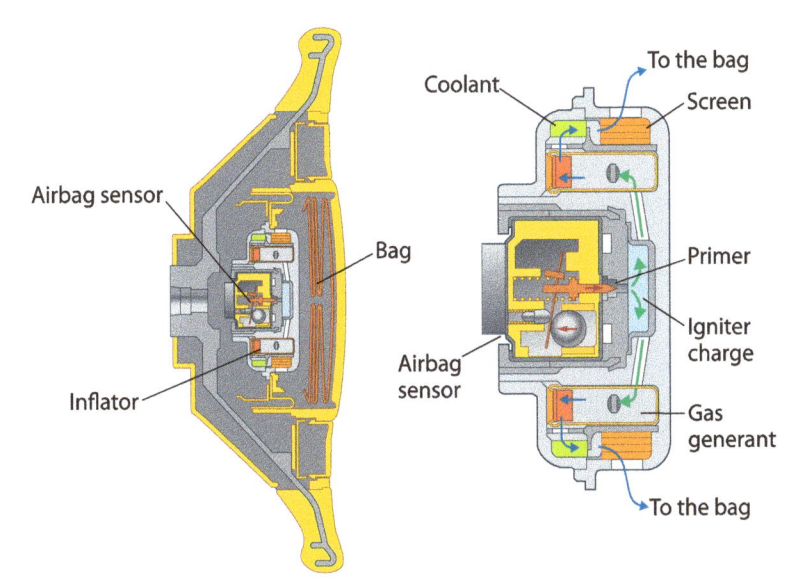

4 Diagnosis & rectification of light vehicle auxiliary electrical faults

Figure 4.66 Mechanical airbag unit

> **Safe working**
>
> Mechanical airbags and pre-tensioners were unstable and prone to accidental **deployment**. For this reason, you must take care when working around a mechanical airbag as sudden shocks could cause the firing mechanism to go off. Always set any safety locks on the airbag units when removing and refitting them.

> **Key terms**
>
> **Pre-tensioner** – mechanism that slightly tightens a seatbelt in the event of an accident.
>
> **Deployment** – putting into action.

Electronic systems

Electronic SRS systems are now the most common, as they give precise control and increased levels of safety. The use of electronic sensing devices and processing computers means that different types of accident can be detected and the action of the SRS components tailored to provide the greatest amount of protection.

Electronic SRS components include:

- **Driver's airbags:** mounted on the steering wheel to protect the head and face in the event of a frontal impact.
- **Passenger's airbags:** mounted in the dashboard to protect the head and face in the event of a frontal impact.
- **Door-mounted airbags:** mounted in the door panel to help protect the occupants in the event of a side impact.
- **Seat-mounted airbags:** mounted in the seat to help protect the occupants in the event of a side impact.
- **Curtain shield airbags:** mounted at the edges of the roof headlining, they deploy across the windows to help protect occupants from broken glass in the event of a side impact.
- **Roll over airbags:** mounted in the headlining to help protect the occupants in the event of the car rolling over during an accident.
- **Lower dash airbags:** mounted in the bottom section of the dashboard to help protect the passengers' legs in the event of a frontal impact.
- **Seatbelt pre-tensioners and force limiters:** working in conjunction with the airbags, these are designed to put occupants in the correct position for safest contact with the airbags.

Airbag operation

Mounted around the car in strategic positions are a number of crash sensors. It is their job to send a signal to the SRS ECU in the event of an impact. An ECU, usually mounted centrally to the floor of the car, contains a second sensing mechanism that detects the rate of deceleration and the angle of impact. This second sensor is often referred to as the safing sensor.

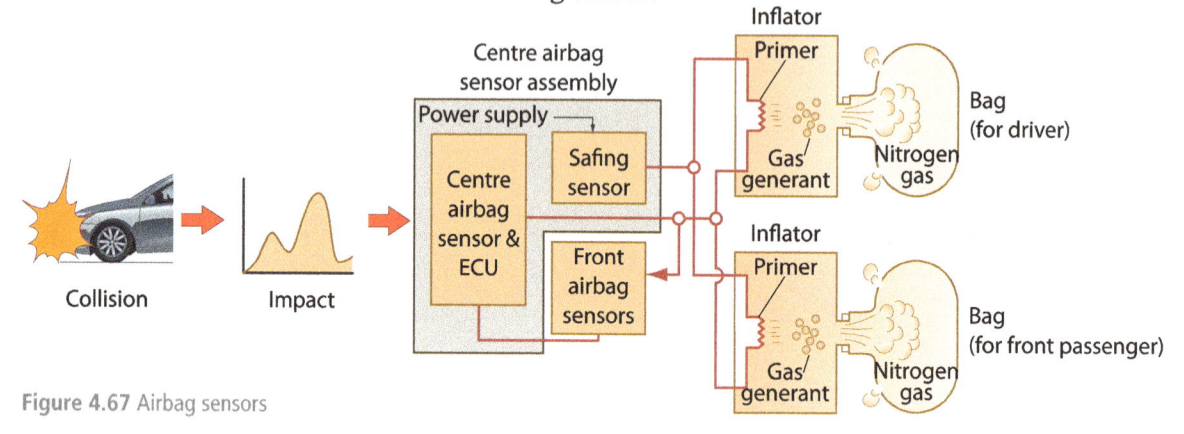

Figure 4.67 Airbag sensors

> **Did you know?**
>
> Front airbags often need an impact of around 30 degrees to the centre line of the vehicle against an immovable object in order to activate. Many airbags are designed not to deploy if the impact occurs below a set speed, typically around 10–20 mph.

In the event of an accident, the crash sensor sends a signal to the ECU. If the safing sensor determines that the impact is happening at above a predetermined speed and within a certain impact angle, it will trigger the seatbelt pre-tensioners and deploy one or more airbags.

Airbag inflation

When an impact occurs which meets the predetermined criteria, the ECU will actuate a small igniter device in the airbag unit called a **squib**. The squib is an explosive detonator which heats up chemicals stored in a gas generator unit. A chemical reaction creates a large quantity of nitrogen and carbon dioxide gas, which is used to inflate

> **Did you know?**
>
> The cover on an airbag is designed with a weak or perforated seam, so that as the airbag is inflated it will tear open easily without obstructing deployment. A powder such as corn starch or talcum powder is used to help lubricate the nylon and keep the bag pliable. This can normally be seen as the airbag deploys; it is often mistaken for smoke created by the explosive or gas generator.

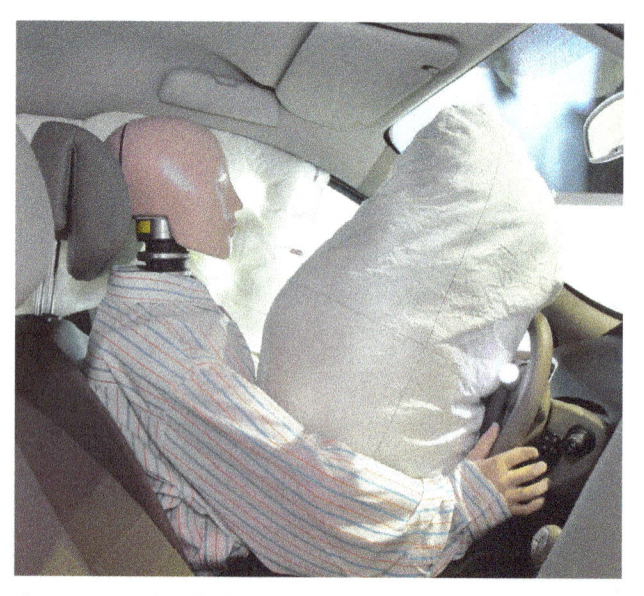

Figure 4.68 Airbag deployment

the airbag. The nylon airbag will then burst out from its cover at over 200 mph.

The airbag will inflate to a pressure of around 0.5 bar and help cushion the impact on the occupant. As soon as the individual has collided with the airbag, large holes in the rear of the bag are designed to allow the gas to escape and the bag rapidly deflates. Passenger airbags normally function in a similar manner to the driver's airbag, although they are generally much larger, having a capacity up to five times that found on a steering wheel airbag. Frontal airbags are designed to work in conjunction with seatbelt mechanisms so that the occupant is in the correct position for their face to impact the bag in the middle.

Other forms of airbag deployment

Additional airbags, such as curtain shield and roll over, may be deployed using a pre-pressurised gas container. In the event of an impact of sufficient speed and direction, a signal from the ECU ignites a squib which punctures a gas container. The pressurised gas is allowed to escape into the nylon airbags, deploying them at high speed.

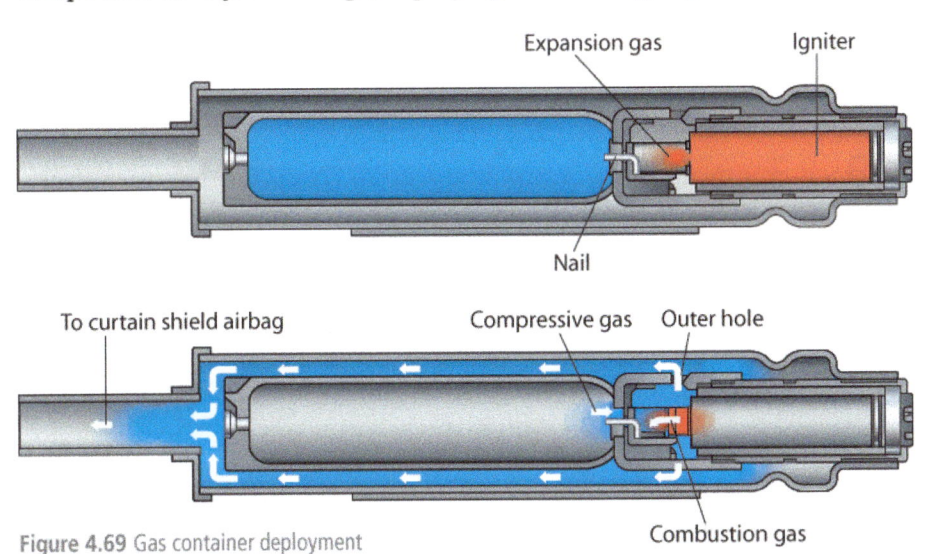

Figure 4.69 Gas container deployment

Clock springs

In order to work correctly, the driver's airbag needs a reliable electrical connection that is able to rotate with steering wheel operation. Conventional wiring and connectors could create a possible problem with premature wear or breakage caused by the constant turning of the steering wheel. To overcome these issues, a special component called a clock spring is fitted between the wiring harness and the airbag module behind the steering wheel. A **Mylar** tape is wound in a similar manner to the spring found inside a clock, and is able to wind up and unwind with steering wheel rotation. One end of the Mylar tape is connected to the fixed wiring on the steering column, and the other end is connected to the steering wheel. This allows a constant electrical connection to be maintained with the airbag unit at all times.

Emergency

If an incorrect seating position is used or seatbelts are not worn, an airbag can cause serious injury or even death. During the vehicle design process, crash test dummies often have lipstick applied to their faces, so that when tested, outlines and marks are left on the airbag following a collision. This allows manufacturers to recommend the best seating positions and seatbelt use for safe operation.

Did you know?

Nylon airbags and gas generators degrade over time. As a result, airbags may be considered service items and many manufacturers recommend that they are replaced approximately every ten years.

Key terms

Squib – a small explosive detonator used to start the deployment of an airbag or seatbelt pre-tensioner.

Mylar – a form of polyester resin used to make heat-resistant plastic films and sheet.

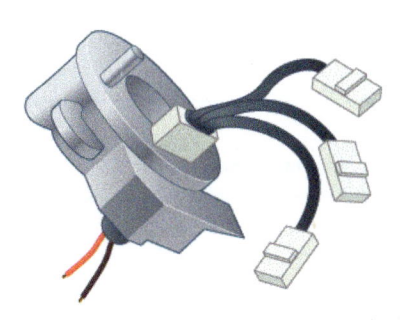

Figure 4.70 Clock spring

237

Level 3 Light Vehicle Technology

Figure 4.71 SRS wiring

Figure 4.72 SRS ECU

> **Safe working**
>
> SRS systems contain a backup power source so that they can still deploy in an accident, even if the battery has been damaged. A number of capacitors can be charged and act as an independent power source for the airbags and pre-tensioners. If you are going to work on an SRS system, you must disconnect the battery and allow time for the capacitors to discharge before you start any work.

> **Safe working**
>
> Mercury is toxic. If an SRS sensor is known to contain mercury it must be treated as hazardous waste.

SRS wiring

To help prevent accidental deployment of airbags and pyrotechnic pre-tensioners, the SRS system usually has its own wiring loom that is separate from the main vehicle wiring. It is connected in parallel to the main wiring and reduces the possibility of stray electrical signals from other systems activating airbag squibs. The SRS wiring is usually covered in bright yellow insulation (see Figure 4.71) in order to differentiate it from other vehicle wiring and to act as a warning that it should be handled/tested with caution.

SRS electronic control units

The SRS ECU receives signals from the crash sensors and safing sensors and processes the information. If the signals meet a preset criteria, the ECU will supply a voltage to the airbag squib and initiate deployment. It is common for the ECU to contain one or more of the sensors, and so its positioning when installed in the car is vital to correct operation. Many ECUs have markings on them to show how they should be fitted and orientated.

SRS ECUs are designed so that in the event of an accident and airbag deployment, codes are stored in the memory that cannot be erased. This safety feature means that a new unit must be fitted when repairing/resetting the system after an accident.

Sensors

There are a number of different crash and safing sensor types used in the construction of an SRS system. Although mechanical sensors are sometimes used, the most common type in modern SRS is electronic.

Mechanical sensors

- **Mass/inertia sensor:** this has a small weight in the form of a roller or ball housed inside an enclosure. During the rapid deceleration created by the impact of an accident, **inertia** causes the ball or roller to move against the force of a contact spring, which will send an electric signal to the ECU.
- **Mercury sensor:** this has a small amount of mercury contained in a tube that is inclined upwards at an angle. It is mounted so that when an accident occurs, the mercury will rise up the tube and bridge two electrical contacts, which sends a signal to the ECU.

> **Safe working**
>
> Many side airbag systems use piezoelectric crash sensors mounted in the doors. You must take great care if you need to strip out the inner door panels, for example to work on the electric window mechanism. If the SRS system is not disarmed, the movement of these sensors may cause accidental deployment of the airbags. Also, incorrect alignment on reassembly may cause the sensors to send the wrong signal in the event of an accident.

4 Diagnosis & rectification of light vehicle auxiliary electrical faults

Electronic sensors

Electronic sensors can work on the principle of **accelerometers**, or can be a strain gauge type device. The sensor is a small plate which has a weighted flap that moves slightly when force is applied to it. A strain gauge and a small integrated circuit are also mounted on the plate.

During the rapid deceleration created in an accident, the flap is forced to move. This creates a signal in the integrated circuit, which can be transmitted to the ECU. This type of sensor is very accurate and it also produces a signal that is proportional to the force of deceleration. Because of this, the ECU is able to calculate the severity of the impact and apply the appropriate safety measures.

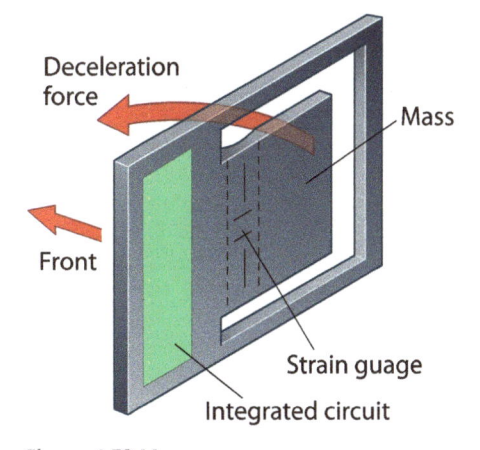

Figure 4.73 Mass sensor

Pre-tensioners and force limiters

During an accident, if the people involved are not in the correct position when airbags are deployed, injury or death may occur. Seatbelt pre-tensioners are designed to work in conjunction with the airbags by **cinching** up the belt components to remove any slack and bring the body and face into line with the airbag.

The pre-tensioner mechanism can be in the inertia reel or the seatbelt stalk buckle mechanism. In a similar manner to airbags, many pre-tensioners use a squib and a gas generator to create pressure on a type of piston that is able to retract the seatbelt slightly when an accident occurs. Because these systems use small explosive charges, they are often called **pyrotechnic** pre-tensioners.

In the moments after deployment has occurred, the force tightening the belts is released as gas pressure is exhausted, in a similar manner to an airbag deflating after an accident. This force limiting reduces the pressures on internal organs, which can rise rapidly following an impact.

Figure 4.74 Seatbelt pre-tensioner mechanism

> **Did you know?**
>
> Seatbelt pre-tensioners are designed to be used only once. Many pre-tensioner systems include an indicator device that appears after the belt mechanism deploys. This makes the occupants aware that the components must be replaced.

> **Skills for work**
>
> You spot that a very skilled technician is about to remove a steering wheel airbag without taking any safety precautions. He is much more experienced than you, and although he probably knows what he is doing, you feel you have a responsibility to intervene.
>
> This situation requires that you use particular personal skills. Some examples of these skills are shown in Table 6.1 at the start of Chapter 6, on pages 304–305.
>
> 1. Using the examples given in Table 6.1, choose one skill from each of the following categories that you think you need to demonstrate in this situation.
>   - General employment skills
>   - Self-reliance skills
>   - People skills
>   - Specialist skills
> 2. Now rank these skills in order of importance, starting with the one that it is most important for you to have in this situation.
> 3. Which of the skills chosen do you think you are good at?
> 4. Which of the skills chosen do you think you need to develop?
> 5. How can you develop these skills and what help might you need?

> **Key terms**
>
> **Inertia** – a force that holds the body steady or keeps it moving at a constant speed and in a constant direction.
>
> **Accelerometer** – an instrument for measuring acceleration.
>
> **Cinching** – tightening up and securing.
>
> **Pyrotechnic** – involving the use of explosives, in a similar manner to fireworks.

Level 3 Light Vehicle Technology

> **Action**
>
> Examine the cars in your workshop and list the number, position and type of all the airbag systems that you find.
>
> Are the pre-tensioner seatbelt systems in the inertia reel or the stem/buckle unit?

Timing of SRS operation

When an accident occurs, the pre-tensioners operate and the airbags will deploy from their housings at over 200 mph. Approximate timings of SRS operation are shown in Table 4.12.

Table 4.12 Approximate timings of SRS operation

Timing of SRS operation	Illustration
0 to 10 milliseconds Crash sensors and safing sensors detect the impact and signal the SRS ECU.	  10 milliseconds
10 to 15 milliseconds The airbag is activated and starts to inflate. The seatbelt pre-tensioners are deployed, taking up the slack in the belt and pulling the occupant into an appropriate position.	  15 milliseconds
20 to 30 milliseconds Seatbelt tensioning is complete and the airbag is fully inflated. Inertia keeps the person moving and they come into contact with the airbag, cushioning the impact.	 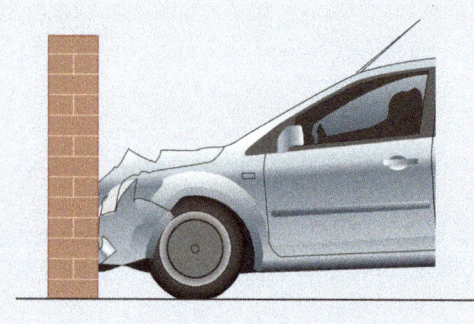 20–30 milliseconds

CONTINUED ▶

4 Diagnosis & rectification of light vehicle auxiliary electrical faults

Timing of SRS operation	Illustration
70 to 90 milliseconds The main impact is over and the force limiters on the seatbelt slacken. As the occupant moves against the airbag, the pressure of gas inside is released through the holes in the back of the bag and it begins to deflate.	70–90 milliseconds
120 to 150 milliseconds The airbag fully collapses and the inertia reel action of the seatbelt returns the occupant to an upright position.	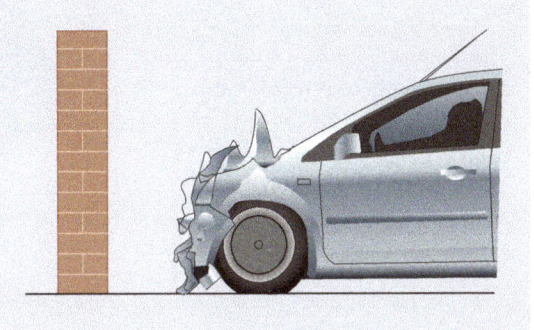 120–150 milliseconds

SRS fault diagnosis

If a fault occurs with a supplementary restraint system, the malfunction warning light should illuminate on the dashboard. This warning light has a self-check procedure: when the ignition is first switched on, it should light up for around 5 to 10 seconds; if no fault exists, it should then go out.

If the light remains illuminated or flashes, this indicates that the system has detected a fault and the airbags and pre-tensioners are disabled and will not deploy in the event of an accident. This should not affect normal seatbelt operation.

In order to diagnose the system, you need to have access to an appropriate fault code reader.

1. Connect the fault code reader to the vehicle's diagnostic socket.
2. Read all fault codes and record them.

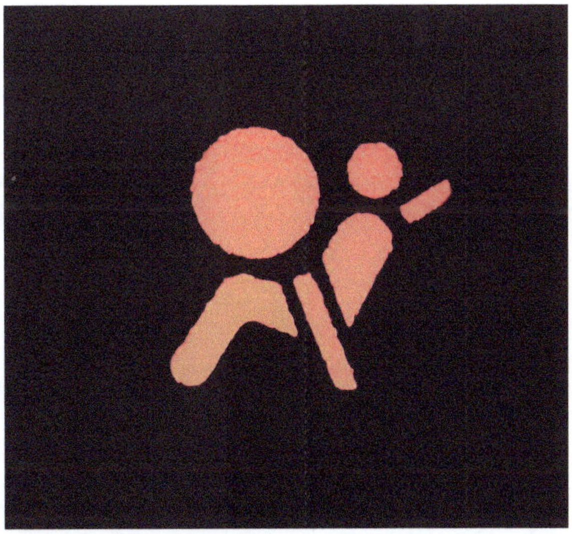

Figure 4.75 SRS warning light

Level 3 Light Vehicle Technology

Did you know?

Some early SRS system fault codes could be accessed by counting 'blink codes'. A connection could be made at the diagnostic socket that allowed the warning light on the dashboard to flash or an LED could be inserted so that it also flashed. By counting the number of long or short flashes produced by the light, a two-digit code could be retrieved and compared to the manufacturer's fault code list.

Safe working

You should not use conventional electronic test equipment to check SRS components, as they sometimes produce small electric currents which could make the SRS system deploy accidentally.

Safe working

Working on or around the components of a supplementary restraint system can be very dangerous. You must always take special care when handling components. Table 4.13 shows some precautions that you should observe, but you should also ensure that all health and safety practices are carried out during repair work. Always follow the manufacturer's recommended procedures.

3. Some SRS ECUs have no memory facility and only display faults that are present at the time of testing. Other ECUs are able to log fault codes, and these will remain in the memory until cleared. To ensure that these codes have not been created during connection and disconnection or testing processes involved with your diagnosis, clear the memory using the code reader and then recheck.

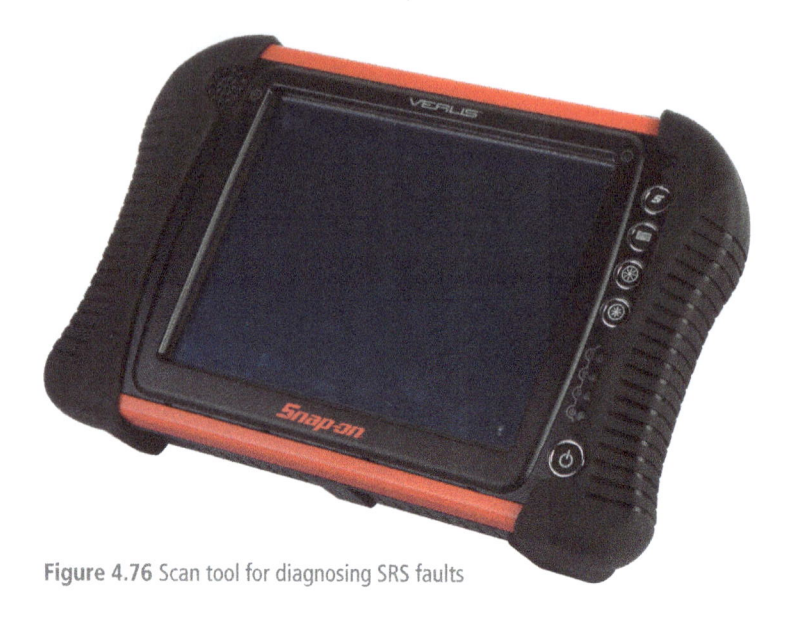

Figure 4.76 Scan tool for diagnosing SRS faults

4. Once you have retrieved the fault codes, carefully inspect components and connections for damage.
5. If a code indicates that an SRS component is faulty, you can often disconnect it and replace it with a special diagnostic resistor, which simulates an operational part.
6. You can now clear the codes and, if the fault has disappeared, you should fit a new SRS component.
7. If the fault code indicates that there is an issue with the wiring, disconnect all SRS components and isolate the wiring, then check with a multimeter.

Table 4.13 SRS dos and don'ts

Dos	Don'ts
Before undertaking any repairs to a vehicle, remove the airbag and pre-tensioner sensors if any shocks are likely to be applied to the sensors or in the vicinity. These shocks may not deploy the SRS, but they can damage the sensors so that they don't work in the event of an accident.	Never use memory keepers when working on SRS airbag systems. These special tools are designed to keep electrical components powered when the battery has been disconnected, so that radio codes, etc., are not lost during repair procedures. Using memory keepers means that the SRS system will remain 'live' and could deploy accidentally during repair work.

CONTINUED ▶

4 Diagnosis & rectification of light vehicle auxiliary electrical faults

Dos	Don'ts
If you are undertaking any electric welding on a vehicle, always disconnect the airbags and seatbelt pre-tensioners. Electrical welding uses the vehicle body to conduct the welding current. The action of welding may cause accidental deployment of airbags and pre-tensioners.	Never use conventional electronic test tools on SRS system components. For example: • ohmmeters create current flow in the circuits they measure • power probes are designed to supply electric current for test purposes • test lights connected in parallel draw extra electric current through a circuit. All of these tools may lead to accidental deployment of airbags or pre-tensioners.
Always check that all fault codes are clear after you have completed the repair work. Fault codes that remain in the memory can mean that the system is disabled and will not deploy in the event of an accident.	Never reuse an SRS ECU if the vehicle has been involved in a collision which has resulted in deployment of airbags or pre-tensioners.
Always treat SRS ECUs as toxic waste when disposing of them if they are known to contain the mercury switch type safing sensor. Mercury is toxic and special controls are needed to ensure safe disposal.	Many SRS ECUs are connected to the vehicle body to provide a safe earthing method. Never disconnect or reconnect the SRS ECU if it is not mounted to the vehicle's body. Static electricity could cause deployment of airbags or pre-tensioners.
If a vehicle has been involved in a collision, even if the airbags or pre-tensioners have not deployed, always inspect them thoroughly and check for system fault codes.	Never use second-hand SRS components from a donor vehicle. They may be faulty, incompatible or beyond their service life and may not function correctly in the event of an accident. Always use new parts.
When disposing of airbag or pre-tensioner assemblies, they must always be deployed first. This includes scrapping a vehicle – end of life (ELV) – with these assemblies still in place. You should only deploy airbags and pre-tensioners if you have been trained in how to do so safely. Special service tools are available to enable safe deployment, and you should always follow the manufacturer's recommended procedures.	Never attempt to repair SRS assemblies, including: • sensors • airbags • pre-tensioners • ECU • loom. This is because they may not function correctly in the event of an accident. Always replace the faulty part.
Always store removed airbags and pre-tensioner mechanisms in a specially designed explosives cabinet and ensure that others are aware of the contents.	Never expose airbag or pre-tensioner components to heat. Never use grease or cleaning agents on SRS components.
Always follow manufacturer-specific instructions when diagnosing SRS airbag faults.	Never place removed airbags deployment side down. If they deploy, they will take off like a rocket and this can be very dangerous.
Always wear appropriate PPE when working on SRS system components.	Do not hang diagnostic scan tools on steering wheels during test procedures. Electronic discharge through exposed contacts or electromagnetic forces may cause accidental deployment. This may result in the scan tool being propelled at high speed into the tester.

Case study

Mr Khan brings his car into your garage and complains that the SRS warning light on the dashboard stays on. He says it's been like that for six days and he's never had a problem before. Your boss asks you to check it out.

Here's what you do:

- ✓ Listen to the customer's description of the fault.
- ✓ Question the customer carefully to find out the symptoms.
- ✓ Test the car to confirm the symptoms described by the customer.
- ✓ Gather information from technical manuals (including how to retrieve fault codes) before you start and take it to the vehicle.
- ✓ Devise a diagnostic strategy.
- ✓ Check the quick things first (10-minute rule) and carry out a visual inspection of airbag components and connections.
- ✓ Ask the customer for authorisation if further testing is required.
- ✓ Conduct as much diagnosis as possible without stripping down.
- ✓ Using a scan tool, read the diagnostic trouble codes. (001C Airbag driver's side, resistance too high)
- ✓ Record the diagnostic trouble codes and clear them from the memory.
- ✓ Road test the car and re-scan to check that the code has returned. (Code 001C Airbag driver's side, resistance too high, has returned.)
- ✓ Concentrate your diagnostic procedure on the driver's side airbag.
- ✓ Record electrical security codes and switch off the ignition.
- ✓ Disconnect the battery and wait the recommended time period to allow the system capacitors to discharge.
- ✓ Wearing appropriate PPE, remove the driver's airbag and store it face up in your company's explosives cabinet.
- ✓ Inform others that there is an undeployed airbag in the cabinet.
- ✓ Test the circuit by installing a diagnostic resistor in place of the airbag and clearing the trouble code. (The code clears and the warning light switches off.)
- ✓ Keep the customer informed of progress and costs and gain authorisation to replace the airbag.
- ✓ Correctly reassemble any dismantled components/systems.
- ✓ Reconnect the battery (negative lead last) and reinstate any security codes.
- ✓ Clear any diagnostic trouble codes.
- ✓ Thoroughly test the system to ensure correct function and operation.
- ✓ Using specialist equipment and all the recommended safety procedures, deploy the old airbag before disposal.

CHECK YOUR PROGRESS

1. In an accident, what is happening with the SRS system 10 to 15 milliseconds after impact?
2. Name two positions for an airbag in an SRS system.
3. What is the purpose of the clock spring in a driver's airbag?

BEFORE YOU FINISH

Recording information and making suitable recommendations

At all stages of a diagnostic routine, maintenance or repair, you should record information and make suitable recommendations. The table below gives examples of how to do this.

Stage	Information	Recommendations
Before you start	Record customer/vehicle details on the job card. Make a note of the customer's repair request and any issues/symptoms. Locate any service or repair history.	Advise the customer how long you will require the car. Describe any legal, environmental or warranty requirements.
During diagnosis and repair	Carry out diagnostic checks and record the results on the job card or as a printout from specialist equipment. List the parts required to conduct a repair. Note down any other non-critical faults found during your diagnosis.	Inform your supervisor of the required repair procedures so that they can contact the customer and gain authorisation for the work to be conducted.
When the task is complete	Write a brief description of the work undertaken. Record your time spent and the parts used during the diagnosis and repair on the job card. (This information should be as comprehensive as possible because it will be used to produce the customer's invoice.) Complete any service history as required.	Inform the customer if the vehicle will need to be returned for any further work. Advise the customer of any other issues you noticed during the repair.

Remember, any work you carry out on a customer's car should be assessed to ensure that it is conducted in the most cost-efficient manner. You should consider reconditioning, repair and replacement of components within units.

FINAL CHECK

1. An inductive ammeter is **not** suitable for conducting which type of test?
   a slow cranking
   b alternator voltage output
   c glow plug operation
   d parasitic drain

2. In which year did diesel cars have to be E-OBD compliant?
   a 2000
   b 2001
   c 2003
   d 2005

3. Which of the following is **not** a multiplex network type?
   a delta
   b star
   c daisy chain
   d bus

4. What does the acronym CAN stand for?
   a controller area network
   b combined area network
   c composite area network
   d common area network

5. Which type of gas is commonly found in an HID bulb?
   a argon
   b hydrogen
   c nitrogen
   d xenon

6. How many satellites are used for worldwide GPS coverage?
   a 36
   b 24
   c 12
   d 6

7. Which of the following is **not** a type of electronic security device?
   a steering lock
   b immobiliser
   c siren/alarm
   d Doppler radar

8. Which of the following is an active safety system?
   a airbags
   b seatbelt pre-tensioners
   c ABS
   d crumple zones

9. Which tool is safe to use for SRS diagnosis?
   a oscilloscope
   b scan tool
   c ohmmeter
   d power probe

10. What is the normal operating voltage of a hybrid vehicle electrical drive system?
    a 0 to 12 volts
    b 24 to 48 volts
    c 100 to 300 volts
    d 500 to 1000 volts

4 Diagnosis & rectification of light vehicle auxiliary electrical faults

PREPARE FOR ASSESSMENT

The information contained in this chapter, as well as continued practical assignments, will help you to prepare for both the end-of-unit tests and the diploma multiple-choice tests. This chapter will also help you to develop diagnostic routines that enable you to work with light vehicle auxiliary system faults. These advanced system faults will be complex and non-routine and may require that you work with and manage others in order to successfully complete repairs.

You will need to be familiar with:

- Diagnostic tooling
- Electrical and electronic principles related to light vehicle electrical circuits
- Diagnostic planning and preparation
- Multiplexing and network systems
- Electrical faults
- Batteries: starting and charging (including hybrid)
- External lighting systems
- Electric window, mirror and screen heating systems
- Electrical comfort and convenience systems
- In-car entertainment (ICE) and satellite navigation systems
- Wiper, washer and central locking systems
- Supplementary restraint systems (SRS)

This chapter has given you an overview of advanced vehicle electrical auxiliary systems and has provided you with the principles that will help you with both theory and practical assessments. It is possible that some of the evidence you generate may contribute to more than one unit. You should ensure that you make best use of all your evidence to maximise the opportunities for cross-referencing between units.

You should choose the type of evidence that will be best suited to the type of assessment that you are undertaking (both theory and practical). These may include:

Assessment type	Evidence example
Workplace observation by a qualified assessor	Carrying out a two-stage diagnosis on an electrical auxiliary system
Witness testimony	A signed statement or job card from a suitably qualified/approved witness stating that you have correctly tested and reset an electric sunroof mechanism
Computer-based	A printout from a diagnostic scan tool showing the results of a test to check the function of a CAN bus system
Audio recording	A timed and dated audio recording of you describing the process involved in setting up and calibrating a satellite navigation system
Video recording	Short video clips showing you carrying out the various stages involved in the removal and safe disposal of an airbag
Photographic recording	Photographs showing you carrying out the stages of an electrical diagnosis when the assessor is unable to be present for the entire observation (because this process may take several days to complete). The photos should be used as supporting evidence alongside a job card.

CONTINUED ▶

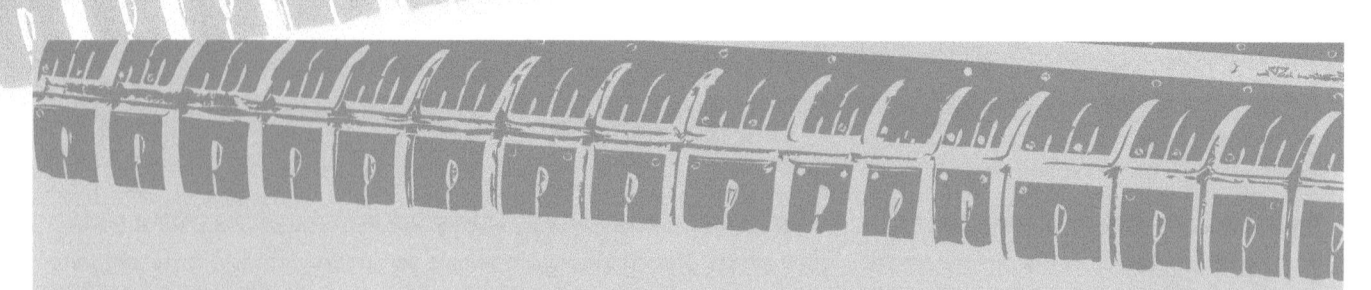

Assessment type	Evidence example
Professional discussion	A recorded discussion with your assessor about how you diagnosed and repaired a central locking system fault
Oral questioning	Recorded answers to questions asked by your assessor, in which you explain how you safely isolated a high voltage hybrid system in order to work on the vehicle's electrics
Personal statement	A written statement describing how you carried out the repair of a vehicle charging system
Competence/skills tests	A practical task arranged by your training organisation, asking you to use a multimeter correctly to conduct a volts drop test
Written tests	A written answer to an end-of-unit test to check your knowledge and understanding of light vehicle electrical auxiliary systems
Multiple-choice tests	A multiple-choice test set by your awarding body to check your knowledge and understanding of light vehicle electrical auxiliary systems
Assignments/projects	A written assignment arranged by your training organisation requiring you to show in-depth knowledge and understanding of a particular electrical system

Before you attempt a theory end-of-unit or multiple-choice test, make sure you have reviewed and revised any key terms that relate to the topics in that unit. Ensure that you read all the questions carefully. Take time to digest the information so that you are confident about what each question is asking you. With multiple-choice tests, it is very important that you read all of the answers carefully, as it is common for two of the answers to be very similar, which may lead to confusion.

For practical assessments, it is important that you have had enough practice and that you feel that you are capable of passing. It is best to have a plan of action and work method that will help you.

Make sure that you have the correct technical information, in the way of vehicle data, and appropriate tools and equipment. It is also wise to check your work at regular intervals. This will help you to be sure that you are working correctly and to avoid any problems developing as you work.

When you are undertaking any practical assessment, always make sure that you are working safely throughout the test. Light vehicle auxiliary electrical systems are dangerous and precautions should include:

- where possible, disconnecting and isolating any electrical power sources before conducting repairs
- observing all health and safety requirements
- using the recommended personal protective equipment (PPE) and vehicle protective equipment (VPE)
- using tools correctly and safely.

Good luck!

5 Diagnosis & rectification of light vehicle transmission & driveline faults

This chapter will help you to gain an understanding of light vehicle advanced transmission technology. It will also give you an overview of diagnosis and diagnostic routines that can lead to the rectification of electrical and electronic transmission control system faults. It explains and reinforces the need to test light vehicle transmission and driveline systems and evaluate their performance. It will support you with knowledge that will aid you when undertaking both theory and practical assessments. It will also help you to develop a systematic approach to light vehicle transmission system diagnosis.

This chapter covers:

- Advanced light vehicle transmission technology
- The operation of light vehicle clutches and fluid couplings
- Manual and automatic gearboxes
- Light vehicle driveline components
- Symptoms and faults in light vehicle transmissions and driveline systems

BEFORE YOU START

Safe working when carrying out light vehicle diagnostic and rectification activities

There are many hazards associated with the diagnosis and repair of advanced transmission systems. You should always assess the risks involved with any diagnostic or repair routine before you begin, and put safety measures in place. You need to give special consideration to the possibility of:

- **Uncontrolled vehicle movement:** Automatic transmissions have a tendency to move or creep when the engine is running and the vehicle is in gear (a drive range is selected). Always make sure that you select neutral or park if you need to run the vehicle while it is stationary during your diagnostic routine.
- **Dermatitis or cancer:** During maintenance and repair procedures, there is the possibility that you will come into contact with waste transmission lubrication oils. Contact with these oils should be kept to a minimum to reduce the risk of skin irritation and disease.

You should always use appropriate personal protective equipment (PPE) when you work on these systems. Make sure that your selection of PPE will protect you from these hazards.

Electronic and electrical safety procedures

Working with any electrical system has its hazards, and you must take safety seriously. When you are working with light vehicle electrical and electronic systems, the main hazard is the possible risk of electric shock. For information on basic first aid for electrical injuries, see Table 1.3 in Chapter 1, page 18. Although most systems operate with low voltages of around 12V, an accidental electrical discharge caused by incorrect circuit connection can be enough to cause severe burns. Where possible, isolate electrical systems before conducting the repair or replacement of components.

If you are working on hybrid vehicles, take care not to disturb the high voltage system. You can normally identify the high voltage system by its reinforced insulation and shielding, which is often brightly coloured. These systems carry voltages that can cause severe injury or death. If you carry out repairs to hybrid vehicles, always follow the manufacturer's recommendations.

Always use the correct tools and equipment. Damage to components, tools or personal injury could occur if the wrong tool is used or a tool is misused. Check tools and equipment before each use.

If you are using measuring equipment, always check that it is accurate and calibrated before you take any readings.

If you need to replace any electrical or electronic components, always check that the quality meets the original equipment manufacturer (OEM) specifications. (If the vehicle is under warranty, inferior parts or deliberate modification might make the warranty invalid. Also, if parts of an inferior quality are fitted, this might affect vehicle performance and safety.) You should only carry out the replacement of electrical components if the parts comply with the legal requirements for road use.

Information sources

The complex nature of advanced light vehicle transmission systems requires you to have a comprehensive source of technical information and data. In order to conduct diagnostic routines and repair procedures, you will need to gather as much information as possible before you start. Sources of information include:

Information source	Example
Verbal information from the driver	A description of the symptoms that occur when driving the car
Vehicle identification numbers	Model year taken from VIN plate
Service and repair history	A check of the service history that shows when the clutch was last replaced
Warranty information	Is the car under warranty and is it valid? (Has the required service and maintenance been conducted?)
Vehicle handbook	To confirm how to correctly choose an automatic gearbox drive range for a certain speed and load (e.g. when towing)
Technical data manuals	To find the recommended oil grade and quantity when changing gearbox oil
Workshop manuals	To find the recommended procedures for adjusting automatic gearbox brake bands
Wiring diagrams	To trace the electrical circuit used for powering the solenoids of a dual clutch system
Safety recall sheets	To confirm which components need to be replaced for safe operation of a paddle shift gear change mechanism
Manufacturer-specific information	Vehicle specific diagnostic trouble codes relating to semi-automatic gearbox control
Information bulletins	Information on a common fault found on dual mass flywheels
Technical help lines	Advice on the correct routine for setting up a limited slip differential
Advice from master technicians/colleagues	An explanation of how to drain the oil from a final drive unit
Internet	An Internet forum page where a number of people who had a similar problem with the failure of a Haldex drive coupling explain how it was resolved
Parts suppliers/catalogues	A cross-reference of automatic transmission fluid (ATF) part numbers, so that you can make sure that the parts comply with the gearbox manufacturer's requirements
Job cards	A general description of the work to be conducted on a customer's driveline
Diagnostic trouble codes	A fault code showing that the vehicle speed sensor needs to be tested to ensure correct operation
Oscilloscope wave forms	A wave form from a vehicle speed sensor to confirm correct function and operation

Remember that no matter which information or data source you use, it is important to evaluate how useful and reliable it will be to your diagnostic routine.

Operation of electrical and electronic systems and components

The operation of electrical and electronic systems and components related to light vehicle transmission and driveline systems:

Electrical/electronic system component	Purpose
ECU	The electronic control unit (ECU) is designed to monitor and control the operation of light vehicle electrical systems. It processes the information received and operates actuators that control gearbox and driveline systems for effective transmission of power to the road wheels.
Sensors	The sensors monitor various vehicle speed and load conditions against set parameters. As the driver and road conditions make demands on the engine and drive, these loads create signals in the form of resistance changes (voltage), which are relayed to the ECU for processing.
Actuators	The actuators are used to control transmission system operation. Motors, solenoids, valves, servos, etc., are operated by the ECU to help control the action of gearboxes and provide the most appropriate gear range for effective vehicle performance.
Electrical inputs/voltages	The ECU needs reliable sensor information in order to correctly determine the needs of the transmission system. If battery voltage was used to power sensors, its unstable nature would create issues (battery voltage constantly rises and falls during normal vehicle operation). Because of this, sensors normally operate with a stabilised 5-volt supply.
Digital principles	Many vehicle sensors create analogue signals (a rising or falling voltage). The ECU is a computer and needs to have these signals converted into digital (on and off) before they can be processed. This can be done using a component called a pulse shaper or Schmitt trigger.
Duty cycle and pulse width modulation (PWM)	Lots of electrical equipment and electronic actuators can be controlled by duty cycle or pulse width modulation (PWM). This works by switching components on and off very quickly so that they only receive part of the current/voltage available. Depending on the reaction time of the component being switched and how long power is supplied, variable control is achieved. This is more efficient than using resistors to control the current/voltage in a circuit. Resistors waste electrical energy as heat, whereas duty cycle and PWM operate with almost no loss of power.
Fibre optic principles	To keep up with developments in vehicle transmission technology, the demand for high speed communication of information and data has increased. Fibre optics use light signals transmitted along thin strands of glass to provide digital data transmission. (The light source is switched on and off.) In this way, information is transmitted essentially at the speed of light.

Electrical and electronic control is a key feature of all the systems discussed in this chapter.

Tooling

No matter what task you are performing on a car, you will need to use some form of tooling. Always use the correct tools and equipment.

The following table shows a suggested list of diagnostic tooling that could be used when testing and evaluating light vehicle transmission and driveline systems. Due to the nature of complex system faults, you will experience different requirements during your diagnostic routines and so you will need to adapt the list shown for your particular situation.

Tool	Possible use
Oscilloscope	To test the signal produced by a vehicle speed sensor (VSS)
Multimeter	To conduct a volt drop test on a clutch actuator
Test lamp/ logic probe	To test the existence of system voltage at the automatic transmission inhibitor switch

(Always use test lamps with extreme caution on electronic systems, as the current draw created can severely damage components.) |
| Power probe | To power an automatic transmission shift solenoid |

CONTINUED ▶

253

Tool	Possible use
Code reader/ scan tool 	To retrieve diagnostic trouble codes (DTCs) related to the transmission system. To clear trouble codes, reset the malfunction indicator lamp, and evaluate the effectiveness of repairs.
Laser thermometer 	A non-contact thermometer, also known as a pyrometer, can help determine the temperature of transmission oil, enabling you to make sure that you take level measurements at the appropriate time

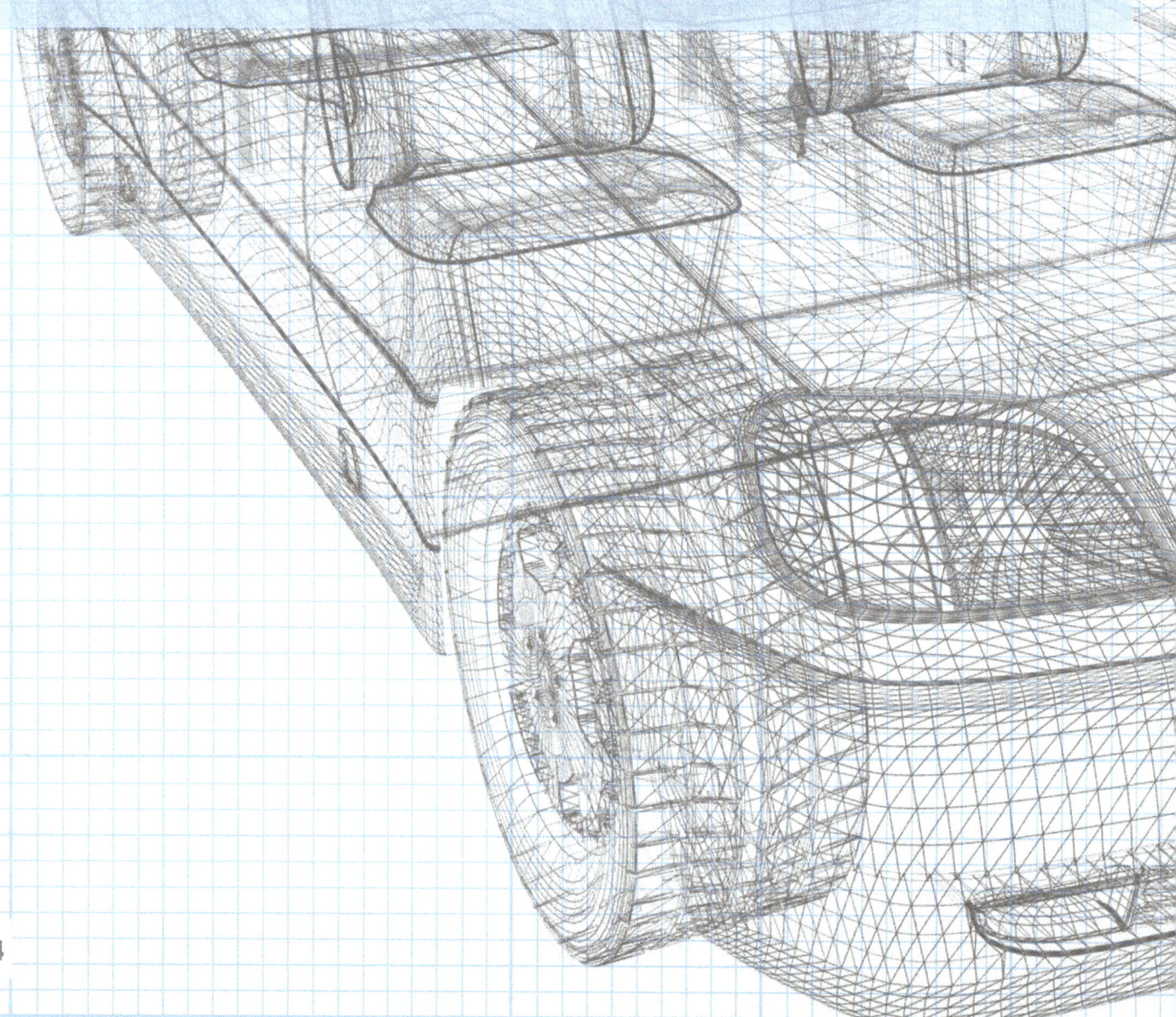

Advanced light vehicle transmission technology

Modern cars are designed so that engine power is delivered to the road wheels as efficiently as possible. To achieve this, advanced engineering techniques and computerised electronic control are used in the design of powertrain operation. Engines and transmission systems are coupled in order to produce low emission output and high fuel economy while maintaining performance and driving pleasure. Manufacturers are continually striving to find new technologies that can be used to enhance overall driving experience. These include:

- semi-automatic transmission
- seamless shift dual clutch transmission
- automatic and continuously variable transmission (CVT)
- paddle shift
- limited slip differentials.

The operation of light vehicle clutches and fluid couplings

The purpose of a clutch is to:

- link up the power from the engine to the driving shafts that turn the wheels
- separate the power from the engine to the driving shaft.

In this way, the clutch allows the drive of the car to be stopped and started.

A clutch provides three main functions:

1. It provides a smooth take-up of drive (going from stationary to moving).
2. It provides a 'temporary position of neutral' (allows the car to come to a stop without taking it out of gear or stalling the engine).
3. It allows the engine to be disconnected from the gearbox so that gear change can take place without the gear teeth of the transmission hitting each other and causing a lot of damage.

Many clutches make use of the principles of friction.

Friction clutches are made up from a number of components. An overview of the main parts are shown in Table 5.1.

Figure 5.1 Engaging a friction clutch

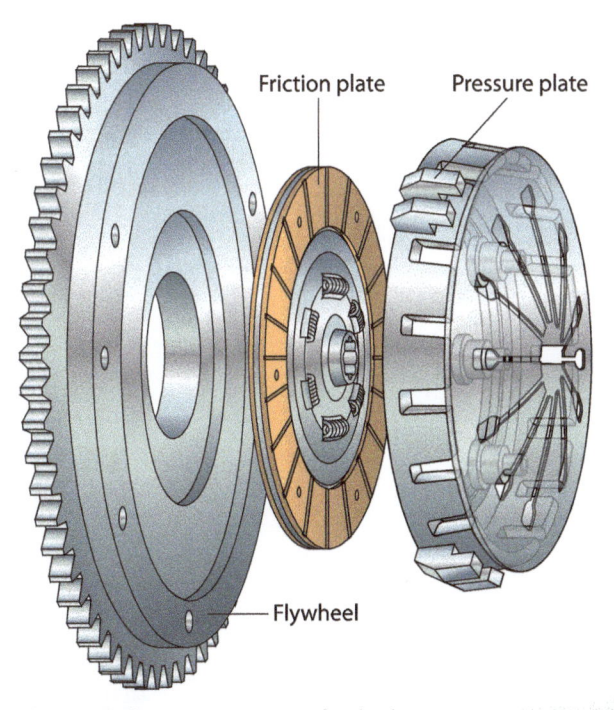

Figure 5.2 The component parts of a clutch

Level 3 Light Vehicle Technology

Table 5.1 Friction clutch components

Friction clutch component	Purpose
Drive surface	This is often the flywheel of an engine which rotates with the crankshaft and forms a flat surface to drive the clutch. In a multi-plate or dual clutch system, the drive surface may be a clutch basket or housing which is driven directly or indirectly by the engine. If a multi-plate clutch is used, a number of **plain plates** are inserted between the friction plates to increase the surface area and number of surfaces in contact.
Clutch cover	The clutch cover houses the components of a **dry clutch** system. It is bolted to the flywheel and rotates at crankshaft speed, transferring engine rotation to the pressure plate.
Pressure plate	The pressure plate provides the clamping surface (operated by springs) to drive the friction plate. In a multi-plate clutch, the pressure plate compresses all of the major drive components (friction and plain plates).
Friction plate	Clamped between the pressure plate and the flywheel, the friction plate transfers drive to the input shaft of the gearbox. In multi-plate and dual clutch systems, a number of friction plates are used to keep the components compact while still transmitting large amounts of torque. Some friction plates are designed to operate in oil to help control friction and heat. These are known as **wet clutches**.
Release bearing	The release bearing operates against the clutch springs while the engine is turning to remove the clamping force on the pressure plate and disengage the clutch. A clutch release bearing is not always needed, as some systems use hydraulic forces to engage and disengage the clutch plates.
Release fork/ servos and valves	The release fork operates against the release bearing when the pedal is pushed to actuate the clutch mechanism. If hydraulic pressure is used to engage and disengage the clutch, valves and servos may be used to control this action.

> **Key terms**
>
> **Plain plates** – a set of smooth metal discs used in a multi-plate clutch to increase surface area.
>
> **Dry clutch** – a clutch where friction surfaces are operated dry (with no lubrication).
>
> **Wet clutch** – a clutch where the friction surfaces operate in oil to control heat and grip.
>
> **Fulcrum** – a pivot point, such as the one on a seesaw.

Coil and diaphragm spring clutches

A number of different construction designs are used in the operating mechanisms of clutches that clamp the pressure plates against other friction components. The design will normally involve the use of coil or diaphragm springs.

Diaphragm spring

Many dry clutch systems use a diaphragm spring. This is a single metal plate made into a series of sprung steel fingers. It is slightly dished in shape. When one end of the fingers is pressed by the clutch release bearing, the fingers pivot about a **fulcrum**. This moves the opposite end of the diaphragm fingers in the other direction. When this happens, the pressure plate is moved away from the friction plate and disengages the clutch.

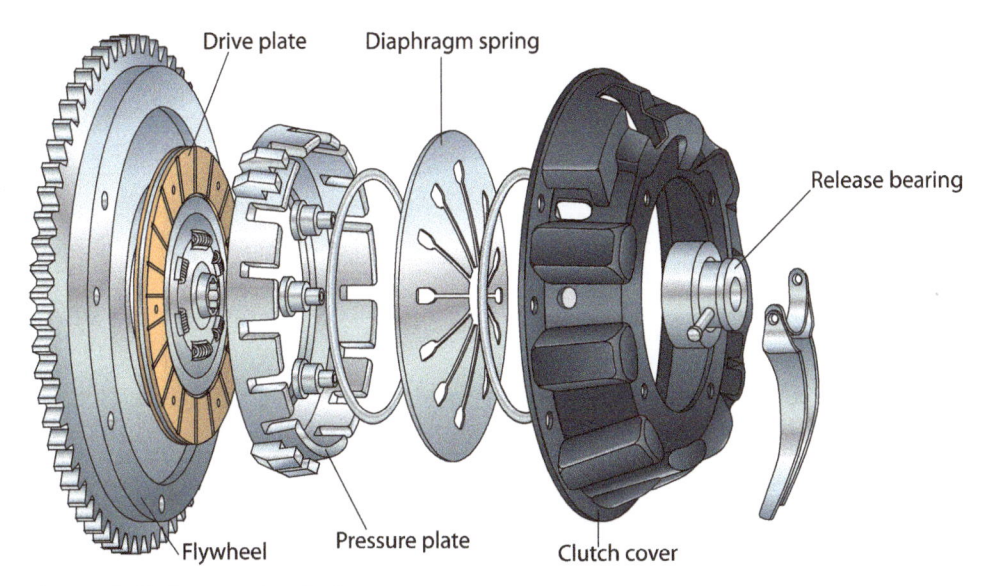

5 Diagnosis & rectification of light vehicle transmission and driveline faults

Figure 5.3 Diaphragm spring clutch

When the driver lifts their foot off the clutch pedal, the ends of the fingers of the diaphragm spring are released. Because the steel fingers are sprung, they return to their original position and reapply pressure to the friction plate. This then reconnects the drive. Because the spring diaphragm fingers are made from a single piece of metal, an even clamping force can be produced.

Self-adjusting clutches

Many manufacturers are now producing self-adjusting diaphragm spring clutches. In this design the main diaphragm spring is not permanently attached to the clutch cover. Instead it rests against another diaphragm spring, which keeps it in tension against the pressure plate. This way, as the clutch friction plate wears, any excessive **free play** is taken up by the second diaphragm spring. This will give a consistent feel to the clutch operation during its normal operating lifespan.

> **Key term**
>
> **Free play** – a small amount of clearance or movement between two components.

It is often necessary to use a specialist tool to align and provide initial tension on the self-adjusting mechanism if clutch units are replaced. If this is not done, judder, premature wear and failure may occur. Always follow the manufacturer's instructions.

Coil spring

A number of early dry clutches used coil springs to provide the clamping effort. Because of wear caused by normal operation, coil spring tension and length will change over time. This can result in an uneven clamping force being produced on the friction plate.

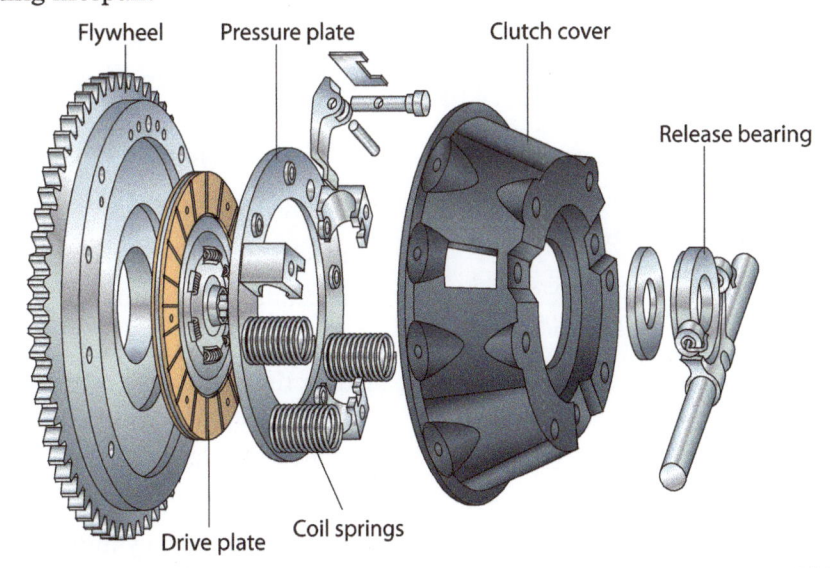

Figure 5.4 Coil spring clutch

This uneven clamping force can lead to clutch **drag**, **slip** or vibrations (often referred to as clutch **judder**) as take-up of drive is required.

> **Skills for work**
>
> During a road test conducted with the customer to diagnose a clutch fault, you notice that their driving style will cause excessive wear of the clutch components and may lead to premature failure.
>
> 1. Describe how you could explain to the customer what they are doing wrong without causing offence.
> 2. This situation requires that you use particular personal skills. Some examples of these skills are shown in Table 6.1 at the start of Chapter 6, on pages 304–305. Using the examples given in Table 6.1, choose one skill from each of the following categories that you think you need to demonstrate in this situation.
>   - General employment skills
>   - Self-reliance skills
>   - People skills
>   - Customer service skills
>   - Specialist skills
> 3. Now rank these skills in order of importance, starting with the one that it is most important for you to have in this situation.
> 4. Which of the skills chosen do you think you are good at?
> 5. Which of the skills chosen do you think you need to develop?
> 6. How can you develop these skills and what help might you need?

Multi-plate clutch systems

With multi-plate clutch systems, the coil springs are sometimes used to provide the tension on the pressure plate. However, it is more common for hydraulic pressure to be used to engage the clutches, and coil springs may be used to help disengage the components.

Clutch engagement

The engagement and disengagement of clutches may be:

- mechanical, using a clutch cable
- hydraulic, using fluid pressure against a **slave cylinder**.

In a hydraulic system, fluid pressure causes a release bearing to act against the mechanism that disengages the pressure plate of the clutch. The slave cylinder can be mounted externally to the gearbox and operate a fork lever against the release bearing, or mounted inside the clutch bell housing and act directly on the release bearing.

Direct acting slave cylinders provide a positive action, with little free play created as the clutch wears. This allows a consistent feel and

> **Key terms**
>
> **Drag** – when the clutch is not fully disengaged and the friction surfaces rub against each other.
>
> **Slip** – when the clutch is not fully engaged and the friction surfaces slide over each other.
>
> **Judder** – a vibration felt during the take-up of drive.
>
> **Slave cylinder** – a hydraulic piston used to operate the clutch release.

steady operation to be achieved. A disadvantage of internal slave cylinders is that if a leak occurs, it may contaminate the clutch friction surfaces with hydraulic fluid and cause clutch slip. Because of its location, in order to inspect and replace a direct acting slave cylinder, you will usually have to remove the gearbox.

Clutch operating mechanisms

In standard systems, movement from the clutch pedal can be transferred to the clutch mechanically or hydraulically.

Mechanical operation

In a mechanical system, a clutch cable is attached to one end of a pivoting clutch pedal. When the pedal is pressed, it pulls on this cable and operates the clutch fork at the gearbox end. The clutch fork also pivots, usually pressing the release bearing against the fingers of the diaphragm spring, pushing them inwards towards the flywheel.

As the diaphragm fingers pivot against their fulcrum the outer end of the diaphragm springs moves the pressure plate away from the friction plate disengaging the clutch. A **mechanical advantage** can be gained by the leverage produced at the clutch pedal and release arm. This leverage means you will notice a difference in the amount of movement at the pedal and release arm. (The pedal moves a long way, while the release arm and bearing only move a short distance.) Mechanical advantage makes it easier for the driver to operate the clutch.

Figure 5.5 Direct acting slave cylinder with thrust bearing attached

> **Key term**
>
> **Mechanical advantage** – where effort is increased by leverage.

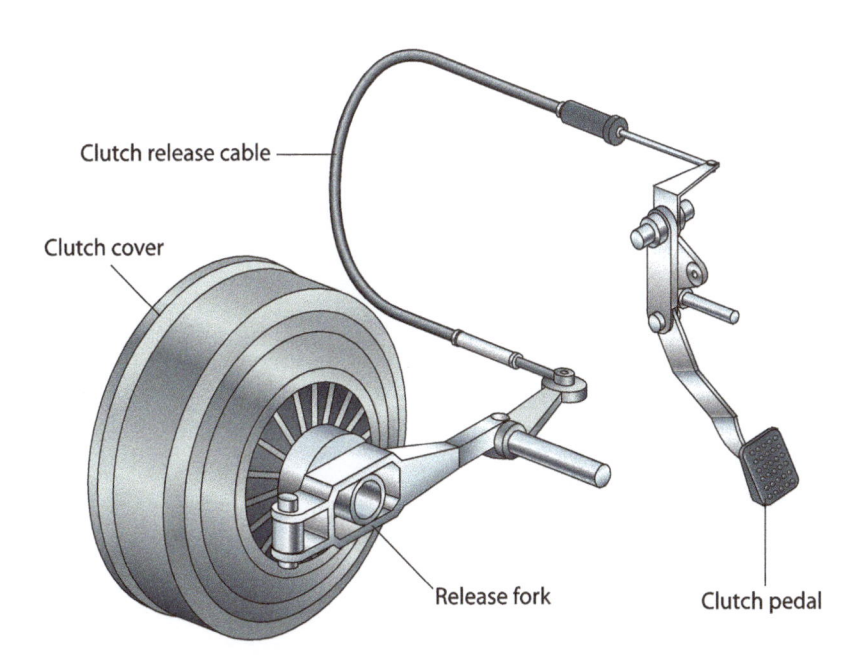

Figure 5.6 Cable-operated clutch assembly

In a mechanical system, the cable has to transmit drive through a number of different angles. This creates drag (making clutch operation harder) and wear (which can lead to premature failure of the clutch cable mechanism as the clutch cable is prone to snapping).

With a mechanical cable-operated system, as the clutch begins to wear a method of adjustment is needed to keep the clutch running and operating properly. This can be done manually using a screw thread or automatically using a pedal ratchet mechanism.

Table 5.2 shows some examples of symptoms created by mechanical clutch cables and their possible causes.

Table 5.2 Examples of symptoms created by mechanical clutch cables and their possible causes

Symptom	Possible cause
Clutch slip	Not enough free play in the clutch cable adjustment (possible faulty self-adjusting mechanism)
Clutch drag	Too much free play in the clutch cable adjustment (possibly stretched cable)
Clutch judder	Cable has become stiff due to lack of lubrication or wear in nylon cable liner
Clutch stiff to operate or cable snaps	Incorrect routing causing the cable to turn too many corners and placing excessive strain on mechanism

Hydraulic operation

A hydraulic system uses the same principle that is used in the braking system. The clutch pedal is attached to a master cylinder. When the pedal is depressed (pushed down), this forces a piston inside the master cylinder to push fluid through a series of pipes and hoses. The fluid is used to operate a slave cylinder piston and the clutch release end.

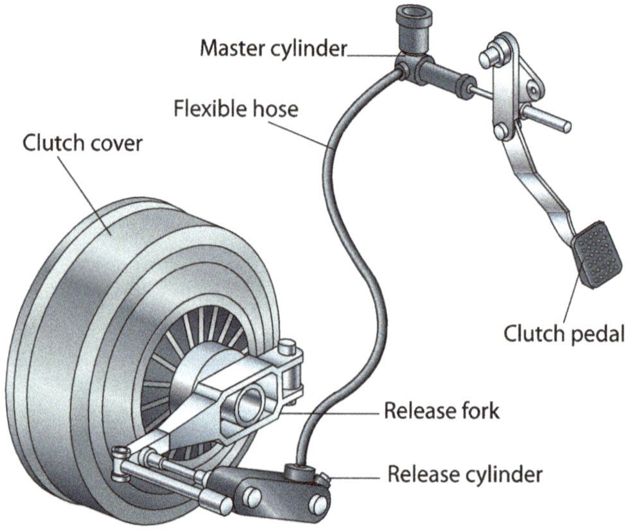

Did you know?

The same type of hydraulic fluid is often used to operate both brake and clutch systems.

Figure 5.7 Hydraulically operated clutch assembly

Table 5.3 shows some examples of symptoms created by hydraulic operating systems and their possible causes.

Table 5.3 Examples of symptoms created by hydraulic operating systems and their possible causes

Symptom	Possible cause
Clutch slip	Fluid contamination of the friction plate from a leaking direct acting slave cylinder
Clutch drag	Air in the hydraulic system, causing sponginess and not allowing complete disengagement of the clutch mechanism

Case study

Mrs Edwards brings her car into your garage, saying that she needs her car servicing. When asked what sort of service she needs, she replies that she doesn't know, but it must need doing because the car has lost power and there is a funny smell when she drives it. Your boss asks you to check it out.

Here's what you do:

- ✓ Listen to the customer's description of the fault.
- ✓ Question the customer carefully to find out the symptoms.
- ✓ Check the vehicle's service history. (A full service was completed four months ago and the car has since done a further 2200 miles.)
- ✓ Carry out a visual inspection to see whether there is anything obviously unusual.
- ✓ Start and run the engine to check operation.
- ✓ Return to reception and ask permission to road test the car with the customer accompanying you so that you are able to confirm the symptoms.
- ✓ When driving the car, you feel that the clutch is not performing as it should.
- ✓ On a clear section of road, with a slight incline, you hold the car in gear at a steady 20 mph, and quickly dip and release the clutch pedal.
- ✓ The clutch is slow to recover, indicating that it is beginning to slip, reducing vehicle performance and producing a burning smell from the friction material.
- ✓ Explain your findings to Mrs Edwards and return to the workshop for further investigation.
- ✓ Gather information from technical manuals (including clutch adjustments and measurements) before you start and take it to the vehicle.
- ✓ Devise a diagnostic strategy. (Start with the clutch system components that it is easiest to access.)
- ✓ Check clutch free play and adjustment. (They are all within tolerance.)
- ✓ Conduct as much diagnosis as possible without stripping down.
- ✓ Inform the customer that you will have to remove the gearbox to check the condition of the clutch and get permission to undertake the work.
- ✓ With the gearbox removed, you find that the direct acting clutch slave cylinder has leaked hydraulic fluid and contaminated the clutch friction material.
- ✓ Check the condition of the dual mass flywheel (DMF) before reporting the faults you have found. (The DMF is in good condition.)
- ✓ Give an estimate for the cost of repair and gain authorisation to conduct the work. (The slave cylinder and clutch unit need to be replaced.)
- ✓ Replace the faulty clutch components and refit the gearbox.
- ✓ Bleed the clutch hydraulic system following the manufacturer's recommended procedures.
- ✓ Correctly reassemble any dismantled components/systems.
- ✓ Thoroughly test the system to ensure correct function and operation.

Clutch by wire (CBW) and semi-automatic gearbox clutch operation

The operation of some standard friction clutches has been designed in order to incorporate electronic control. Two main systems exist.

In a clutch by wire (CBW) system, the physical connection between the pedal and the clutch components has been removed. The pedal is attached to a sensor mechanism that simulates the feel of a standard clutch. As the driver moves the pedal to engage and disengage the clutch, an ECU calculates the best possible operation of the clutch mechanism depending on engine and vehicle load and speed.

CBW still allows the driver to control when the clutch is engaged and disengaged, but can overcome many of the problems caused by poor manual clutch control. It means that vibrations created when pulling away are eased, and strain and component wear are reduced. The integration of other system information, such as engine management data, ABS and traction control, work in conjunction with movement sensors on the clutch to provide the best possible delivery of power through the clutch.

Some systems have removed the clutch pedal completely. Various engine and transmission sensors work together and control hydraulic fluid pressure to the clutch slave cylinder. The fluid pressure is created by a pump and stored in an accumulator unit for use at a moment's notice. Depending on the actions of the driver and the requirements of the vehicle, the clutch can be operated electronically in a fully automatic manner. As the driver applies or releases the brakes and moves the gear shift, sensors decide if the clutch should be engaged or disengaged. The main actions of the clutch (providing a smooth take-up of drive, providing a temporary position of neutral and disengaging the engine from the gearbox to allow for gear change) are all controlled by the ECU, giving better vehicle control and driver comfort.

Dual mass flywheels (DMF)

As engine technology and driver comfort have advanced, many manufacturers now use dual mass flywheels (DMF) in the design of their transmission systems to help reduce driveline vibrations. The flywheel forms the driving surface for most clutch systems, although the main purpose of a flywheel is to store **kinetic** energy and keep the crankshaft turning during the engine's non-power strokes.

The delivery of power strokes to the crankshaft is not smooth and can create **pulsations** in the transmission system. These pulsations can normally be seen as vibrations at the gearstick. Over time, the shaking created through the transmission will lead to premature wear and damage of gearbox components.

To help reduce judder during the take-up of drive, many clutch friction plates have **torsional** damping springs incorporated in their design. A DMF works on a similar principle. It is essentially two separate

---

Did you know?

If the car is run in semi-automatic mode, a system is normally incorporated to inform the driver if they have selected the wrong gear. Many systems have a gear number indicator on the dashboard or may cut fuel injection in and out so the engine hesitates, prompting the driver to change to a more appropriate gear ratio.

Key terms

Kinetic – movement energy.

Pulsation – a rhythmic vibration or oscillation.

Torsional – using a twisting action.

Radial direction – moving in a circular motion.

flywheels connected by a series of torsion springs and dampers. A friction ring between the two main flywheel sections allows them to slip across each other in a **radial direction**. Both flywheel sections are supported in the centre by a bearing that carries most of the load. During normal operation, pulsations from the crankshaft are smoothed out, giving a more comfortable ride and increasing the overall lifespan of transmission components.

The main benefits of a DMF include:

- elimination of gear rattle
- reduced drivetrain noise
- less need to change up and down gear so often
- less synchroniser wear
- lower engine operating speeds can be used, saving fuel and reducing emissions
- less drivetrain torque fluctuation.

A disadvantage of DMFs is that, if the damping springs and connecting components between the two flywheels wear out, pulsations from the crankshaft are exaggerated. This will create rapid wear in other transmission components. This means that when you are changing a clutch, you need to examine DMFs carefully and replace them if necessary.

Fluid couplings

An automatic transmission system also needs a method of transferring drive from the engine to the input shaft of the gearbox. However, a friction clutch is not always appropriate. A fluid coupling is a component that uses hydraulic forces to create drive. It is still able to provide a temporary position of neutral so that the vehicle can be held stationary while in gear without stalling the engine. One of the most common fluid couplings is the torque converter.

A torque converter is mounted at one end of the engine crankshaft, in a similar position to a standard friction clutch. It consists of three main components (as shown in Figure 5.8 overleaf):

- impeller
- turbine
- stator.

These components are sealed inside the torque converter casing. The casing is pressurised with automatic transmission fluid (ATF) from a crankshaft-driven oil pump. When the torque converter is spun by the engine crankshaft, fluid is taken into the impeller blades and thrown outwards by **centrifugal force**.

Torque converter operation

Fluid exiting the impeller at the outer edge strikes the blades of the turbine, making it spin. The centre of the spinning turbine is connected to the input shaft of the gearbox, which now also turns.

The hydraulic fluid (ATF) now leaves the turbine and strikes the blades of the stator, which direct it back into the impeller at high speed.

Action

Research vehicles that use dual clutch, seamless shift transmission and name two models that use this system.

Research vehicles that use launch control and name two models that use this system.

NEW TECH

Launch control

Some manufacturers are now producing road cars with a system of launch control. This is a system that has its origins in racing and motor sports. To achieve the quickest possible start and acceleration from rest, the engine and transmission ECUs work together to control the throttle and clutch. Once the driver has set a switch inside the vehicle, the throttle is held wide open. When the driver releases the brake, the car pulls away in the most efficient manner, without spinning the wheels or over-revving the engine and damaging the clutch or gearbox. Once the vehicle is moving, the system is disabled.

CHECK YOUR PROGRESS

1. Name three components of a clutch.
2. What does the acronym CBW stand for?
3. Why are dual mass flywheels used?

Level 3 Light Vehicle Technology

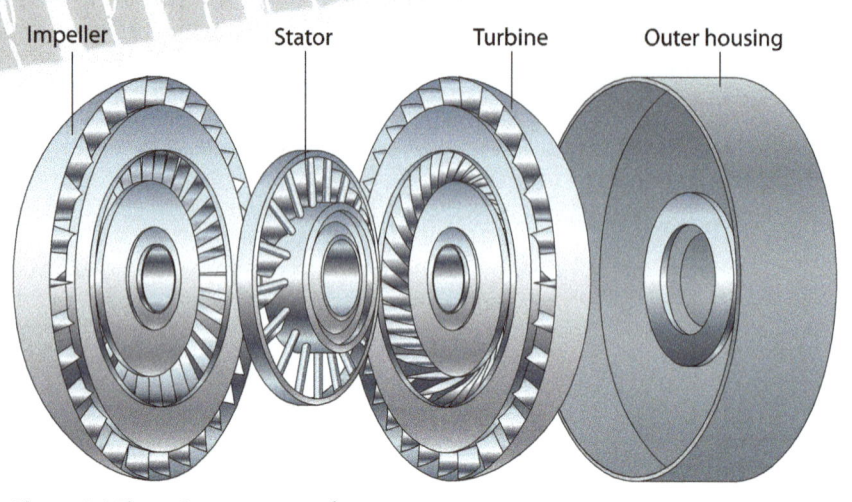

Figure 5.8 The main components of a torque converter

Labels: Impeller, Stator, Turbine, Outer housing

Did you know?

A feature of the torque converter is that when the brake pedal is released and fluid from the impeller begins to react against the turbine, drive to the gearbox begins. This means that the vehicle will start to move unless it is held stationary by the brake. This movement is often referred to as **creep**.

Key terms

Centrifugal force – force that makes rotating objects move outwards.

Torque multiplication – an increase in engine turning effort.

Stall – the point of greatest torque multiplication, when the impeller and turbine are moving at different speeds (usually moving away from rest).

Coupling point – when the impeller and turbine are turning at the same speed and torque multiplication falls to zero.

Freewheel – to spin freely with no drive connection.

Creep – movement of the vehicle as the brakes are released, caused by drag inside the torque converter.

This force helps to multiply the torque provided by the crankshaft and leads to the name torque converter.

The largest amount of **torque multiplication** happens when there is the greatest difference in speed between the impeller and the turbine. This is usually when the vehicle is starting to pull away and is sometimes known as **stall**. The torque multiplication at this time can be around 2.2 : 1.

As the speed of the impeller and the turbine begin to synchronise, torque multiplication falls to zero. This is called **coupling point**. At coupling point, fluid is leaving the centre of the turbine blades with such speed that it would create drag as it strikes the blades of the stator. The stator is mounted on a one-way clutch and, as coupling point is reached, the action of the hydraulic fluid striking the blades makes the stator **freewheel** and prevents drag.

Slip between the turbine and impeller can reduce performance, so as the torque converter reaches coupling point, a hydraulically operated lock-up clutch can be used. This holds all internal components together and prevents slip. This helps improve fuel economy, reduce emissions and maintain engine performance. For most efficient operation, many modern systems use electronic control to activate the lock-up clutch inside a torque converter.

If the car is held stationary, using the brakes, the turbine is also held still while the impeller spins. As the brakes are released, the hydraulic action of the transmission fluid striking the turbine blades makes them turn and provides a smooth take-up of drive.

Stall testing

Stall testing is used on automatic transmissions as a diagnostic method that can be used to:

- check the operation of the automatic transmission clutch
- check the operation of the torque converter clutch
- check engine performance.

To conduct a stall test:

1. Check that throttle valve opens fully.
2. Check that engine oil level is correct.
3. Check that coolant level is correct.
4. Check that ATF level is correct.

5. Check that differential gear oil level is correct.
6. Ensure that the ATF is at the correct operating temperature by running the engine for approx 30 minutes (with the gear selector lever set to 'N' or 'P').
7. Place wheel chocks at the front and rear of all wheels and apply the parking brake.
8. Move the gear selector lever to the drive range.
9. While forcibly depressing the foot brake pedal, gradually depress the accelerator pedal until the engine operates at full throttle.
10. When the engine speed has stabilised (the engine speed should not reach full revs), record the maximum RPM reading and release the accelerator pedal.
11. Move the gear selector lever to 'N' range and cool down the engine by idling it for more than one minute.
12. Compare the engine stall speed with the manufacturer's recommendations.

Note: Do not conduct the stall test for more than five seconds at a time as this can cause overheating and damage to the automatic transmission.

Advantages and disadvantages of torque converters

Table 5.4 lists some of the advantages and disadvantages of torque converters.

Figure 5.9 Torque converter lock-up clutch

Table 5.4 Advantages and disadvantages of torque converters

Advantages	Disadvantages
Multiplication of engine turning effort, which is not accomplished using other forms of clutch	Drag and slip created inside a torque converter reduce overall performance, making it inefficient
Comfort and convenience, as no driver interaction is required to operate the smooth take-up of drive	Creep created by the torque converter when the vehicle is placed in gear can allow the car to move unexpectedly and cause an accident
Normally very long lasting as little wear takes place in the main torque converter components	If the torque converter goes wrong, it can be a very expensive component to replace

Safe working

If an automatic vehicle is to be left stationary for any period with the engine running, you must take it out of gear and select the neutral or park position. The neutral position will simply remove the connection to the gearing inside the box, while the park position has the added advantage of locking a lever mechanism into the gearing to physically prevent any movement.

Did you know?

Early fluid flywheels did not contain a stator to help redirect fluid flow inside. As a result, the drag created when pulling away made them very inefficient.

CHECK YOUR PROGRESS

1 What is creep?
2 What is coupling point?
3 What is the purpose of a lock-up clutch?

Manual and automatic gearboxes

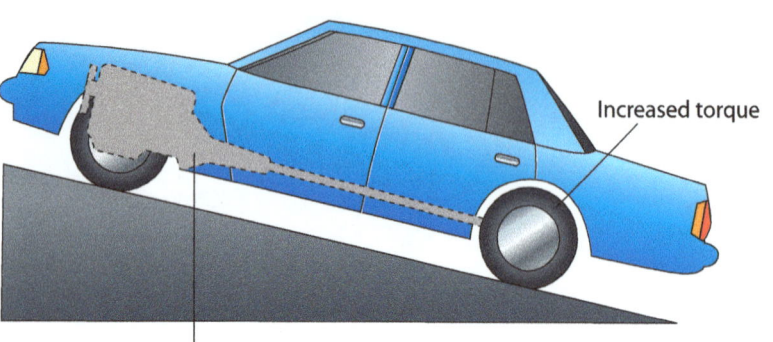

Figure 5.10 The need for a gearbox

Depending on the design of an engine, it delivers torque in a very narrow rev band. This means that the greatest effort being produced by the engine is only available when the engine is running at certain speeds. When moving from rest, climbing a hill or transporting heavy weights, the vehicle will be under load and require large amounts of torque. To overcome this, the gear ratios need to be raised to a point where torque is increased and speed is decreased.

When travelling at speed under light loads, very little torque is required to maintain **momentum**. To ensure good fuel economy and low exhaust emissions, the gear ratios need to be reduced to a point where speed is increased and torque is decreased. This is known as **overdrive**. In reality, a combination of these conditions exist during normal driving. A gearbox is needed in order to make the vehicle driveable in all situations.

Gearbox requirements for different engine types

A diesel engine produces torque low down in its rev range. This means that the crankshaft is turning more slowly when the greatest amount of torque is created. As a result, the gearbox has to be designed so that this torque can be transmitted to the road with sufficient speed for general use.

A petrol engine produces torque higher in its rev range. This means that the crankshaft is turning fast when the greatest amount of torque is created. As a result, so that it is suitable for general use, the gearbox has to be designed so that speed is reduced and torque is multiplied and transmitted to the road under different load conditions.

A hybrid vehicle that combines a petrol engine and an electric motor produces a combination of torque. An electric motor gives its greatest amount of torque when starting. A petrol engine gives its greatest amount of torque at speed. Many manufacturers use a continuously variable transmission (CVT) (see pages 280–283) with hybrid vehicles to deliver the most efficient amount of torque and speed to the road wheels no matter which motor is driving at the time.

The construction and operation of manual gearboxes

A transmission casing mounted between the engine and final drive unit contains gears of varying sizes. When engaged with each other, these gears multiply torque through the principle of leverage.

- When a small drive gear is connected to a large driven gear, torque is multiplied and speed is reduced.
- When a large drive gear is connected to a small driven gear, torque is reduced and speed is increased (overdrive).

> **Key terms**
>
> **Momentum** – movement created by the speed and weight of the vehicle.
>
> **Overdrive** – when the output speed of a gearbox is higher than the input speed.

> **Did you know?**
>
> A series hybrid car doesn't need a gearbox. The internal combustion engine acts as a generator to charge batteries, which operate the electric motors to drive the wheels. As there is no direct connection between the engine and wheels, the engine doesn't need to have the torque multiplied (as this is supplied directly by the electric drive motors).

5 Diagnosis & rectification of light vehicle transmission and driveline faults

Gears of varying sizes are normally arranged on two or three shafts, supported on bearings to allow low friction rotation.

- The input shaft receives drive from the engine through the clutch unit.
- The output shaft is connected to the final drive, which through a number of other shafts and couplings/joints will rotate the wheels.
- In a rear-wheel drive car, a third shaft called a counter shaft or layshaft is often used to help transfer movement through the gears.

Most gearboxes operate on a principle known as constant mesh. This means that all of the gear teeth of all of the gears are always in contact with each other, whether a gear ratio is selected or not.

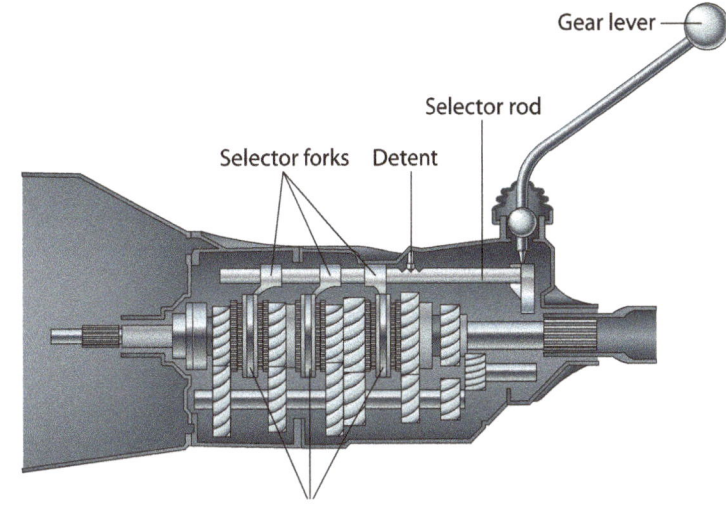

Figure 5.11 Manual gearbox

Gear types

Two main types of gear are used in a standard manual transmission system: spur gears and helical gears.

Spur gears

A spur gear has straight-cut teeth, as shown in Figure 5.13. These are direct-acting and create low amounts of drag, which helps improve overall performance, but they can be noisy. A spur-cut gear can slide in and out of mesh with another spur-cut gear, making them ideal for use as a reverse gear idler. The characteristic whine when reversing a car is caused by the design of the teeth on the spur-cut gear.

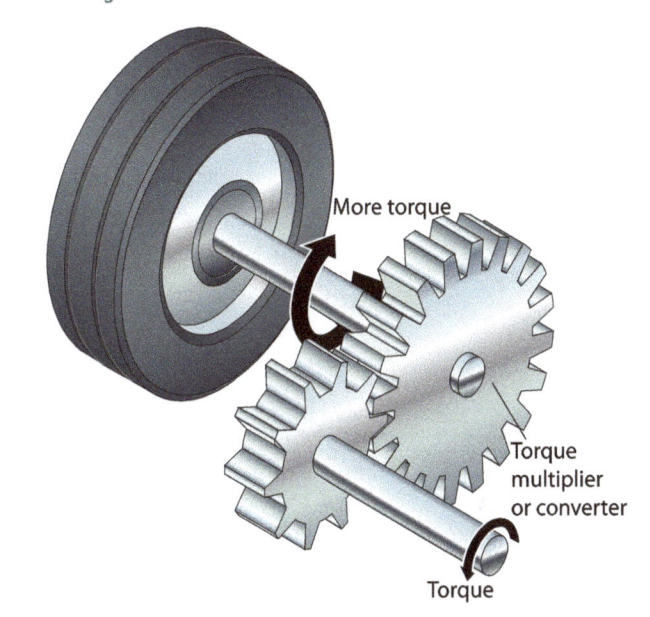

Figure 5.12 Using gears to multiply torque

Helical gears

A helical gear has teeth which are cut on an angle, as shown in Figure 5.14. The word 'helical' comes from the word 'helix', which is a form of spiral (think of the spiral of a coil spring). You can see from Figure 5.14 that the teeth are not just cut on a diagonal; if they were extended around a cylinder they would actually be shaped like a screw thread.

Figure 5.13 Spur gears – with straight-cut teeth

Figure 5.14 Helical gears – with teeth cut at an angle

Helical-cut gear teeth have a large surface area, which makes them very strong, less prone to wear and quiet in operation. Most light vehicle transmission systems use this design as the main type of gearing for transmission of drive.

Because of the design and shape of helical-cut gears, they cannot be slid in and out of mesh like spur gears. This means that they must be used in a constant mesh type gearbox.

Gear ratios

Gear ratios give an indication of how much torque is multiplied inside the gearbox. The gear ratio is a comparison between the number of teeth on the driving gear and the number of teeth on the driven gear (input and output).

To calculate a gear ratio, use this equation:

$$\text{Ratio} = \frac{\text{Driven}}{\text{Driver}}$$

You need to divide the number of teeth on the output (driven) gear by the number of teeth on the input (driver) gear.

A simple example:

If the input gear has 16 teeth and the output gear has 32 teeth, the input gear will need to turn two complete revolutions to make the output gear turn one complete revolution. This gives a gear ratio of 2:1.

- This 2:1 gear ratio means that speed is reduced by half. For example, an input turning at 200 rpm will give an output of 100 rpm.
- On the other hand, this gear ratio of 2:1 doubles the output torque (turning effort). For example, an input torque of 200 **Newton metres** will give an output torque of 400 Newton metres.

Gear ratios of compound gear sets

In practice, a gearbox uses a combination of gears to achieve torque multiplication. When drive is sent through a number of gears, this is known as a compound gear set.

The equation shown in the previous example will only give you the gear ratio for one set of gears. To calculate the overall gear ratio of a compound set, calculate the ratio of each set of gears individually and then multiply them together to get the overall gear ratio.

$$\text{Ratio} = \frac{\text{Driven}}{\text{Driver}} \times \frac{\text{Driven}}{\text{Driver}}$$

Gears are arranged so that the different sizes produce a usable range of gear ratios that are suitable for the type of vehicle and how it will be used under all driving conditions.

- Low gears produce large amounts of torque for situations when high demands are placed on the transmission system. These gears are usually numbered one to three.

Did you know?

Some racing cars use spur-cut gears as part of their main transmission design. The low amounts of drag created make them an efficient design for the best transmission of engine power without loss of performance through the gearbox. They are also an ideal gear for use in a sequential racing gearbox design (see pages 272–273).

Key term

Newton metre – the unit of measurement for torque (turning effort).

Did you know?

In the example described above, an even number of teeth are used (16 and 32), but in reality odd numbers of teeth are used. When you use the equation given to calculate a gear ratio, you will get an answer containing a decimal point.

The reason for using odd numbers of teeth is that it avoids the same teeth coming into contact with each other too often as the gears rotate against each other. This reduces overall wear and improves the lifespan of the individual gears.

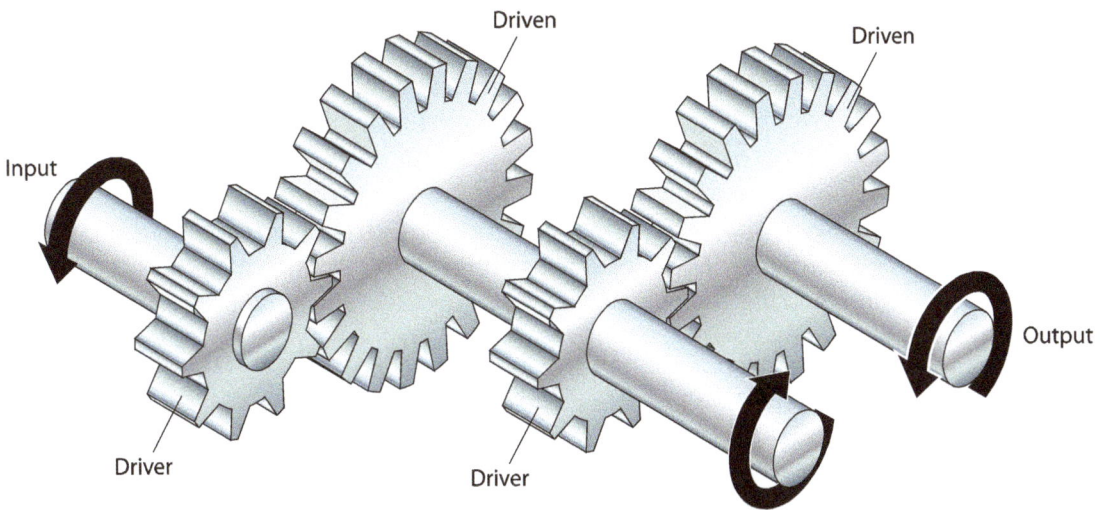

Figure 5.15 Compound gear train ratios

- Fourth gear usually produces a ratio of approximately 1:1, meaning that input and output speed and torque are normally equal.
- Some gearboxes have five or even six gears, and these produce ratios of lower than 1:1. This results in overdrive, where the speed output of the gearbox is higher than the input, but the amount of torque is reduced.

For more information on calculating gear ratios including the multiplication of torque and speed, see *Level 2 Principles of Light Vehicle Maintenance & Repair Candidate Handbook,* Chapter 5, page 339.

Reverse gear

Another function of the gearbox is to provide a method of reversing the car. To do this, the output of the transmission must be rotated in the opposite direction. Reverse gear is achieved by placing an intermediate gear, called an idler gear, between two of the drive gears. This reverses the direction of the output – if the input gear turns clockwise, the idler gear turns anticlockwise and the output gear turns clockwise, as shown in Figure 5.16. The idler gear has no overall effect on the gear ratio between the input and the output, but simply changes the direction of rotation.

Action

With the gearbox removed from a vehicle, select all gear ratios and count the difference in the number of turns required at the input shaft to those produced at the output shaft.

With the gearbox stripped down, count the number of teeth on each gear set and calculate the individual gear ratios.

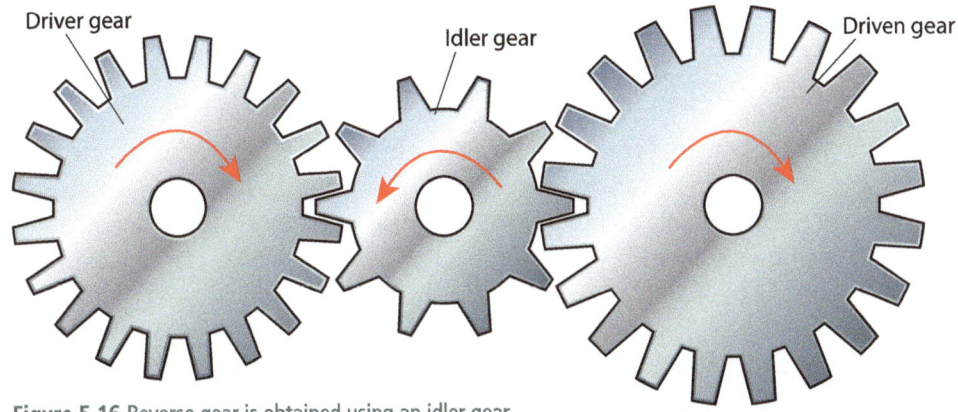

Figure 5.16 Reverse gear is obtained using an idler gear

269

Level 3 Light Vehicle Technology

Selector hubs and synchromesh mechanisms

Because helical gears cannot slide in and out of mesh, a selector hub is used, which is **splined** to the output shaft of the gearbox. The selector hub sits between the helical-cut gears and, when a gear ratio is selected by the driver, a selector fork slides the hub towards an appropriate gear setting, as shown in Figure 5.11 on page 267. **Dog teeth** on the side of the selector hub locate with dog teeth on the side of the appropriate gear, locking the two components together, engaging the gear and producing drive.

Synchromesh

Since gears of different sizes are used inside the gearbox, they will all be travelling at different speeds. Because dog teeth or dog clutches are used to provide the positive engagement between input and output, when the driver wants to select a certain gear, the speed of these teeth must be **synchronised** before they can be engaged. (This means that they must be sped up or slowed down so they are all turning at the same speed.) Otherwise, severe noise and damage may occur.

To achieve this, a system sometimes called a synchromesh (Figure 5.17) is used in combination with the selector hub. As the selector hub is moved towards the gear to be used, two surfaces come into contact with each other and act as a friction clutch. This provides grip to bring the speed of the selector hub either up or down, so that it is spinning at the same speed as the gear (synchronising their speed).

> **Key terms**
>
> **Splined** – attached using splines. Splines are a series of grooves cut into the outside of a shaft and inside a circular housing. When slotted together, the splines prevent rotational movement.
>
> **Dog teeth** – teeth formed on the edge of a gear that act as a positive engagement clutch.
>
> **Synchronised** – operating together at the same speed and time.

> **Action**
>
> A five-speed manual gearbox has the following gear ratios (input:output):
> - First gear = 16:1
> - Second gear = 12:1
> - Third gear = 8:1
> - Fourth gear 1:1
> - Fifth gear 0.8:1
>
> With an input speed of 4800 rpm, what is the output speed of each gear?
>
> How many rpm does the synchroniser need to speed up or slow down when changing from:
> - first to second?
> - second to third?
> - third to fourth?
> - fourth to fifth?

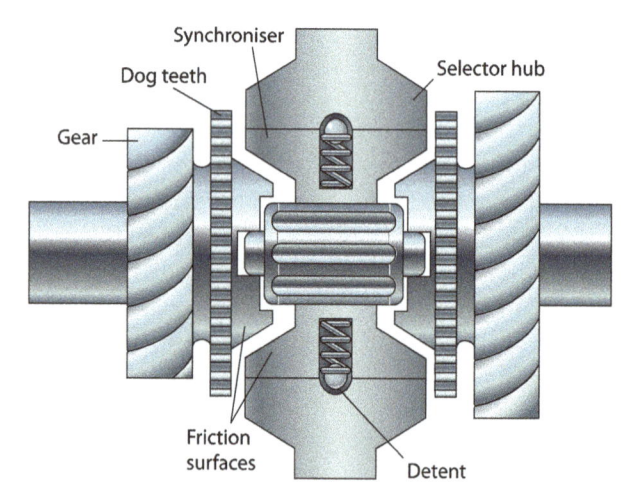

Figure 5.17 Synchromesh selector hub

Now that the gears are spinning at the same speed, when the selector hub dog teeth and gear dog teeth come into contact with each other, they are able to lock together without noise or damage and provide a positive engagement.

Baulk ring

Because a synchromesh unit relies on friction between two surfaces to equalise the speeds of both the hub and gear, it is possible to outrun the synchronisation by forcing the gear selector lever too quickly. In order to overcome this problem, a component called a **baulk ring** is used to prevent gear selection if the hub and the chosen gear speeds are not synchronised.

A baulk ring has teeth around its outer edge that are the same size and shape as the dog teeth on the gear and selector hub. When gear speeds are not synchronised, the baulk ring is able to move slightly in a radial direction when compared to the gear and selector hub, and is said to be 'out of register'. This means that if the teeth on all three components (selector hub, baulk ring and gear) don't line up, then gear selection is blocked.

As the driver pushes the gear lever, the baulk ring acts like the drive plate of a friction clutch. It is sandwiched between the selector hub and the gear. As pressure is applied, friction makes it grip and bring all three components up to the same speed. Once all three components are travelling at the same speed, the teeth on the baulk ring line up and move 'into register'. This allows the selector hub to slide across, locking the dog teeth to those on the side of the chosen gear.

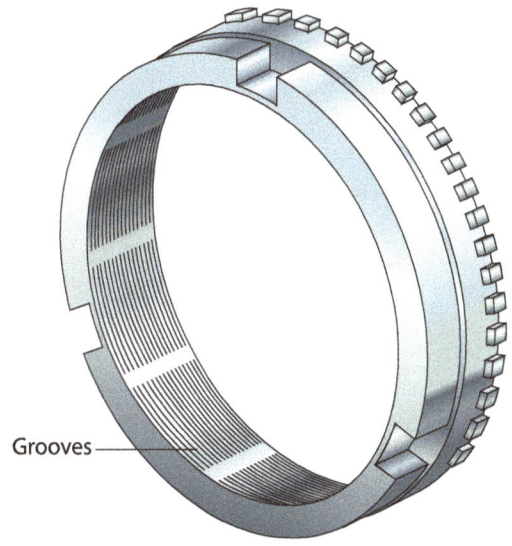

Figure 5.18 Synchromesh baulk ring

The components inside a manual gearbox are lubricated by **splash feed**. This means that as the gears rotate, they scoop up oil and drag it around the gears, bearings and shafts, etc. Because friction is needed between the selector hub, the baulk ring and the appropriate gear, a system is needed to allow small amounts of friction to exist (cutting through the lubricating oil).

Many baulk rings are manufactured with small grooves in the friction surface that act like the tread found on tyres (see Figure 5.18). Just as tyre tread is designed to cut through water tension to prevent aquaplaning, the grooves that are found on the synchromesh baulk ring perform a similar function with the lubricating oil in the gearbox.

Over a period, the grooves on the baulk ring begin to wear. This can lead to slippage of the friction surfaces, which might mean that complete synchronisation of the gear speeds is not possible. If this happens, the dog clutch teeth may be travelling at different speeds and will strike each other, leading to noise and gear teeth damage. If it is difficult to engage a specific gear without a crunching noise being heard, you should carefully inspect the baulk ring next to the chosen gear and replace it if necessary.

Interlock mechanism

It is possible that when the gear selector is operated by the driver, two selector rods are moved at the same time. If this happens, the selector forks inside the gearbox may try to lock two gears to the output of the transmission simultaneously. Because the gears inside the gearbox are different sizes, they are turning at different speeds. If both gears were selected at the same time, the gearbox would lock solid.

> **Key terms**
>
> **Baulk ring** – a blocking ring that sits between the selector hub and the gear to be selected.
>
> **Splash feed** – a method of lubrication used in some gearboxes that drags oil around with the movement of the gears.

> **Action**
>
> Examine the vehicles in your workshop and, using technical data, find the manufacturer's specification for transmission oil used in three of the cars.

Level 3 Light Vehicle Technology

To prevent this, a system called an interlock is provided in the selection mechanism. A number of different interlock mechanisms are used by manufacturers; two common designs are described below.

- **Ball and plunger:** a series of balls, plungers or rods that lock the selector shafts when operated, so that only one shaft is able to move at any one time (see Figure 5.19).

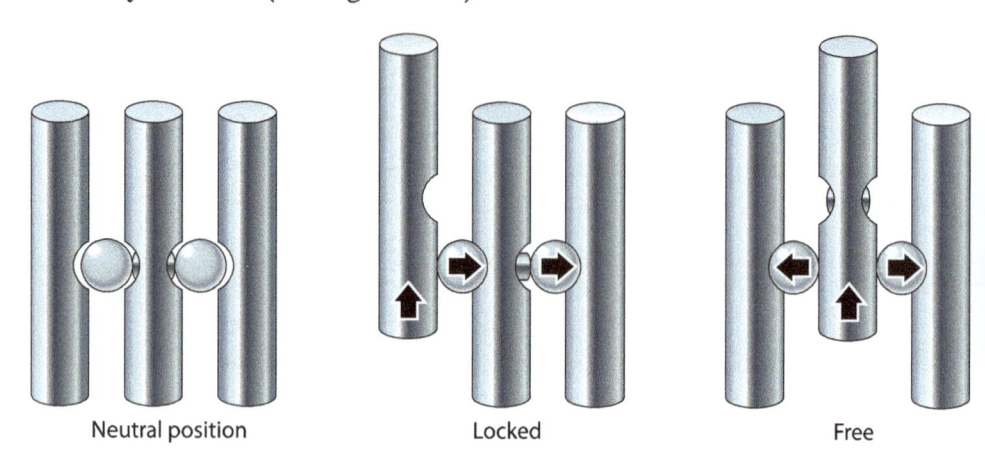

Neutral position   Locked   Free

Figure 5.19 Ball and plunger interlock

- **Plate:** a movable locking plate that is able to pivot from side to side through a groove machined in the selector rod mechanism. It has an opening on one side that is the same width as one of the selector rods. As the gearshift mechanism is operated, the slot in the locking plate only allows one selector rod to move at a time.

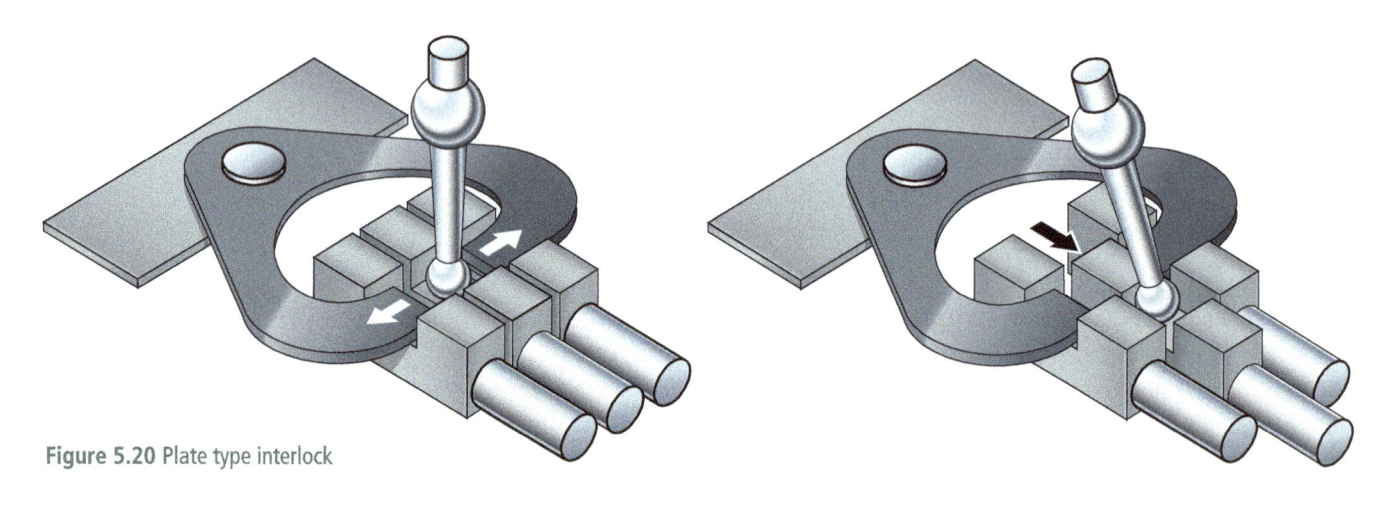

Figure 5.20 Plate type interlock

Key term

Sequential – in sequence, one after another, following a set pattern.

Sequential manual gearbox operation

In a **sequential** manual gearbox (SMG) the gears must be selected in order, one after another. To move up and down the gears, the driver operates a gear lever, but instead of acting on a series of selector rods, a ratchet mechanism rotates a selector drum. The selector drum is a cylindrical component with angled grooves machined in the outer surface. A selector fork is located by a peg in each of the machined grooves. As the drum rotates, the selector forks are forced to move in the direction of the machined grooves, causing the synchromesh selector hub to engage a single gear. The grooves on the surface of the selector drum are designed so that as one gear is disengaged, another is engaged. This removes the need for an interlock device.

An overshift limiter mechanism is often incorporated in the design of a sequential gearbox. It only allows the driver to change up and down the box one gear at a time in sequence, so gears cannot be skipped.

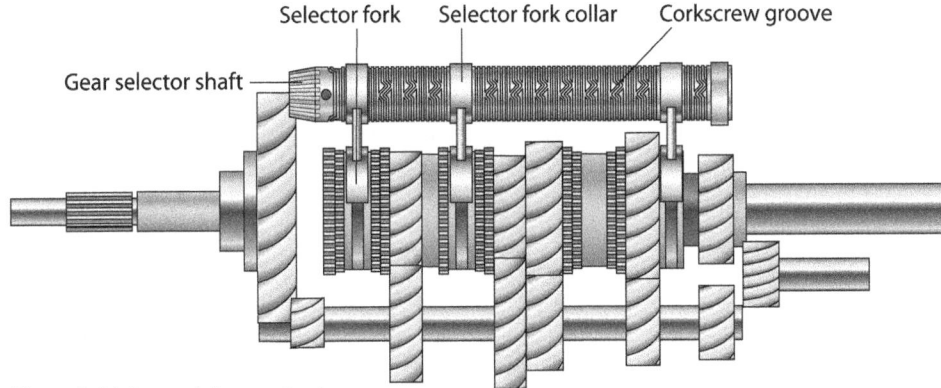

Figure 5.21 Sequential gear selection

Seamless shift dual clutch systems (DCS)

Some manufactures now use a paddle shift gear change mechanism. Instead of the traditional gear lever and clutch pedal, a pair of flat levers or 'paddles' is mounted behind the steering wheel. When the driver wants to select a different drive ratio, they simply operate one of the paddles to change up or down. The paddles send a signal to a transmission control ECU, and electronics, actuators and hydraulics then perform the actual gear change in the gearbox.

To improve the speed and smoothness of the gear change in this type of transmission, a dual clutch, seamless shift design is used. The gearbox has two different shafts to support the drive gears. One shaft is hollow, with the other shaft running through the middle. The shafts can rotate independently of each other, although from the outside it may look like a single shaft. The odd-numbered gears are mounted on one shaft and the even-numbered gears are mounted on the other. Each shaft has its own multi-plate clutch pack, leading to the term 'dual clutch'.

- When the driver selects first gear, actuators in the gearbox move the selector hub, which locks first to the drive of the gearbox. The multi-plate clutch connected to the odd-numbered shaft is engaged by hydraulics and drive is taken up.
- When the driver wants to change up to second gear, they operate a paddle behind the steering wheel. Actuators in the gearbox move the hub to select second gear. Once second gear is selected, the multi-plate clutch pack on the odd-numbered shaft is disengaged and the multi-plate clutch pack on the even-numbered shaft is engaged hydraulically.

Level 3 Light Vehicle Technology

The smooth changeover between the clutch packs and gear shafts gives an almost seamless gear change. This process continues for all other gears in a sequential pattern, up and down the gearbox.

CHECK YOUR PROGRESS

1. Name one advantage of a spur gear and one advantage of a helical gear.
2. What is the purpose of a baulk ring?
3. Name two types of interlock mechanism.

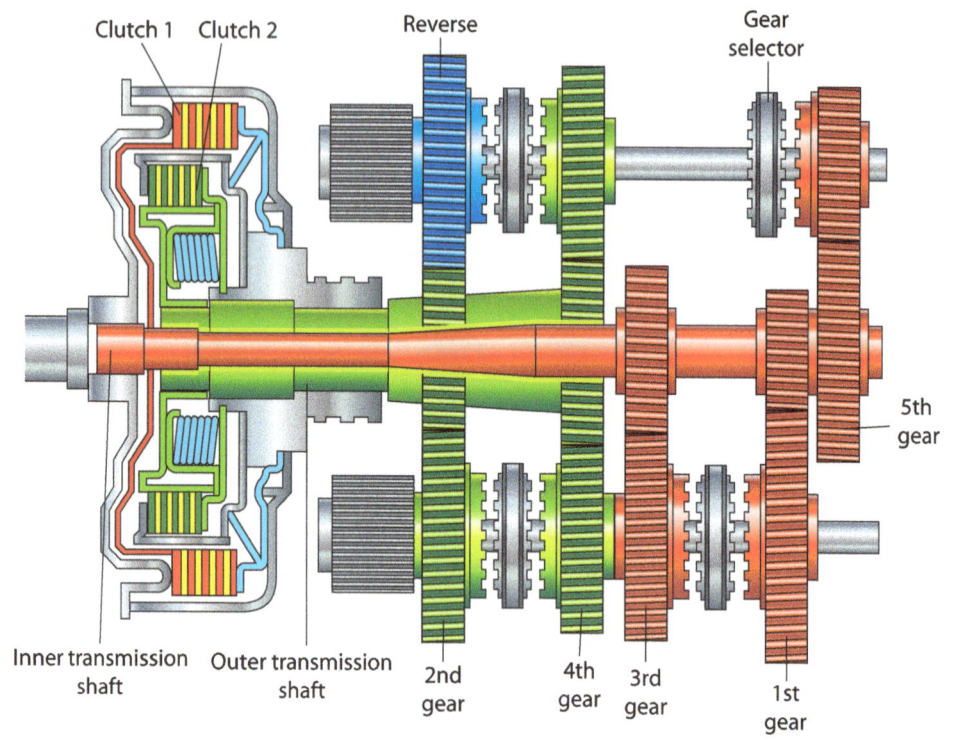

Figure 5.22 Dual clutch seamless shift gearbox

For further information on manual transmission symptoms and possible faults, see Table 5.7 on page 297.

Automatic transmission

In a manual gearbox, the driver selects the gear ratio for the driving situation. In an automatic gearbox, engine speed and load are detected by the transmission, and the system itself chooses the most appropriate gear.

Figure 5.23 Automatic gearbox

Many automatic transmission systems use a different type of gearing method from that of manual transmission systems. Automatic transmissions still need to achieve the appropriate ratios required for the multiplication of torque. In a standard automatic transmission, instead of spur or helical gears that are engaged and disengaged, a system called an epicyclic gear train is used.

Epicyclic gear train

An epicyclic gear train (see Figure 5.24) uses gears that are constantly in mesh and consist of:

- a large outer ring gear, often called the 'annulus'
- a central gear, often called the 'sun gear'
- a series of intermediate gears (that sit between the sun gear and the annulus) called the 'planet gears'. These are supported on spindles attached to a planet carrier.

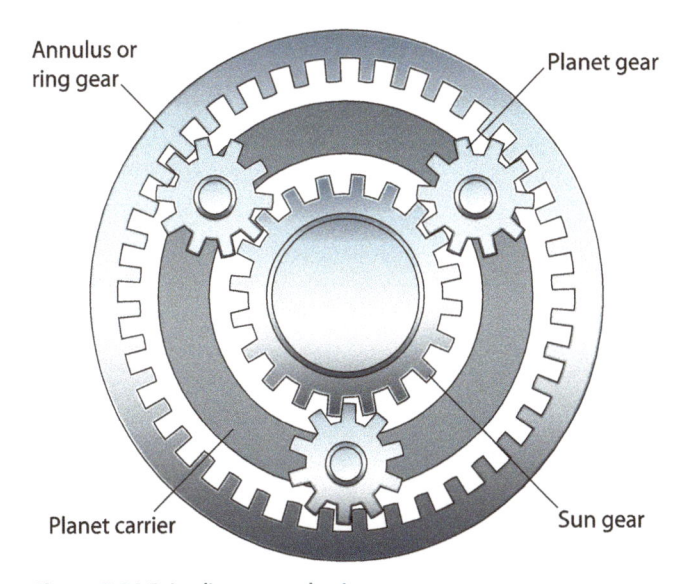

Figure 5.24 Epicyclic gear mechanism

To select an appropriate gear ratio, one section of the epicyclic gearing will be locked to another part of the transmission. This section of gearing then becomes an idler gear with no direct effect on the gear ratio, meaning the remaining two gears become input and output. The differing numbers of teeth on the input and output are now able to provide various gear ratios, including reverse. The gear ratios show how much the torque is increased (multiplied).

Gear ratios are calculated in the same way as for a manual gearbox:

$$\text{Ratio} = \frac{\text{Driven}}{\text{Driver}}$$

(Since one component of the epicyclic gear mechanism is operating as an idler, this can be ignored during your calculation of gear ratio.)

A single epicyclic gear mechanism is able to provide three forward gear ratios and one reverse, as shown in Table 5.4.

> **Did you know?**
>
> The number of planet gears used in an epicyclic gear mechanism varies depending on the manufacturer. To work out the number of planet teeth to use in the calculation of gear ratio, subtract the number of teeth on the sun gear from the number of teeth on the annulus:
>
> Annulus teeth – Sun gear teeth = Planet teeth

Table 5.4 Example of epicyclic gear selection, torque increase and direction of travel

Stationary	Input	Output	Ratio and direction
Annulus	Sun gear	Planet carrier	3.4:1 (forward)
Sun gear	Planet carrier	Ring gear	0.71:1 (forward)
Planet carrier	Sun gear	Ring gear	2.4:1 (reverse)
When annulus, planet carrier and sun gear are locked together			1:1 (forward)

These gear ratios are not suitable for all driving conditions. As a result, many systems use at least two epicyclic gear sets joined together to

Level 3 Light Vehicle Technology

form a compound gear train. Two main types of compound gear set are common:

- **Simpson gear set:** A single long sun gear is used between the two epicyclic gear sets to join the gear sets together.
- **Ravigneaux gear set:** In this design, it is the planet carriers that are connected to join the gear sets together.

Most automatic gearboxes use hydraulic fluid pressure to control the operation and selection of gears. Automatic transmission fluid (ATF) under pressure is directed through a series of channels and galleries by valves. The hydraulic fluid will then operate brake bands or multiplate clutches and provide engagement of a particular gear ratio. Early systems relied on fluid pressure created by an engine-driven oil pump to sense load and speed and initiate gear change. Modern systems now use engine management and transmission sensor data to control the operation of solenoid valves in a valve block. The sensor information is processed by a transmission ECU, which actuates the solenoid valves controlling the hydraulic system of the automatic gearbox. In this way, gear ratio and shift timing can be accurately matched to all road situations and driver demands.

Figure 5.25 Automatic transmission valve block

Electronic control system

As with all electronic control systems, the management of transmission gear selection involves three main processes:

- **Input:** This comes from various sensors, normally related to load and speed.
- **Processing:** The ECU takes the sensor information and calculates the best gear ratios for a given driving situation, pre-programmed into its memory.
- **Output:** Signals are sent from the ECU to various actuators, **servos** and solenoids which control the operation of clutches and brake bands and allow gear change to take place.

Key term

Servo – a mechanism that converts a small mechanical motion into a larger movement with greater force.

Did you know?

The transmission control ECU will have a number of different shift patterns programmed into its memory. This allows the driver to select an economy or sport mode, for example. Some transmission ECUs are able to learn a driving style and adapt the shift pattern to suit the driver.

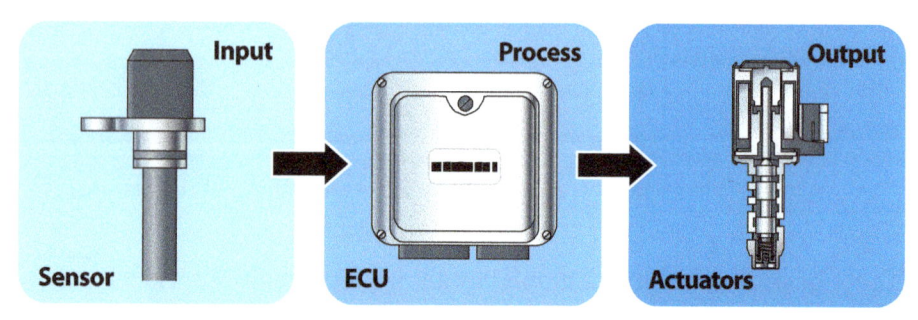

Figures 5.26 Transmission electronic control

276

Three main methods are used to control the selection of gearing inside an automatic transmission system:

- brake bands
- multi-plate clutch packs
- unidirectional clutches.

Brake band

Brake bands are an actuator system that hold a section of the epicyclic gearing stationary by anchoring it against the transmission casing. A brake band is similar to a belt that is wrapped around the outside of one of the gearing components. One end of the brake band is fixed to the transmission casing by an adjustable mounting. The other end of that brake band is connected to a hydraulic servo, which when operated will try to squeeze the ends of brake band together. The inside of the brake band usually has a friction material attached, so that when it is pinched together it provides grip and stops part of the gear set rotating (see Figure 5.27).

Over time the friction material will wear and grip on the gear set will be reduced. You can usually access the fixed end of the brake band from the outside of the gearbox casing to adjust it during maintenance procedures.

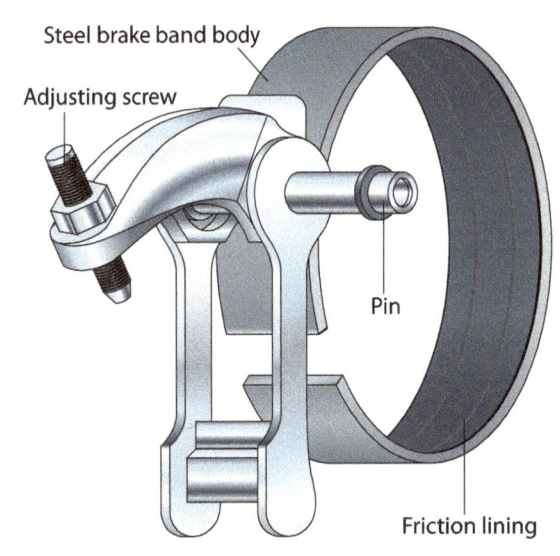

Figure 5.27 Automatic transmission brake band

Multi-plate clutch packs

To connect two rotating gear sets in an epicyclic gear train, a hydraulically operated multi-plate clutch pack can be used. A number of friction plates, interspaced with plain plates, are housed in a clutch basket. A pressure plate is mounted on the outside end of the clutch basket. Hydraulic pressure acts on the pressure plate, squashing the component parts together to create friction. The friction provides the grip that locks parts of the rotating gear set together.

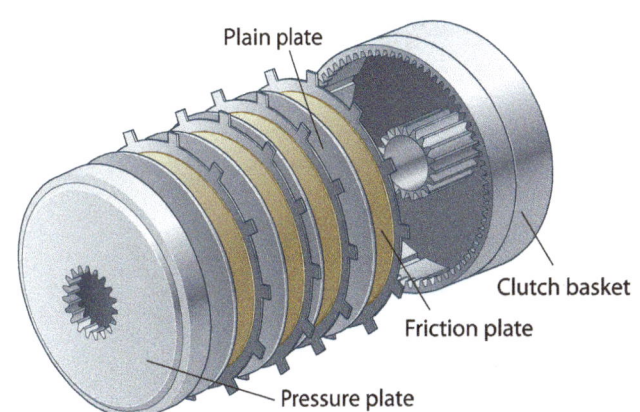

Figure 5.28 Multi-plate clutch pack

Unidirectional clutches

To prevent parts of an epicyclic gear set rotating in the wrong direction, a unidirectional (one-way) clutch can be used. This is similar in construction to ball bearing race, but instead of balls a set of sprags are used.

A sprag is a small 'S'-shaped wedge held between an inner and outer bearing race. When rotated in one direction, the 'S'-shaped sprag tilts to one side, allowing the two bearing races to turn freely. When rotated in the opposite direction, the 'S'-shaped sprag wedges itself tightly between the two bearing surfaces and stops it from turning.

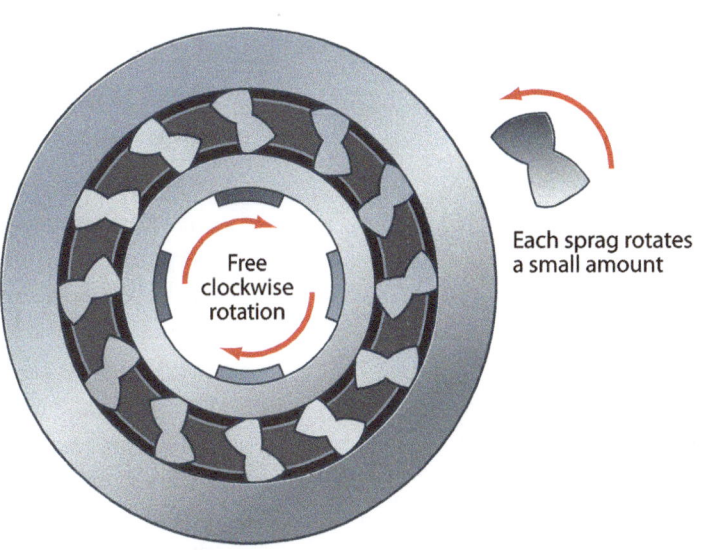

Figure 5.29 Unidirectional sprag clutch

Clutch and brake band operation for automatic transmissions

Table 5.5 gives an example of how brake bands, multi-plate clutches and unidirectional clutches are combined to connect epicyclic sets for gear selection.

Table 5.5 Clutch and brake band operation for automatic transmissions

Gear	C0	U0	C1	C2	B0	B1	B2	U1	B3	U2
Park	Active									
Reverse	Active			Active					Active	
Neutral	Active									
First	Active	Active	Active							Active
Second	Active	Active	Active				Active	Active		
Third	Active	Active	Active	Active			Active			
Overdrive			Active	Active	Active		Active			
Low Range Second	Active	Active	Active			Active	Active	Active		
Low Range First	Active	Active	Active						Active	Active

Key: C = Multi-plate clutch pack   B = Brake band   U = Unidirectional (one-way) clutch

Diagnosis and repair

When a car is presented with an automatic transmission fault, your diagnostic routine should be broken down into three areas:

- **Mechanical system:** Thoroughly road test the car, using all of the gear settings and positions under different driving conditions. Depending on the symptoms produced and the gear operation available, you can use Table 5.5 above to help you identify which clutch or brake band may be at fault.

- **Hydraulic system:** Thoroughly assess the level and condition of the hydraulic fluid in the automatic gearbox. A low fluid level will give incorrect shift patterns or may not allow the correct operation of brake bands and clutches. If the fluid is dirty or discoloured, and sometimes has a slightly burnt smell, this can be an indication that the clutches have worn out.

5 Diagnosis & rectification of light vehicle transmission and driveline faults

- **Electrical and electronic system:** Many automatic transmission systems incorporate a self-diagnostic facility. You should attach a suitable scan tool then record and clear any stored fault codes. Next, take the car for a full road test, re-scan and focus your diagnostic routine around any codes that have returned. If a code indicates a problem in an electrical part, you should test the component and circuit to confirm the fault before replacing it.

The overhaul and repair of many automatic transmission systems may require specialist knowledge and tooling. Many garages simply do not have the resources available to undertake repairs to these systems. However, general maintenance procedures, such as fluid levels and brake band adjustment, should be possible if you follow the manufacturer's guidelines.

For further information on automatic transmission system symptoms and possible faults, see Table 5.8 on page 297.

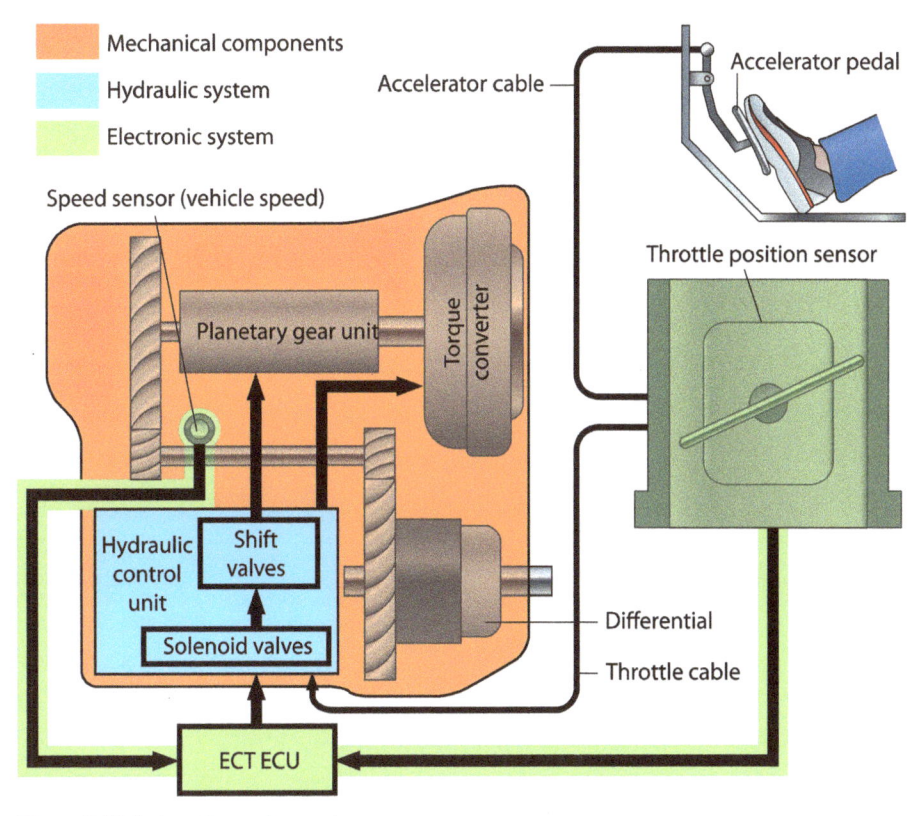

Figure 5.30 Automatic gearbox systems

Skills for work

Following the diagnosis of an automatic transmission fault, you realise that you don't have the specialist tools required to conduct the overhaul of the gearbox. You need to explain this to your customer in a way that will not discourage them from coming back to your garage in the future.

This situation requires that you use particular personal skills. Some examples of these skills are shown in Table 6.1 at the start of Chapter 6, on pages 304–305.

1. Using the examples given in Table 6.1, choose one skill from each of the following categories that you think you need to demonstrate in this situation.
   - General employment skills
   - Self-reliance skills
   - Specialist skills
   - People skills
   - Customer service skills

2. Now rank these skills in order of importance, starting with the one that it is most important for you to have in this situation.
3. Which of the skills chosen do you think you are good at?
4. Which of the skills chosen do you think you need to develop?
5. How can you develop these skills and what help might you need?

> **Safe working**
>
> You must always check the fluid level and condition of an automatic transmission system before you carry out a road test. A level that is too high or too low could cause unnecessary damage to the gearbox during your road test.

Case study

Mr Simpson brings his car into your garage and complains that his automatic transmission has stopped changing gear. Your boss asks you to check it out.

Here's what you do:

- ✓ Listen to the customer's description of the fault.
- ✓ Question the customer carefully to find out the symptoms. (You discover that the engine management light is illuminated and that the speedometer stopped working about a month ago.)
- ✓ Carry out a visual inspection. (Check transmission fluid levels, etc.)
- ✓ Gather information from technical manuals (including diagnostic trouble codes) before you start and take it to the vehicle.
- ✓ Devise a diagnostic strategy.
- ✓ Use a scan tool to see if any diagnostic trouble codes have been stored. (P0721 Output Speed Sensor Circuit Range/Performance)
- ✓ Record and clear the diagnostic trouble codes.
- ✓ Conduct a road test to confirm the symptoms.
- ✓ Re-scan for any fault codes that have been stored. (P0721 Output Speed Sensor Circuit Range/Performance)
- ✓ Check the quick things first (10-minute rule) and carry out a visual inspection of connections, fuses, etc.
- ✓ Conduct as much diagnosis as possible without stripping down.
- ✓ Connect an oscilloscope to the output speed sensor and check for correct function. (No wave form is produced, meaning that the sensor is faulty.)
- ✓ Strip out and replace the faulty speed sensor.
- ✓ Correctly reassemble any dismantled components/systems.
- ✓ Clear all fault codes and reset the ECU adaptions.
- ✓ Thoroughly road test the system to ensure correct function and operation. (Check gear change, engine management light and speedometer.)

> **Did you know?**
>
> Although the CVT gearbox design has been around for over 100 years, many manufacturers are now offering this type of gearbox as an option because of its efficient delivery of torque and power.

Continuously variable transmission (CVT)

A continuously variable transmission (CVT) gearbox is a form of automatic transmission. Instead of using a fixed set of five or six gear ratios, it is able to offer a stepless ratio between an upper and lower limit. This means that, when coupled to an engine, a CVT gearbox is always able to run within its optimum range.

Two main types of CVT are used:

- variable diameter pulley (VDP)
- toroidal CVT.

Variable diameter pulley (VDP)

Instead of using mechanical gear sets, as in an epicyclic gear train, VDP uses a drive belt held between two pulleys similar to the chain and sprockets on a bicycle. Originally this belt was made of rubber, but as technology has developed, a steel drive belt has replaced the original design.

A bicycle is able to vary its gearing by changing the size of the sprocket on which the drive chain runs. VDP operates by changing the size of the drive pulleys, which allows different gear ratios to be created. To do this, the drive pulleys expand and contract. In this way, the drive belt is able to ride up and down within the pulleys, varying their size and therefore the gear ratio. As one pulley expands, the other will contract equally.

The steel drive belt is made up of many small links held together on a metal band. As the drive pulleys rotate, the metal links are forced into compression, causing the belt to push rather than pull.

Because these pulleys do not rely on fixed gear sizes, a stepless gear ratio can be achieved that maintains optimum efficiency for any engine speed or load. The output from the drive pulleys is normally transmitted through a further reduction gear, which can be of epicyclic design. This allows a reverse gear to also be included.

Toroidal CVT

A toroidal CVT has a tapered input disc and output disc, which are placed face to face to form a **concave** driving surface. The input and output discs are able to turn independently of one another and are connected using **torus**-shaped rollers. The rollers are able to ride up and down against the concave surface of the input and output drive discs and transfer turning effort between the two.

- When the roller is touching a low point on the input disc curve and a high point on the output disc curve, a low gear ratio is achieved.
- When the roller is touching a high point on the input disc curve and a low point on the output disc curve, a high gear ratio is achieved.

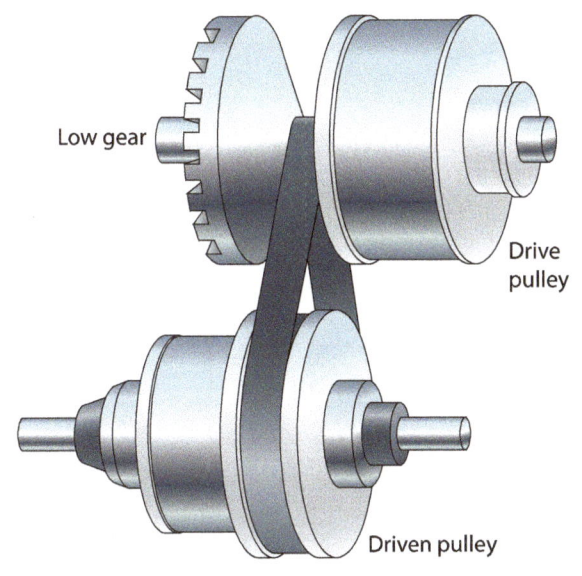

Figure 5.31 CVT low gear range

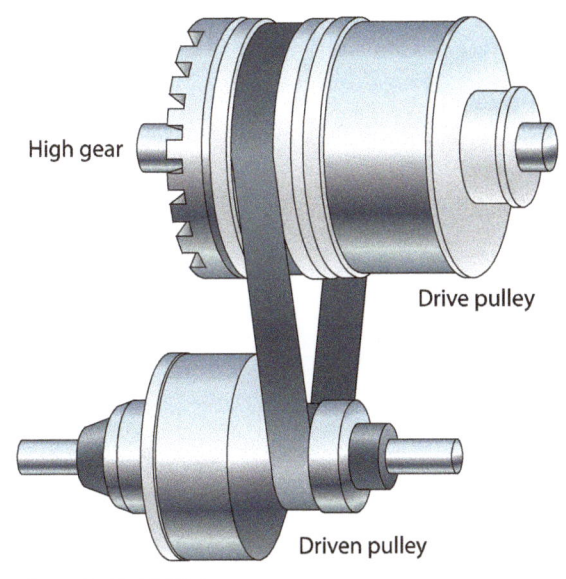

Figure 5.32 CVT high gear range

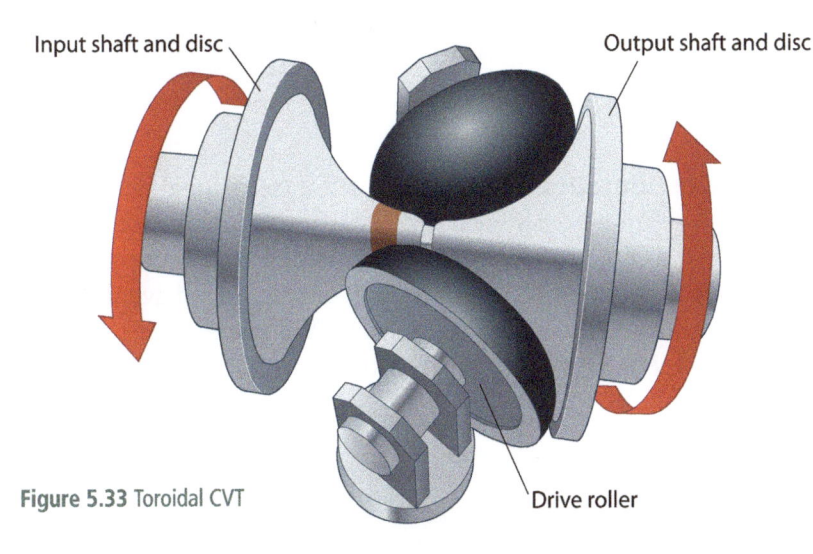

Figure 5.33 Toroidal CVT

Key terms

Concave – curved inwards.

Torus – ring-shaped like a doughnut.

Level 3 Light Vehicle Technology

By moving the rollers across the surfaces of the input and output discs, a continuously variable transmission (CVT) ratio can be achieved. The output from the drive disc is normally transmitted through a further reduction gear, which can be of epicyclic design. This allows a reverse gear to also be included.

Action
Research manufacturers that offer CVT as a gearbox option when purchasing a new vehicle. List three makes and models.

NEW TECH
Continuously variable transmission

Because a continuously variable transmission is able to transmit turning effort to the road wheels in a very efficient manner while keeping the engine at its optimum speed, this design has become very popular with manufacturers of hybrid vehicles.

Drive connection

To join the engine to a continuously variable transmission system and allow a smooth take-up of drive and a temporary position of neutral, some form of clutch or torque converter is required. Three main methods are used to achieve this:

- **Torque converter:** If a CVT is connected to the engine via a standard torque converter, some efficiency can be lost due to the drag created by the automatic transmission fluid and its overall weight.

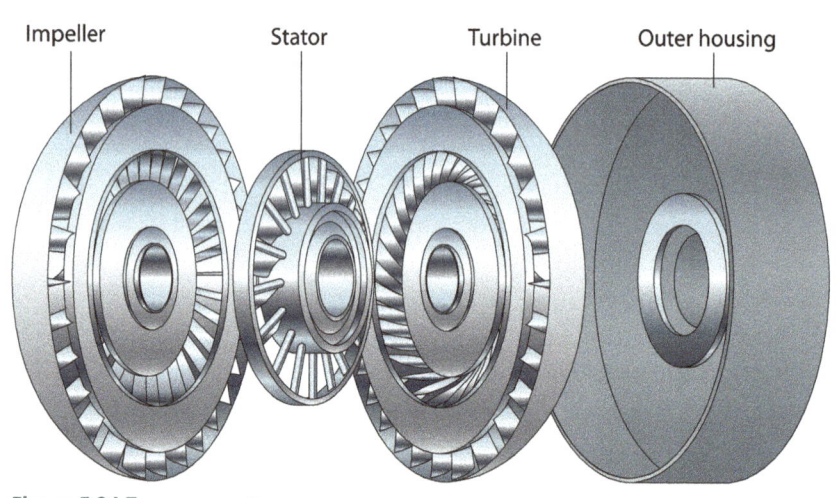

Figure 5.34 Torque converter

- **Centrifugal clutch:** Early systems used a centrifugal clutch in which a set of clutch shoes, similar in construction to brake shoes, were rotated by the engine. Centrifugal force acting on the shoes moved them outwards until they contacted a drive drum. This drive was then transmitted to the CVT system. This was also a very inefficient drive connection between the engine and the gearbox.

Figure 5.35 Centrifugal clutch

- **Electromagnetic clutch:** In this system, a housing of similar size and shape to a standard clutch is mounted on the end of the engine crankshaft. The housing contains a metallic powder, which when energised by an electromagnet, bonds together to provide drive to the CVT. The electromagnet is managed by a transmission ECU, which is able to vary the strength of the magnetic field and therefore control the take-up of drive.

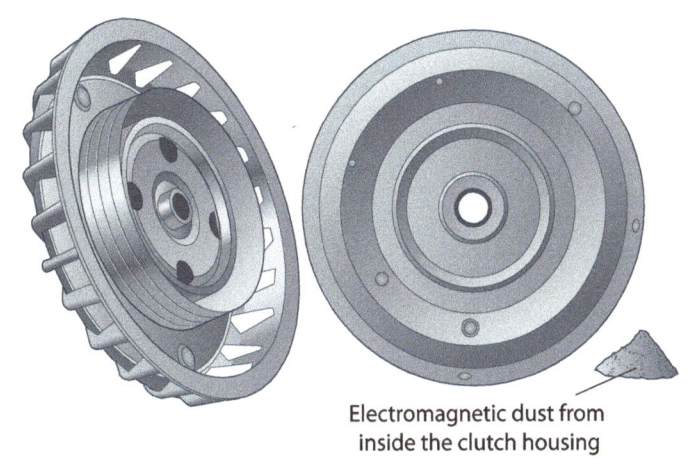

Electromagnetic dust from inside the clutch housing

Figure 5.36 Electromagnetic clutch

CHECK YOUR PROGRESS

1 Name two types of compound epicyclic gear train.
2 Name two types of CVT.
3 Name the 'S'-shaped component in a unidirectional clutch.

Light vehicle driveline components

Once the gearbox has multiplied the torque from the engine, the turning effort must be transferred to the road wheels. Depending on transmission layout (front-wheel drive, rear-wheel drive or four-wheel drive), drive shafts or propeller shafts are used.

Propeller shaft

The propeller shaft is usually called the prop shaft. It is used on front engine rear-wheel drive vehicles and four-wheel drive vehicles. It is simply a metal tube, strong enough to transmit the full power of the engine and the torque multiplied by the gearbox. The prop shaft is connected to the back of the gearbox and runs underneath the floor to join it to the back axle, as shown in Figure 5.37 overleaf.

Universal joints

At each end of the prop shaft, a **universal joint (UJ)** is needed. As the suspension moves up and down, a difference in height occurs

Key term

Universal joint (UJ) – a mechanism that allows drive to be transmitted through an angle.

Level 3 Light Vehicle Technology

Figure 5.37 Vehicle driveline

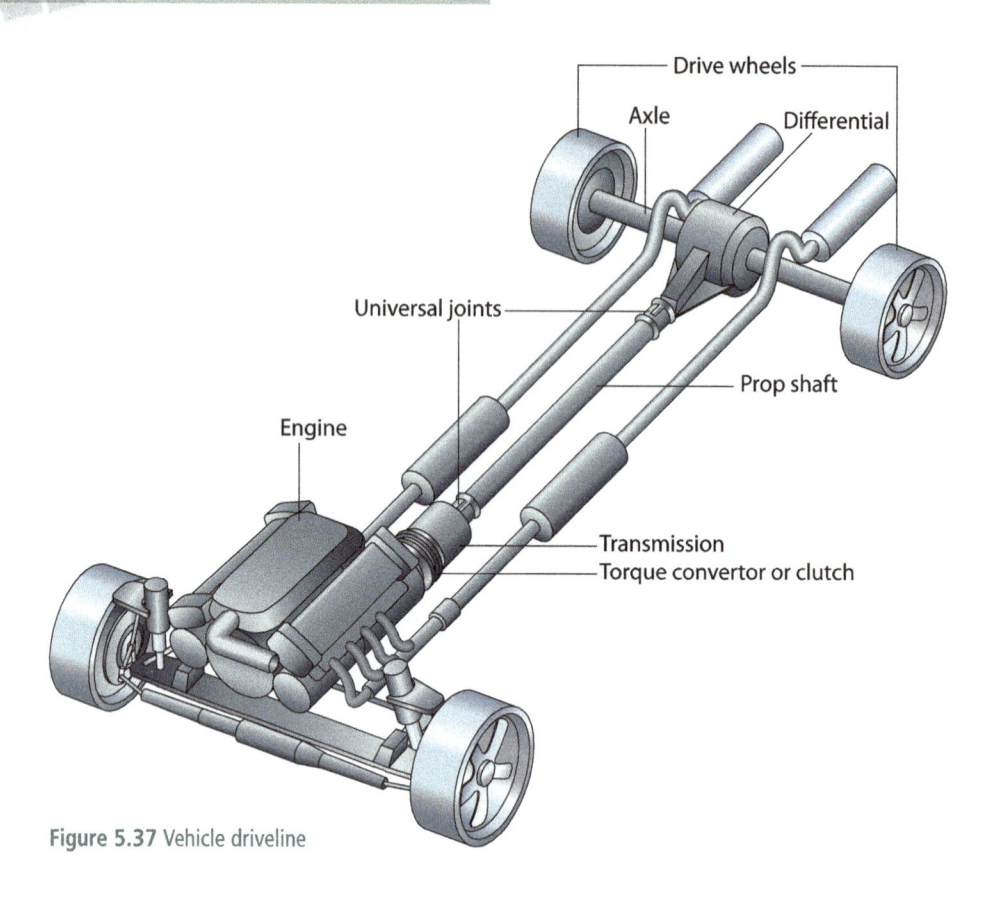

between the rear axle and the gearbox. Using universal joints allows the drive to be transmitted without bending the prop shaft. (Some prop shafts are split in two with an additional universal joint mounted in the middle.)

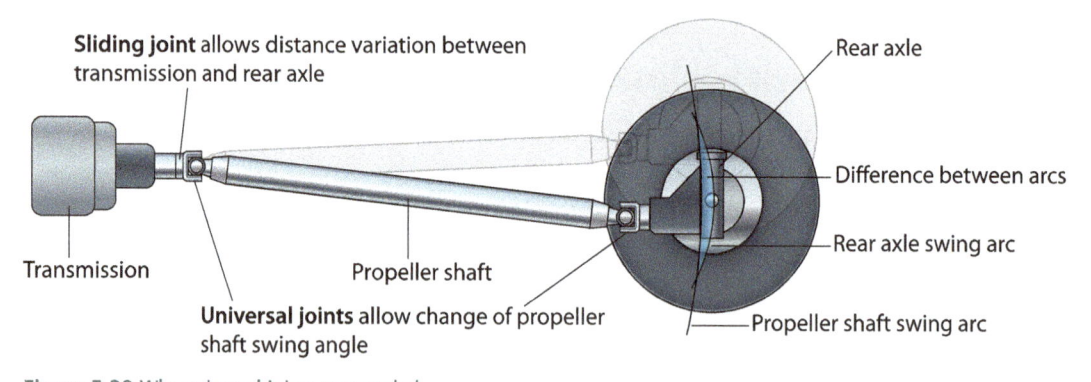

Figure 5.38 Why universal joints are needed

The most common type of universal joint is Hooke's UJ (see Figure 5.39). This is made up of two **yokes** pivoted on a central crosspiece, sometimes called a spider. The spider is formed by two pins crossing over each other at right angles. The yokes, one on the input shaft and the other on the output shaft, are connected to the spider so they are at right angles to each other. This arrangement allows the input and output shafts to rotate together even when their axes are at different angles.

> **Key term**
>
> **Yoke** – a connection between two components so that they move together.

284

5 Diagnosis & rectification of light vehicle transmission and driveline faults

As a universal joint turns through an angle, it speeds up and slows down. Universal joints need to be synchronised because they speed up and slow down as they travel through 90 degrees of revolution. The wave form shown in Figure 5.40 represents the speeding up and slowing down, which cancel each other out. When the two universal joints work together like this it is known as synthesis.

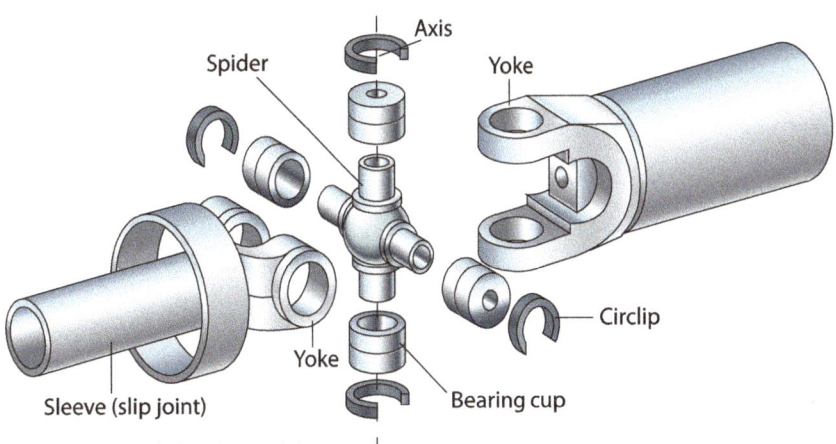

Figure 5.39 Hooke's universal joint

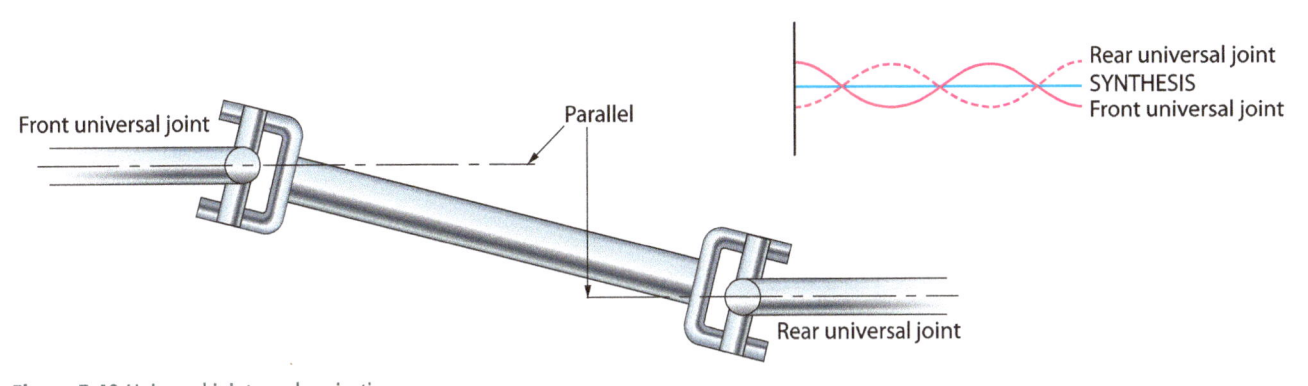

Figure 5.40 Universal joint synchronisation

Sliding joint

As the rear suspension moves up and down, this movement tries to stretch or compress the prop shaft. Because of this, a sliding joint is included to allow the prop shaft to get longer and shorter.

Drive shafts

Drive shafts can be found on both front and rear axles, depending on the drive layout type. Two shafts are placed across the width of the car and connect to the **hub** of the driving wheels. As with a prop shaft, these shafts must be able to cope with suspension movement and, in the case of front-wheel drive, steering movement as well.

CV joint

In a front-wheel drive layout, steering and suspension movement cause large angular movements. A universal joint would be unsuitable for use with front-wheel drive shafts, as the speeding up and slowing down of the UJ would cause transmission vibrations.

Did you know?

As they operate, very long prop shafts have a tendency to flex in the middle and cause a whipping action. To avoid this problem prop shafts are split into two sections and supported in the middle by a centre bearing. By having two shorter prop shafts, the amount of vibration caused by the rotating action can be reduced.

Key term

Hub – an assembly at the outer end of the suspension that carries the wheel, tyre and brake components.

285

Level 3 Light Vehicle Technology

> **Key term**
>
> **Constant velocity (CV) joint** – the joint at the end of the drive shaft on a front-wheel drive vehicle that is able to transmit drive with no variation in speed.

Instead of universal joints, a special type of coupling called a **constant velocity (CV) joint** is used. Different manufacturers use a number of different designs and styles of CV joint, but their job remains the same – to transmit drive at a constant speed regardless of steering or suspension movement and angle.

The most common CV joint is the Birfield joint, which is based on a design patented in 1935 by the Ford engineer Alfred Hans Rzeppa. It transmits drive through a series of ball bearings housed in a cage (see Figure 5.41). Because ball bearings are spheres, it doesn't matter at what angle or at which point the drive is transmitted through these balls – speed (velocity) is kept constant.

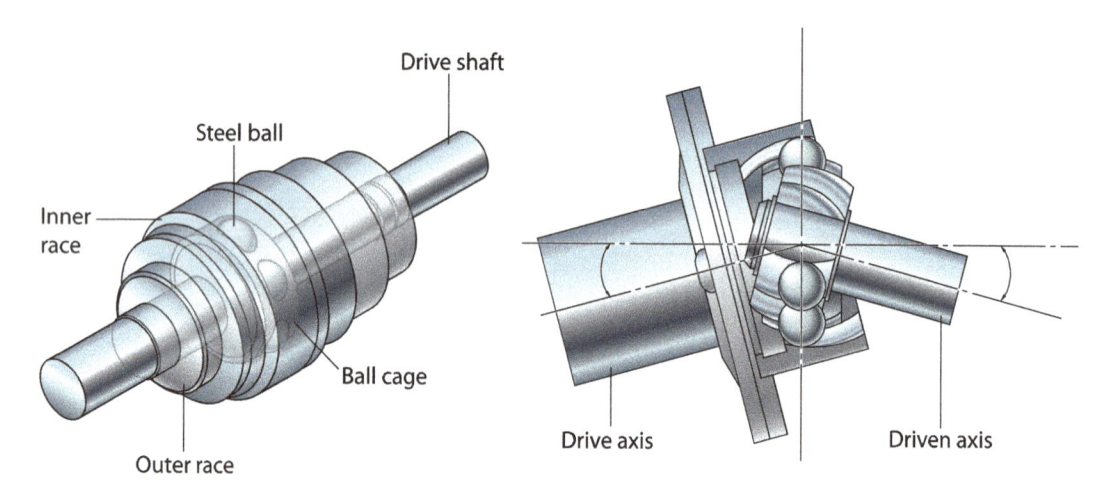

Figure 5.41 Constant velocity (CV) joint

> **Did you know?**
>
> A number of different designs of CV joint are available. Manufacturers tend to produce their own design to overcome copyright issues.

CV joints are prone to wear because of the forces they undergo. They must be kept well lubricated with grease; a rubber boot is used to keep the grease in and water and dirt out. These rubber boots are often called 'drive shaft gaiters' or 'CV gaiters'.

You can often assess wear in CV joints by driving the car in a tight circle. If you can hear a clicking or knocking noise from the area of the CV joint, it may need to be replaced.

Final drive

Having completed its journey through the gearbox, the torque is now transmitted to the **final drive** unit. This represents the last stage in the transmission of power from the engine to the road wheels.

> **Did you know?**
>
> If the final drive and gearbox are incorporated into a single unit, this is often called a **transaxle**.

A final drive unit usually consists of two gears called the crown wheel and pinion. This provides a fixed final gear ratio to increase torque. With a rear-wheel drive, it will also turn the rotation from the prop shaft through right angles to drive the wheels.

The final drive also includes a **differential** unit, which allows one wheel to travel faster than the other when turning a corner (see next page).

5 Diagnosis & rectification of light vehicle transmission and driveline faults

Final drive gear ratios

As with the gears in a gearbox, the final drive reduction depends on the number of teeth on the crown wheel and pinion. A typical figure for final gear ratio is approximately 4:1. It is calculated by using the formula:

$$\text{Ratio} = \frac{\text{Driven}}{\text{Driver}}$$

Once you have worked out the final gear ratio, you need to multiply this figure by the total gearbox ratio to represent the vehicle's true overall gear ratio.

If the crown wheel and pinion are used on a front wheel drive, they are normally standard helical gears.

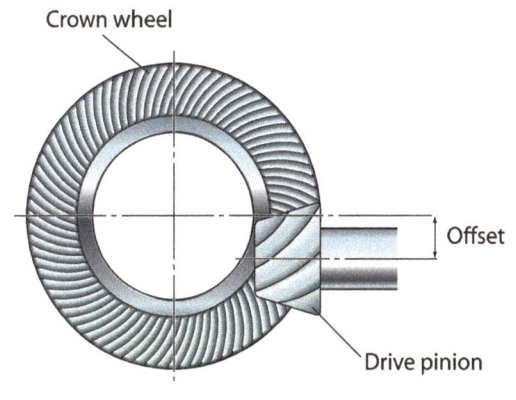

If the crown wheel and pinion are used on a rear-wheel drive, a special type of gearing called a **bevel** gear is used. Most rear-wheel drive axles combine this bevel gear with a spiral **hypoid** design, as shown in Figure 5.42.

In a hypoid design, the gear teeth are curved in such a way that the pinion can be positioned below the centre line of the crown wheel. This means that the prop shaft can be set lower down. Because of this, the transmission tunnel in the floor that houses the prop shaft can be made lower or may not be used at all. The resulting flatter floor area inside the car creates more space for passengers, lowers the centre of gravity of the vehicle and makes it more stable on the road.

Figure 5.42 Hypoid rear-wheel drive crown wheel and pinion

Did you know?

The fixed final gear reduction reduces the output speed of the gearbox. This is because, even when a car is travelling at 70 miles an hour, the road wheels are only turning between 700 and 1200 rpm (depending on their size). Compare this with the speed of the engine crankshaft, which may be doing around 4000 rpm.

Differentials

When a car is travelling along a straight stretch of road, both of the driven wheels cover the same amount of ground at the same speed. When the car takes a bend, the inner driven wheel doesn't have to travel as far as the outer one, so it needs to travel more slowly. If both wheels rotated at the same speed when trying to turn the corner, the inner wheel would be forced into a skid.

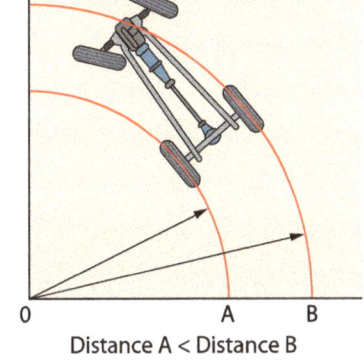

On a bend, drive must be transmitted at varying speeds, and this is done by using a differential unit housed inside the final drive casing. When needed, some of the driving force from the inner wheel is transferred to the outer wheel. This speeds up the outer wheel and slows down the inner wheel.

A differential unit allows this because of its internal gearing system. The turning effort taken from the crown wheel is transmitted to the differential casing, where a metal drive pin is fixed. As the differential casing turns, the drive pin moves end over end. On the drive pin are mounted two small gears, often called 'planet gears', which are in constant mesh with two side gears, often called 'sun gears' (see Figures 5.44 and 5.45).

Distance A < Distance B

RPM of inside wheel < RPM of outside wheel

Figure 5.43 The need for a differential

Key terms

Final drive – a transmission unit containing the differential, crown wheel and pinion.

Transaxle – a unit in which the gearbox and final drive are contained within one casing.

Differential – a mechanism that allows one driven wheel to travel faster than the other when the car goes around a bend.

Bevel – gears that mesh at an angle.

Hypoid – describes the pinion mounted in an offset position against the crown wheel in a final drive unit (usually set below the centre line).

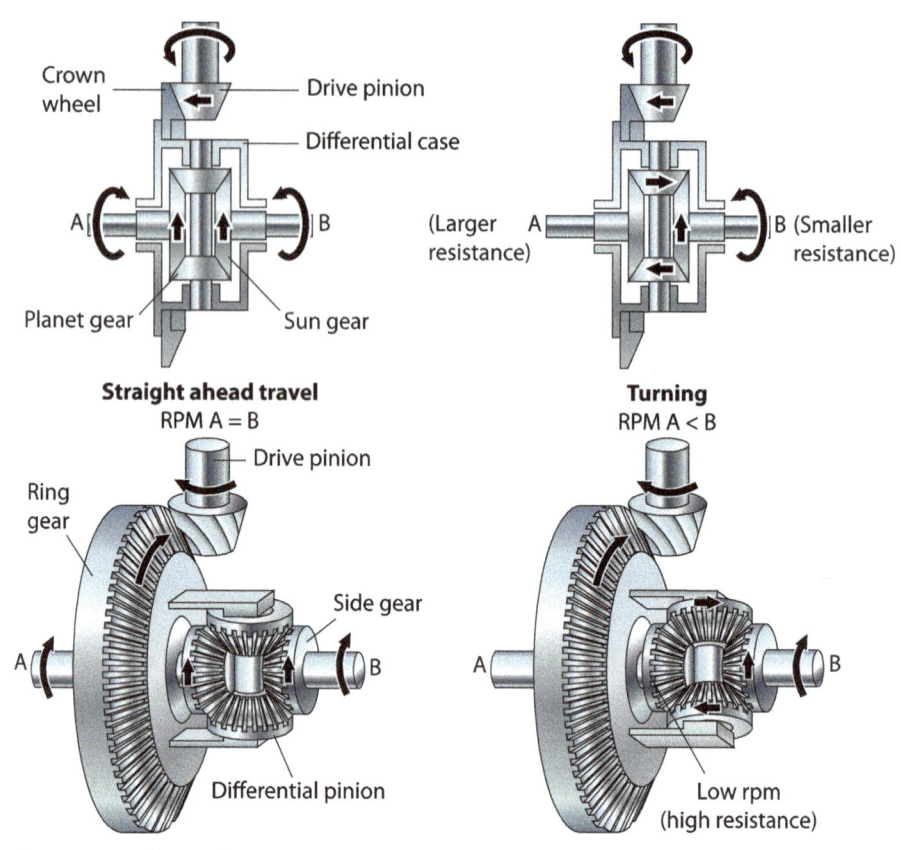

Figure 5.44 Differential operation when travelling straight

Figure 5.45 Differential operation when cornering

Driving in a straight line

When the car is travelling in a straight ahead direction, the drive pin turns end over end. This locks the planet gears directly to the side gears, driving them all at the same speed, as shown in Figure 5.46.

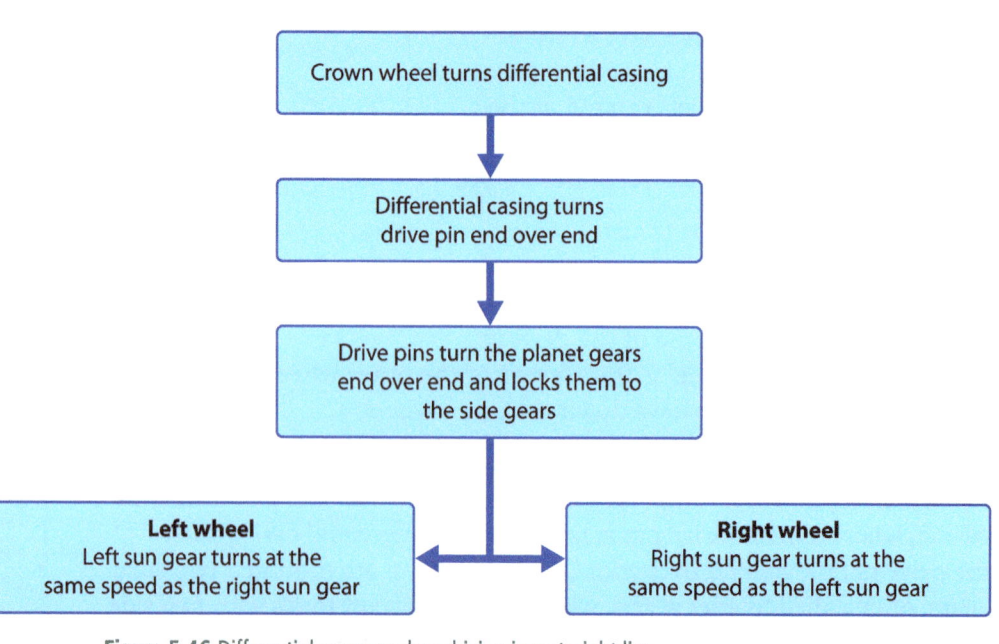

Figure 5.46 Differential process when driving in a straight line

Turning a corner

As the vehicle turns a corner, the extra load tries to slow down one wheel and reduces the speed at one sun gear. The drive pin still turns end over end, providing torque or turning effort to the sun gears, but the planet gears now rotate on the pin, allowing more drive to be transmitted to one wheel than the other. This allows one wheel to travel faster than the other while continuing to transmit drive with the same torque.

Figure 5.47 shows the process involved when turning a left-hand bend.

```
                Crown wheel turns differential casing
                              │
                              ▼
                Differential casing turns
                drive pin end over end
                              │
                              ▼
                Drive pins turn the planet gears
                end over end but the planet gears
                rotate against the side gears
                              │
              ┌───────────────┴───────────────┐
              ▼                               ▼
         Left wheel                      Right wheel
   Left sun gear turns at a         Right sun gear turns at a
 slower speed than the right sun gear   faster speed than the left sun gear
```

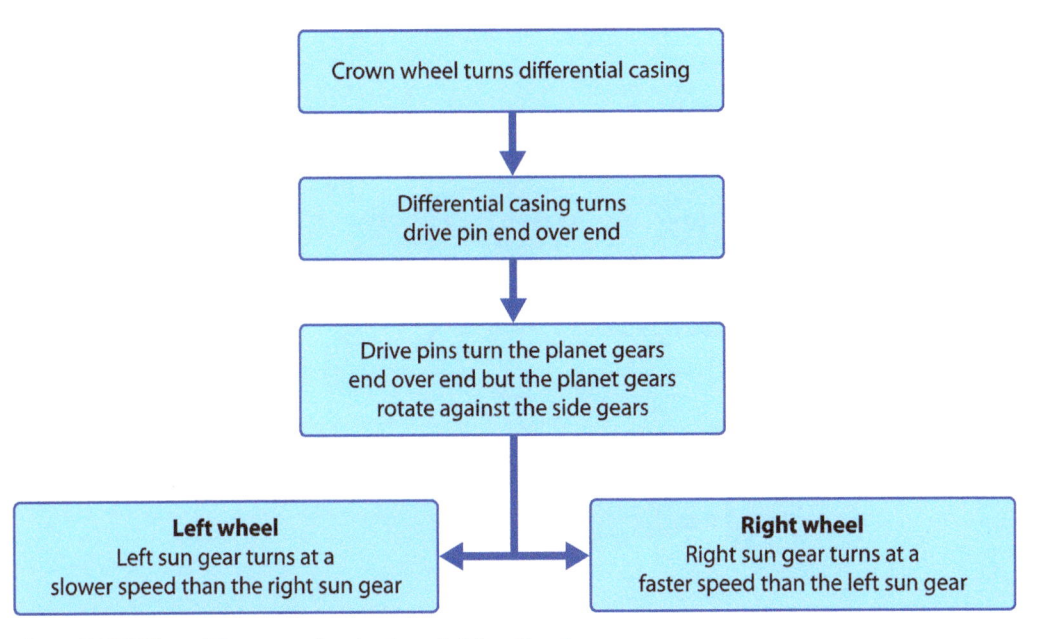

Figure 5.47 Differential process when turning a left-hand bend

Limited slip differentials (LSD)

A disadvantage of a standard differential is that if one wheel loses traction, nearly all of the turning effort will be directed away from the wheel with grip. This means that one wheel will spin uncontrollably and no drive will be transmitted.

A limited slip differential (LSD) is designed to transmit an equal torque to both driving wheels when the car is travelling in a straight ahead direction while continuing to allow standard differential action when going round a corner. Because of this, if one wheel loses traction, the other will still have some drive transmitted, and this will give greater vehicle control.

Three main methods of limited slip differential are in common use:

- clutch type LSD
- viscous coupling type LSD
- torsion wheel type LSD.

Clutch operation

A series of multi-plate clutches are included in the design of the clutch type limited differential unit. Four planet gears are mounted on two drive pins. The ends of the drive pins are **tapered** and sit in corresponding tapered grooves in the differential casing.

When travelling in a straight ahead direction, forces acting on the planet gears cause the drive pins to move up and outwards in the tapered grooves of the casing. This movement creates pressure on the clutches, clamping them together to create friction and transmit drive to both road wheels.

As the vehicle turns a bend, the load on the clutches is reduced as the drive pins move back down into the grooves. Pressure is reduced on the clutches, allowing them to slip; this lets normal differential action take place.

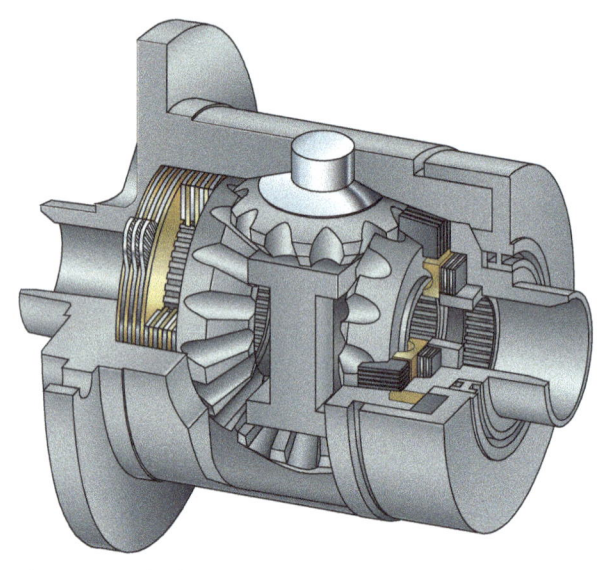

Figure 5.48 Clutch type LSD

Figure 5.49 shows how a clutch type differential operates in a straight ahead direction.

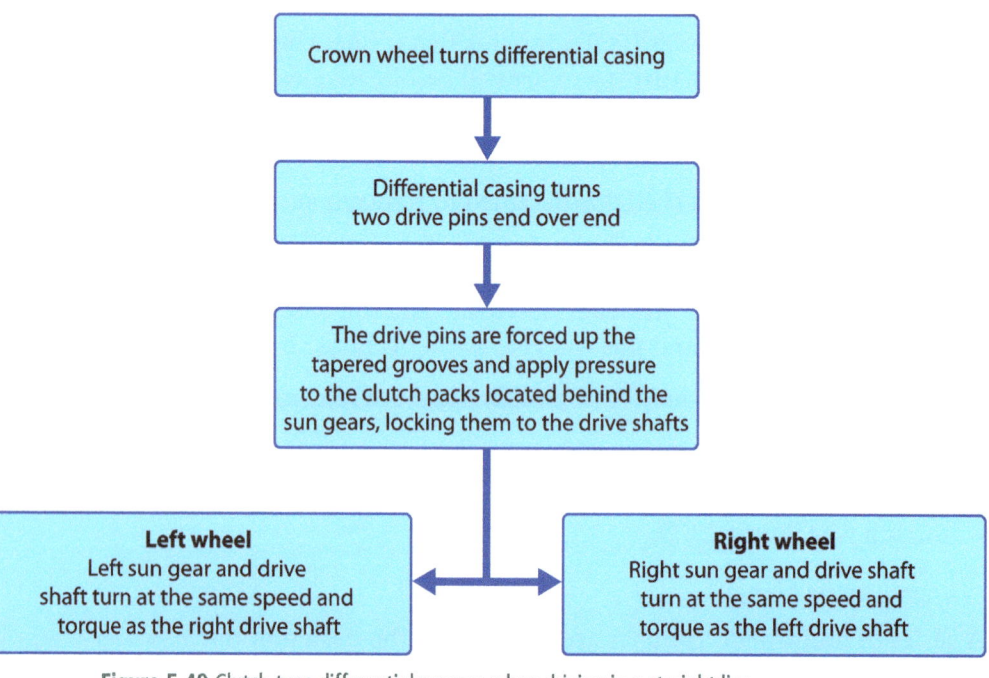

Figure 5.49 Clutch type differential process when driving in a straight line

Viscous coupling operation

A chamber inside the differential contains a **viscous** (thick) liquid and a series of rotor blades. When travelling in a straight ahead direction, the viscous liquid creates **drag**. This reduces slip in the differential gears and transmits an even torque to both drive shafts. As the car turns a bend, normal differential action can take place because the rotor blades are able to **shear** through the viscous liquid, creating an acceptable amount of slip.

> **Key terms**
>
> **Tapered** – angled inwards slightly.
>
> **Viscous** – describes a liquid with a high resistance to flow (a thick liquid).
>
> **Drag** – force on an object that resists its movement through a fluid.
>
> **Shear** – to slice through something with a similar action to scissors.

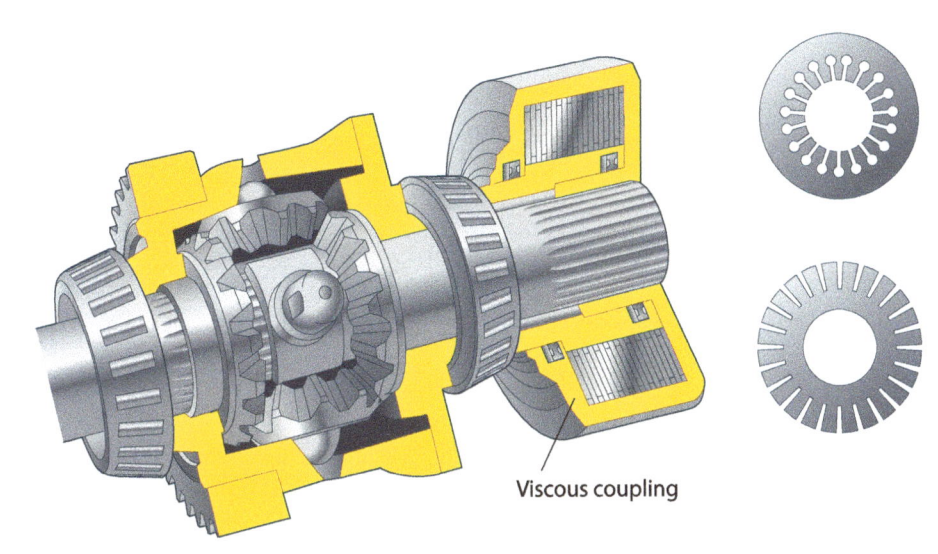

Figure 5.50 Viscous coupling type LSD

Figure 5.51 shows how a viscous coupling type differential operates in a straight ahead direction.

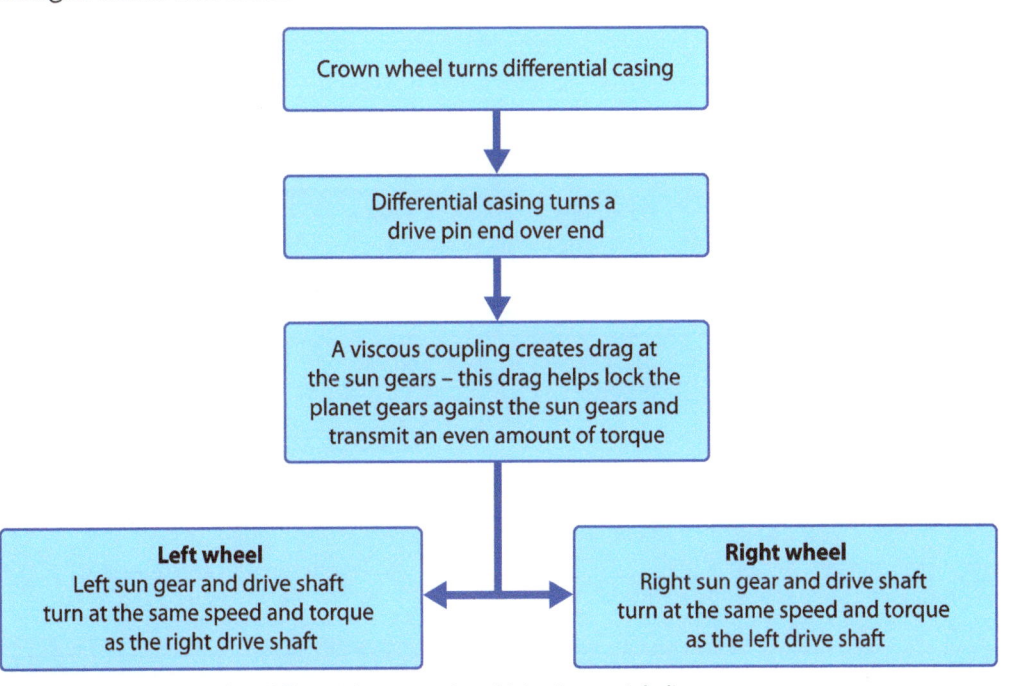

```
Crown wheel turns differential casing
                │
                ▼
Differential casing turns a
drive pin end over end
                │
                ▼
A viscous coupling creates drag at
the sun gears – this drag helps lock the
planet gears against the sun gears and
transmit an even amount of torque
                │
        ┌───────┴───────┐
        ▼               ▼
   Left wheel      Right wheel
```

Left wheel – Left sun gear and drive shaft turn at the same speed and torque as the right drive shaft

Right wheel – Right sun gear and drive shaft turn at the same speed and torque as the left drive shaft

Figure 5.51 Viscous coupling differential process when driving in a straight line

Level 3 Light Vehicle Technology

Key term

Worm gear – a gear consisting of a spiral threaded shaft and a wheel with teeth that mesh into it.

Torsion wheel operation

A series of **worm gears** are included in the design of this differential unit. In a worm and wheel gear set up, the worm can easily drive the gear, but the gear is unable to drive the worm (this allows drive in one direction only). When travelling in a straight ahead direction, gear teeth lock against the worm drive, providing equal torque to both wheels at the same time. As the car turns a bend, one wheel slows down while the other speeds up. When this happens, the worm turns the gear, allowing one wheel to travel faster than the other and provide normal differential action.

Safe working

It is advisable not to test the brakes of a vehicle with a limited slip differential in a brake rolling road. If you are operating or testing wheels individually, you may obtain inaccurate readings or damage the LSD.

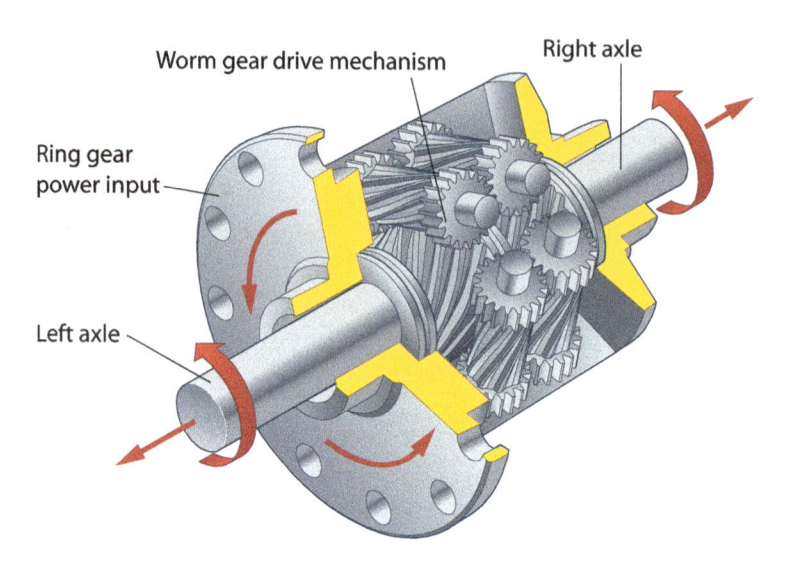

Figure 5.52 Torsion wheel type LSD

Figure 5.53 shows how a torsion wheel type differential operates in a straight ahead direction.

Action

Using research sources available to you, name a vehicle (make and model) that uses the following types of LSD:
- clutch type
- viscous coupling
- torsion wheel.

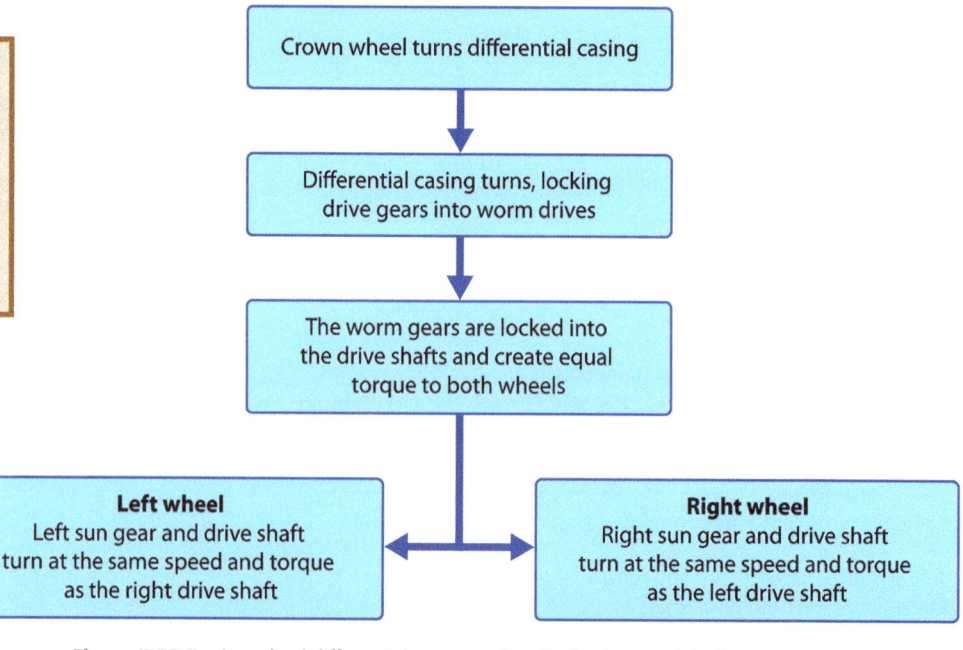

Crown wheel turns differential casing
↓
Differential casing turns, locking drive gears into worm drives
↓
The worm gears are locked into the drive shafts and create equal torque to both wheels
↓
Left wheel
Left sun gear and drive shaft turn at the same speed and torque as the right drive shaft

Right wheel
Right sun gear and drive shaft turn at the same speed and torque as the left drive shaft

Figure 5.53 Torsion wheel differential process when driving in a straight line

Automatic brake differential (ABD)

An automatic brake differential (ABD) is a method of creating the operating characteristics of a limited slip differential, but with the ability to add electronic control. A series of multi-plate clutches are mounted in a similar position to those found in a clutch type limited slip differential. When engaged, these clutches are able to transmit driving torque to the wheel that has the greatest amount of grip.

Pressure to activate the clutches in an ABD system is normally created by an electromagnet. The current flowing in the electromagnet can be varied by an ECU, to manage the amount of slip needed at the clutches and provide the best delivery of torque to the most appropriate wheel.

An advantage of this type of system, when compared to a standard limited slip differential, is that it can work in harmony with other active safety systems to provide dynamic vehicle control.

Other systems that ABD is able to interact with include:

- anti-lock braking systems (ABS)
- traction control systems (TCS)
- electronic stability programs (ESP)
- electronic brake force distribution (EBD)
- active yaw control (AYC).

> **Did you know?**
>
> It is sometimes possible to determine whether a car has a limited slip differential by raising the driven wheels off the ground and placing the transmission in gear. If one of the driven wheels is turned in a particular direction and the opposite wheel rotates in the reverse direction, a standard differential has been used. If both wheels rotate in the same direction, a limited slip differential has been used.

Final drive lubrication

Lubrication of final drive systems is usually achieved using splash feed. To stop the oil being squashed out from between the gear teeth, you need to use a special type of oil. These specialist transmission and final drive oils are often known as extreme pressure (EP) oils and they usually have a high viscosity grade.

Final drive in four-wheel drive vehicles

Transfer box

In a four-wheel drive vehicle, the engine can be mounted transversely (sideways) or longitudinally (in a straight line) depending on the manufacturer's design. As drive exits the gearbox, an extra unit called a **transfer box** is often used (see Figure 5.54 overleaf). This splits the drive so that it can be used by the front and rear axles. A four-wheel drive vehicle has at least two differential assemblies, one for the front wheels and one for the back. In addition, vehicles with permanent four-wheel drive often have a central differential that splits the drive from front to rear.

When a vehicle is travelling with forward motion, the front axle reaches a bend first. Because of this, the front axle needs to be travelling at a different speed from the rear axle. A central differential is sometimes used. This operates in the same way as a standard differential, but instead of allowing a difference in speed between the right and left wheels, it allows a difference in speed between the front and rear axles.

> **Key term**
>
> **Transfer box** – a mechanism used on four-wheel drive vehicles to split the drive between the front and rear axles.

Level 3 Light Vehicle Technology

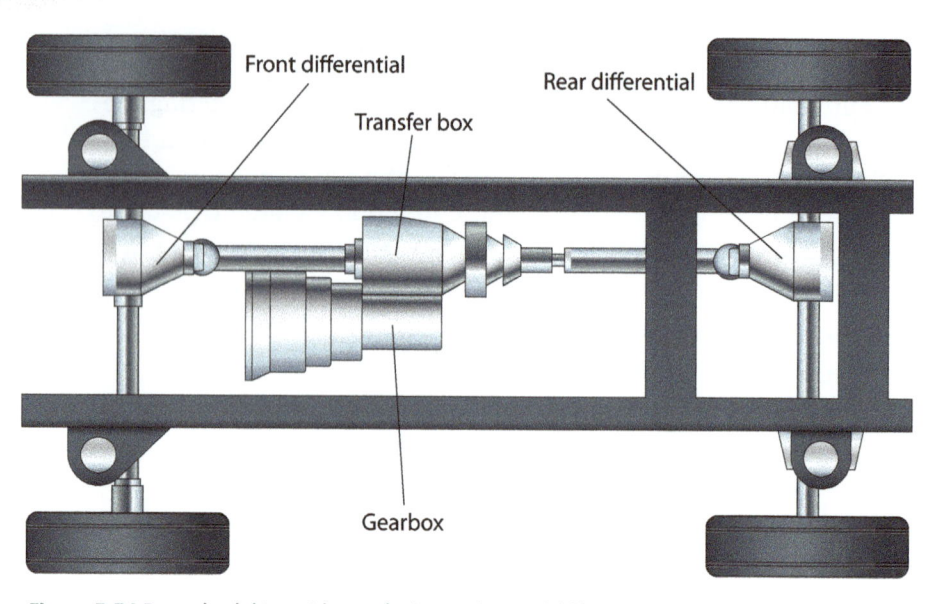

Figure 5.54 Four-wheel drive with transfer box and central differential

Haldex coupling

Some manufacturers do not use a central differential in the design of their four-wheel drive vehicles. A method is still needed to control the distribution of torque between the front and rear axles. This can be achieved using a Haldex coupling.

Did you know?

If a vehicle is to be used off road, some manufacturers include a mechanical locking mechanism in the design of their differential. This is a means of eliminating differential action on loose and slippery surfaces and improves overall traction.

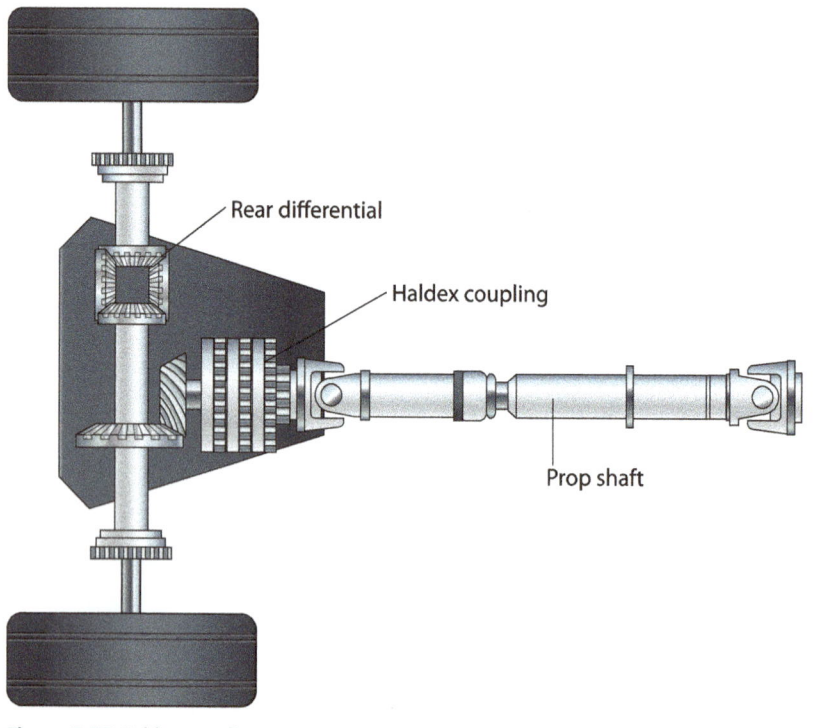

Figure 5.55 Haldex coupling

Normally mounted on or near the rear axle, the Haldex coupling is a multi-plate clutch unit that is able to control the amount of torque delivered to the rear axle by managing slip or drag created at the clutch plates. Pressure to engage the clutches relies on hydraulic forces created by the Haldex coupling's own internal oil pump. During vehicle operation, fluid pressure is controlled by a series of electronic valves activated by an ECU that processes dynamic information from various chassis and drive sensors. This means that the vehicle can instantly respond to changes in road surface grip, giving the best amount of traction for different driving situations.

> **Safe working**
>
> A differential lock should not be used when driving on a normal road surface. Grip created at the road wheels can cause transmission 'wind-up' as a car goes round a bend. A reverse torque is created in the transmission parts, which places undesirable stress on shafts and gears and may cause components to break.

CHECK YOUR PROGRESS

1 What does the term hypoid mean?
2 In four-wheel drive vehicles, why is a central differential required?
3 What is transmission wind-up?

Symptoms and faults in light vehicle transmissions and driveline systems

The symptoms caused by transmission system faults present themselves under different driving and load conditions. When diagnosing transmission-related faults, make sure you follow a systematic routine to avoid wasted time, frustration and misdiagnosis. Your diagnostic routine should include the following:

1. Fully question the driver about the symptoms relating to the fault. The driver will know their vehicle better than anybody else, so allow them to explain the problems.

2. Following an initial visual inspection to ensure that the vehicle is safe to drive, fully road test the vehicle to confirm the symptoms. It is sometimes advisable to take the driver with you on a road test so they can point out problems as the vehicle is driven, or answer questions relating to issues created during the investigation.

3. Once the symptoms have been confirmed, gather information relating to the vehicle and transmission system, including:
   - vehicle details
   - service history
   - warranty period
   - manufacturer recall information
   - system information and technical data
   - diagnostic trouble codes (if applicable)
   - technical advice (from manufacturers, helplines or master technicians).

4. Examine the systems and undertake as much diagnosis as possible before you dismantle components, so that you do not restrict the tests available to you. Check:

    - mechanical systems and components
    - hydraulic and pneumatic systems and components
    - electrical systems and components.

5. Strip down and repair, replace or adjust components as required.

6. Reassemble systems and check for correct function and operation.

7. Road test and evaluate the repairs for safety and to ensure that the problems have been cured and the repairs have not affected any other system. This should include resetting any electronic adaptions (see Chapter 1, page 26) and diagnostic trouble codes.

8. Finally, it is important that you ensure the customer is satisfied with the repairs that have been done.

Transmission faults and symptoms can be wide and varied. Tables 5.6 to 5.10 give some examples of symptoms, faults, possible causes and recommendations for repair.

Table 5.6 Clutch and coupling faults

Symptom	Fault	Possible cause	Recommendation for repair
Abnormal noises	Rattle from dual mass flywheel	Worn damping springs in dual mass flywheel	Replace flywheel
Vibrations	Crankshaft pulsations transferred to transmission	Seized dual mass flywheel	Replace flywheel
Fluid leaks	Internal slave cylinder leaking fluid onto friction material	Seal failure in slave cylinder	Replace slave cylinder Replace friction plate if required
Slip	Poor transmission of torque on acceleration	Worn clutch friction material	Replace clutch
Judder	Vibration as clutch is engaged	Worn clutch pressure plate springs	Replace clutch
Grab	Sudden snatch as clutch is engaged (automatic clutch system)	Incorrect information from clutch position sensor	Re-program clutch position with scan tool/specialist equipment
Failure to release (drag)	Car creeps forward in neutral	Too much free play on clutch cable	Adjust clutch cable

5 Diagnosis & rectification of light vehicle transmission and driveline faults

Table 5.7 Manual gearbox faults

Symptom	Fault	Possible cause	Recommendation for repair
Abnormal noises	Rumbling noise as the vehicle is driven	Worn main shaft support bearings	Overhaul gearbox bearings
Vibrations	Vibration when driving in third gear	Broken gear teeth on third driven gear	Replace third gear
Loss of drive	Car will not move, gearbox jammed	Interlock failure, allowing two gears to be engaged at the same time	Strip out gearbox and disengage jammed gears. Replace interlock mechanism
Difficulty engaging or disengaging gears	Even-numbered gears unavailable (dual clutch transmission)	Hydraulic pressure to one clutch lost due to failed control solenoid valve	Test and replace solenoid valve

Table 5.8 Automatic gearbox faults

Symptom	Fault	Possible cause	Recommendation for repair
Abnormal noises	Rumbling noise as the vehicle is driven	Worn planet gear support bearings	Replace epicyclic gearing
Vibrations	Vibration created on overrun/deceleration	Damaged sprag in one-way clutch	Replace one-way clutch unit
Loss of drive	Car will not move when placed in gear	Low automatic transmission fluid level	Check for repair leaks. Top up ATF and test
Failure to engage gear	Gearbox will not change up properly	Worn brake band	Adjust brake bands and test
Failure to disengage gear	Gearbox stays in third gear	Clutch pack 2 hydraulic pressure not releasing	Check for blockage and solenoid operation
Leaks	ATF dripping from the bottom of the gearbox	Gearbox sump pan gasket failure	Replace sump gasket
Failure to operate	No drive when the gear selector is moved	Shift position sensor failure	Test sensors and circuit. Replace faulty component
Incorrect shift patterns	Gear change happening too late	Incorrect signal from vehicle speed sensor	Test sensors and circuit. Replace faulty component
Electrical and electronic faults	Incorrect gearbox operation	CAN bus network failure	Scan for fault codes, isolate and replace damaged component

Table 5.9 Final drive faults

Symptom	Fault	Possible cause	Recommendation for repair
Abnormal noises	Whine on overrun/deceleration	Incorrect tooth contact between crown wheel and pinion	Adjust final drive
Vibrations	Seized LSD clutch pack	Differential action ineffective, making road wheels slip when cornering	Overhaul limited slip differential
Loss of drive/failure to operate	Car will not move	Broken planet gear drive pin	Overhaul differential unit
Oil leaks	Oil dripping from final drive unit	Pinion oil seal leaking	Replace pinion gear oil seal
Electrical and electronic faults	No differential control (ABD)	Electromagnet high resistance	Conduct a volt drop test and repair resistance

Table 5.10 Driveline and coupling faults

Symptom	Fault	Possible cause	Recommendation for repair
Abnormal noises	Clicking noise when cornering	Worn CV joint	Replace CV joint
Vibrations	Vibration when moving at speeds above 20 mph	Bent drive shaft	Measure drive shaft run-out and replace
Loss of drive	Car will not move	Broken prop shaft universal joint	Replace prop shaft universal joint

BEFORE YOU FINISH

Recording information and making suitable recommendations

At all stages of a diagnostic routine, maintenance or repair, you should record information and make suitable recommendations. The table below gives examples of how to do this.

Stage	Information	Recommendations
Before you start	Record customer/vehicle details on the job card. Make a note of the customer's repair request and any issues/symptoms. Locate any service or repair history.	Advise the customer how long you will require the car. Describe any legal, environmental or warranty requirements.
During diagnosis and repair	Carry out diagnostic checks and record the results on the job card or as a printout from specialist equipment. List the parts required to conduct a repair. Note down any other non-critical faults found during your diagnosis.	Inform your supervisor of the required repair procedures so that they can contact the customer and gain authorisation for the work to be conducted.
When the task is complete	Write a brief description of the work undertaken. Record your time spent and the parts used during the diagnosis and repair on the job card. (This information should be as comprehensive as possible because it will be used to produce the customer's invoice.) Complete any service history as required.	Inform the customer if the vehicle will need to be returned for any further work. Advise the customer of any other issues you noticed during the repair.

Remember, any work you carry out on a customer's car should be assessed to ensure that it is conducted in the most cost-efficient manner. You should consider reconditioning, repair and replacement of components within units.

FINAL CHECK

1. A CBW system operates using:
   a cables
   b rods
   c electronics
   d pneumatics

2. Which of the following is **not** a function of a dual mass flywheel (DMF)?
   a eliminates gear shift rattle
   b reduces gear change frequency
   c lower tickovers can be used
   d a clutch is no longer needed

3. The maximum torque multiplication in a torque converter is around:
   a 2.2:1
   b 1:1
   c 4:1
   d 14.7:1

4. In a manual gearbox, as speed increases:
   a torque increases
   b torque reduces
   c torque remains the same
   d there is no torque

5. In a hybrid vehicle, the greatest amount of torque from the electric motor occurs:
   a starting from rest
   b at top speed
   c at the same time as in a petrol engine
   d at the same time as in a diesel engine

6. In a sequential gearbox, what prevents two gears being selected at once?
   a detent mechanism
   b hydraulics
   c interlock mechanism
   d selector drum

7. A DCS system has:
   a one clutch pack
   b two clutch packs
   c three clutch packs
   d four clutch packs

8. When working out gear ratios, to calculate the number of teeth on the planet gears of an epicyclic gear train, you use the formula:
   a Annulus − Sun gear = Planet gear
   b Annulus + Sun gear = Planet gear
   c Annulus ÷ Sun gear = Planet gear
   d Annulus × Sun gear = Planet gear

9. Which of the following are used in an automatic transmission to select a gear ratio?
   a brake bands
   b multi-plate clutch packs
   c unidirectional sprag clutches
   d all of the above

10. Which of the following is **not** a type of LSD?
    a clutch type
    b viscous coupling type
    c Haldex type
    d torsion type

5 Diagnosis & rectification of light vehicle transmission and driveline faults

PREPARE FOR ASSESSMENT

The information contained in this chapter, as well as continued practical assignments, will help you to prepare for both the end-of-unit tests and diploma multiple-choice tests. This chapter will also help you to develop diagnostic routines that enable you to work with light vehicle transmission and driveline system faults. These advanced system faults will be complex and non-routine and may require that you work with and manage others in order to successfully complete repairs.

You will need to be familiar with:

- Diagnostic tooling
- Electrical and electronic principles
- Diagnostic planning and preparation
- Clutch systems
- Manual gear change systems
- Automatic gear change systems
- Torque converters
- Epicyclic gear trains
- Continuously variable transmission (CVT)
- Dual clutch seamless shift transmission (DCT)
- Final drive and limited slip differentials.

This chapter has given you an overview of advanced vehicle transmission systems and has provided you with the principles that will help you with both theory and practical assessments. It is possible that some of the evidence you generate may contribute to more than one unit. You should ensure that you make best use of all your evidence to maximise the opportunities for cross-referencing between units.

You should choose the type of evidence that will be best suited to the type of assessment that you are undertaking (both theory and practical). These may include:

Assessment type	Evidence example
Workplace observation by a qualified assessor	Carrying out the replacement of a baulk ring synchromesh component
Witness testimony	A signed statement or job card from a suitably qualified/approved witness, stating that you have correctly overhauled a manual gearbox
Computer-based	A printout from a diagnostic scan tool showing the results from a test to check the function of an automatic transmission gear shift map/timing
Audio recording	A timed and dated audio recording of you describing the process involved in road testing a car to check for symptoms produced by a transmission fault
Video recording	Short video clips showing you carrying out the various stages involved in the electrical diagnosis of an automatic clutch system
Photographic recording	Photographs showing you carrying out the stages of a final drive set up and adjustment, when the assessor is unable to be present for the entire observation (this process may take a considerable time to complete). The photos should be used as supporting evidence alongside a job card.
Professional discussion	A recorded discussion with your assessor about how you diagnosed and repaired a vehicle with transmission whine

CONTINUED ▶

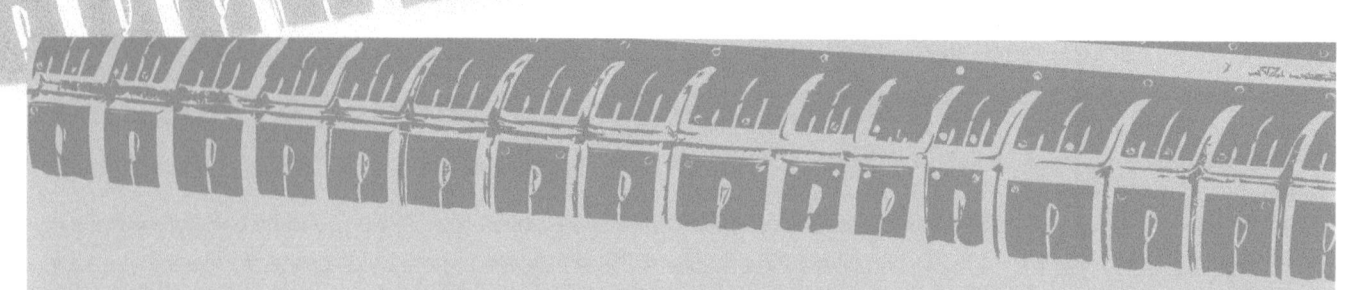

Assessment type	Evidence example
Oral questioning	Recorded answers to questions asked by your assessor, in which you explain how you safely isolated a high voltage hybrid system in order to work on the vehicle's CVT transmission
Personal statement	A written statement describing how you accomplished the repair of a vehicle clutch system
Competence/skills tests	A practical task arranged by your training organisation, asking you to use a multimeter correctly to conduct a volts drop test on an inhibitor switch
Written tests	A written answer to an end-of-unit test to check your knowledge and understanding of light vehicle transmission and driveline systems
Multiple-choice tests	A multiple-choice test set by your awarding body to check your knowledge and understanding of light vehicle transmission and driveline systems
Assignments/projects	A written assignment arranged by your training organisation requiring you to show an in-depth knowledge and understanding of a particular driveline system (e.g. limited slip differentials)

Before you attempt a theory end-of-unit or multiple-choice test, make sure you have reviewed and revised any key terms that relate to the topics in that unit. Ensure that you read all the questions carefully. Take time to digest the information so that you are confident about what each question is asking you. With multiple-choice tests, it is very important that you read all of the answers carefully, as it is common for two of the answers to be very similar, which may lead to confusion.

For practical assessments, it is important that you have had enough practice and that you feel that you are capable of passing. It is best to have a plan of action and work method that will help you.

Make sure that you have the correct technical information, in the way of vehicle data, and appropriate tools and equipment. It is also wise to check your work at regular intervals. This will help you to be sure that you are working correctly and to avoid any problems developing as you work.

When undertaking any practical assessment, always take care to work safely throughout the test. Light vehicle transmission systems are dangerous and precautions should include making sure that:

- the vehicle you are working on is not in gear before the engine is started, and that park is selected when running an automatic in the workshop
- you observe all health and safety requirements
- you use the recommended personal protective equipment (PPE) and vehicle protective equipment (VPE)
- you use tools correctly and safely.

Good luck!

6 Identifying & agreeing motor vehicle customer service needs

This chapter will introduce you to the subject of customer service within an automotive setting. No matter what your job role involves, when you are working in a customer service industry, you will have contact with vehicle owners and drivers. You need to be aware of how your actions when dealing with customers will impact on the company's image and therefore the amount of business that it receives. You will be expected to work in a professional manner at all times and provide sound customer service in line with legal and organisational requirements. This chapter will support you with knowledge that will aid you when undertaking both theory and practical assessments. It will help you develop routines that can promote good service while working in the automotive industry.

This chapter covers:

- Organisational requirements
- Principles of customer communication and care
- Obtaining and providing relevant information to customers
- Company products and services
- Agreeing and undertaking work for customers
- Consumer legislation

BEFORE YOU START

Skills for work

In order to be successful, the motor industry relies on a strong reputation for its customer service. According to many surveys, the general view of automotive maintenance and repair is good and the majority of customers feel that they are dealt with in a professional manner. Unfortunately, a small number of garages do provide very poor customer service, and some of these have been exposed very publicly on television and in the media. This has led to stereotyping of garages and technicians, which can often cloud a customer's opinion about the service they have just received.

As an employee, you have a key role to play in maintaining the standards of customer care and the perception of the automotive industry in the eyes of the public.

Working for any company requires that you are able to demonstrate a series of personal skills that ensure you meet an employable standard and are able to deliver outstanding customer service, no matter what your job role. Your employer will expect you to already have some of these skills when you first start with an organisation, and you will be able to develop more of the skills during your working life.

Some examples of these personal skills are described in Table 6.1.

Table 6.1 Personal skills required at work

Skill	Examples
General employment skills	*Time keeping:* making sure that you stick to agreed timescales
	Problem solving: having a practical, logical approach to your work that gives results
	Flexibility: being versatile, willing and multi-skilled
	Business acumen: having an entrepreneurial attitude, being competitive and a risk taker (within limits)
	IT/computer literacy: good office skills, keyboard skills and a knowledge of software packages
	Numeracy: the ability to make accurate calculations, measurements and estimates
	Literacy: the ability to produce accurate documents, such as records of work done
	Commitment: showing that you are dedicated, trustworthy and conscientious

CONTINUED ▶

Skills for work

Skill	Examples
Self-reliance skills	*Self-awareness:* showing that you are purposeful, focused and have self-belief, but are realistic about what you can achieve
	Proactivity: the ability to take action to prevent a potential problem from occurring
	Resourcefulness: the ability to find out what you need to do to carry out a job
	Willingness to learn: being inquisitive, motivated and enthusiastic
	Self-promotion: showing that you are positive, persistent and ambitious
	Networking: the ability to initiate and build relationships
	Planning action: the ability to make decisions, to plan your work and to prioritise
People skills	*Team working:* being supportive to colleagues and working cooperatively to produce results
	Interpersonal skills: the ability to listen, give advice and cooperate with colleagues
	Assertiveness: the ability to express yourself confidently, without being aggressive
	Oral communication: being a good communicator, presenter and influencer
	Leadership: the ability to motivate others, be energetic and visionary
Customer service skills	*Customer orientation:* showing that you are friendly, caring and diplomatic
	Patience: the ability to be patient and listen attentively
	Tact: the ability to deal with customers tactfully, never being aggressive or defensive or blaming the customer
	Empathy: the ability to understand your customers' needs and moods
	Assessment: the ability to ask questions and gather customer-related information in order to gauge the needs of your customer
Specialist skills	*Specific occupational skills:* specialist relevant knowledge of motor vehicles
	Technical skills: automotive hand skills and tool use
	Diagnostic ability: the skills to accurately diagnose and repair customers' vehicles

Organisational requirements

To promote customer service, every good garage will have a method or policy in place that describes how customers should be treated. Depending on your position within the company, the amount of contact that you have with customers will vary. However, even if you have very little contact with customers, you need to be aware of the customer service requirements of your organisation. You also need to find out the limitations of your authority when dealing with customers (what you are allowed to say, do or offer).

The garage's customer service policy should cover:

- How the customer is greeted, either in person or on the telephone
- How the goods or services offered by your garage should be promoted
- How to record vehicle symptoms and faults
- How to discuss issues and repairs in a non-technical manner
- How to give estimates for work to be conducted and explain the limitations of vehicle and component warranties
- How to book vehicles in for work to be conducted
- How to accept customer vehicles and what information needs to be gathered at the time of acceptance
- How to conduct a pre-work inspection (preferably with the customer present)
- How to obtain the customer's agreement for the work to be conducted and obtain signatures for the scope of the work to be completed
- How to keep the customer aware of progress
- How to inform the customer if work, time or costs are likely to exceed those agreed
- How to conduct a post-work inspection to ensure customer satisfaction
- How to inform the customer that the work has been completed
- How to explain invoicing, including payment methods, and obtain customer signatures when necessary
- How to explain any further work that needs to be conducted to satisfy the conditions of guarantees or warranties
- How to accomplish the final handover and return of the vehicle to the customer
- How to conduct follow-up inquiries to ensure customer satisfaction

Skills for work

When a customer collects their car following some repairs, they notice a small dent in the passenger's door that they say wasn't there when they dropped it off.

This situation requires that you exhibit particular personal skills. Some examples of these skills are shown in Table 6.1 on pages 304–305.

1. Using the examples given in Table 6.1, choose one skill from each of the following categories that you think you need to demonstrate in order to deal with this situation.
   - General employment skills
   - Self-reliance skills
   - People skills
   - Customer service skills
   - Specialist skills

2. Now rank these skills in order of importance, starting with the one that it is most important for you to have in this situation.
3. Which of the skills chosen do you think you are good at?
4. Which of the skills chosen do you think you need to develop?
5. How can you develop these skills and what help might you need?

Principles of customer communication and care

Many customers will make up their minds about your organisation within the first few seconds, so first impressions are vital. Effective communication and showing that you care about your customers are key to creating a good impression of your garage and the work you do. If you can manage to create a **rapport** with your customer when you first speak to them, you are well on the way to gaining their respect and trust and ensuring their future business with your company.

Effective communication involves more than just the words you use. It involves the ability to listen and use positive body language, your attitude to your work, and having a professional appearance (both your own and the appearance of the working environment). All of these will combine to create the impression of your organisation that the customer will take away with them.

Presenting a professional image

In order to present a professional image, a company will often create a sense of identity among its employees. This may involve wearing a uniform or having a **dress code**, so that customers can instantly recognise members of staff. A dress code can also help to encourage teamwork. If no company dress code is required, make sure that your

Key terms

Rapport – a sense of trust and understanding between people, whether or not they know each other well.

Dress code – a company's policy regarding the style and type of clothing that you are required to wear.

appearance is presentable, as this will make a difference to how people react to and interact with you.

Taking care of your appearance is not the only thing you need to do. Consider the impact of your behaviour too. You need to be confident in your actions and behaviour. When speaking to customers, consider what you want to say. Remember that it's not only what you say but how you say it that will create an impression with the customer. Be positive with your choice of words. Nobody wants to hear what cannot be done, so say what you can do and offer options and alternatives.

Listening skills

When a customer brings their car to your garage, you need to find out from them what work needs to be done. To do this, you need to develop good listening skills. Make sure that you show interest by actively listening to your customers and not just waiting for your turn to speak. When dealing with customers, you should use a ratio of approximately 80 : 20 of listening to speaking. Actively listening will help you get things right first time. It will prevent misunderstandings occurring and potential complaints being made.

Active listening involves using your eyes as well as your ears. You need to listen for content as well as underlying emotions. Respond to the customer's feelings; sometimes the true meaning is in their emotion rather than the words they use. Tune in to your customer's body language: what are the non-verbal clues telling you?

Figure 6.1 Presenting a company image

- Content is the information about what needs to be done to the vehicle.
- Underlying emotions are how the customer feels about the situation. (For example, are they upset, worried, cheerful, etc.?)
- The true meaning is what the customer really wants, not what they might be saying. (For example, they may say that everything is fine, but really they are unhappy with the service provided.)

Active listening involves:

- *Giving your customer your full attention:* Make sure you maintain enough eye contact to develop a rapport. Make encouraging noises, for example saying 'okay' to indicate that you are listening to the customer. This is especially important when talking on the telephone as the customer cannot see your responses. Customers will take this as an acknowledgement that you have heard and understood them.
- *Managing your reactions:* You are listening, so try not to interrupt. The person speaking the most should be your customer.
- *Looking for signs of the true meaning behind the words:* Does the customer's facial expression match the words they are using?

Figure 6.2 Active listening means using your eyes as well as your ears

Does their tone of voice tell you anything about what they are really feeling?

- *Giving feedback:* Confirm that you have understood by nodding or using phrases such as 'I see', 'I understand', 'Oh yes' and 'okay'.
- *Asking questions:* If you do not understand, ask questions to find out more information.
- *Listening to tone of voice:* You can still be an active listener on the telephone because your customer's tone of voice can give an indication of their needs. It is not only what your customers say, it is how they say it. Listen carefully for changes in tone of voice and how each word is stressed. Tone is how things are said, for example, loudly, softly, impatiently, quickly, with sensitivity, with respect.
- *Reading body language:* The ability to read a customer's **body language** can make all the difference between just hearing and actively listening. The signals given out through body language can be positive or negative (see below).

> **Key terms**
>
> **Active listening** – paying careful attention to understand more than just the words being spoken.
>
> **Body language** – non-verbal communication shown through conscious or unconscious gestures and movements.

Body language

To gain the best understanding, you need to consider the person's body language together with the words that they use. Try not to interpret one without taking the other into account. For example, it is often said that if a person has their arms tightly crossed, this shows that they are angry, confused or unhappy with the situation. In fact, he or she may just be comfortable in that position or they may be cold. The words and tone of voice that the person uses will help you work out what this body language means.

Figure 6.3 Remember to consider words and body language together

Level 3 Light Vehicle Technology

Case study

You are working in the reception area of your garage one day when a young lady comes in to the office to ask for some advice about her car. She has only just passed her driving test; this is her first car and she doesn't know how to check her tyre pressures. You can see she looks nervous and slightly embarrassed.

Here's what you do:

- ✓ Make her feel welcome by using positive body language (smiling with your mouth and eyes).
- ✓ Give her your full attention, with enough eye contact to develop rapport.
- ✓ Listen carefully to her enquiry and try not to interrupt.
- ✓ Give feedback to confirm you have understood by saying 'okay'.
- ✓ Go with her to the car and demonstrate how to check the tyre pressures.
- ✓ Be positive and ensure that she has understood your instructions.
- ✓ Make no charge for your time and encourage her to return to your garage if she needs any further help or assistance with her car.

Action

Working with a partner, see if you can convey the following emotions and feelings using body language alone:

- happiness
- anger
- confusion
- frustration
- boredom
- satisfaction
- understanding
- assertiveness
- rapport.

Key terms

Gesture – a movement of part of the body, especially a hand or the head, to express an idea or meaning.

Empathy – understanding and sharing the feelings of others.

You should always try your best to use positive body language in order to show your customers that you are willing to help and are respectful of their wishes. This will also give them a positive impression of the organisation. Some examples of positive and negative body language are listed in Table 6.2.

Table 6.2 Positive and negative body language

Positive body language	Negative body language
Smiling with your mouth and eyes	Pursed lips
Leaning forward (but not too much)	Invading personal space
Maintaining eye contact	No eye contact/looking around
Relaxed facial expression	Frowning
Head nodding occasionally	Staring
Head leaning to one side	Yawning
Steady breathing	Rapid breathing
Gesturing with your hands while speaking	Pointing with your hands or fingers
Raised eyebrows while smiling	Tapping your fingers

Gestures include movements of the head and hands – they are used to emphasise what you are saying. If you understand the gestures you tend to use and can recognise them in other people, you can use them successfully to help achieve customer satisfaction.

Positive or supportive gestures will help create **empathy**. For example, your customer will want to know you are listening. You can show that you are by leaning your head to one side and nodding occasionally. Make sure your eyes also really show you are listening. Do not grin from ear to ear, but the occasional smile will help with rapport.

6 Identifying & agreeing motor vehicle customer service needs

Figure 6.4 Always try to show empathy when dealing with customers

Dealing with customers' questions and comments

In your work, you may need to communicate with customers face to face, on the telephone, by electronic means, or using a combination of these methods. Whatever the form of communication, you will need to be respectful, helpful and professional. You can achieve this by being polite and confident and making it very clear what you can and cannot do for your customer.

Communicating with customers in a professional way involves many different factors in order to create a positive impression. These factors include:

- appearance (both of you and your surroundings)
- appropriate behaviour
- product or service knowledge.

Your appearance

When dealing with customers face to face, your appearance and that of your surroundings need to convey the right professional impression. You can achieve this by following your organisation's dress code and keeping your working environment clean and tidy. You will also find it easier to locate the information that you need to help customers if you are well organised, so keep your workspace free of clutter.

Level 3 Light Vehicle Technology

Hang on please. The information is here somewhere!

Figure 6.5 In most cases, being untidy means you are disorganised

Even if you are not dealing with a customer face to face, your colleagues can see you. You need to create the right impression with them too. A tidy appearance and a clean and tidy working area will demonstrate to your colleagues that you care about your work. This will lead to your colleagues trusting in you and your abilities and will help you to gain their respect.

The way you conduct yourself

Effective handling of questions and comments depends to a large extent on your behaviour towards your customers. Equally, you will be affected by your customers' behaviour towards you. If a customer behaves in an angry or disrespectful manner towards you, you may be tempted to react in a confrontational manner. But remember to keep your responses professional at all times.

Behaviour refers to everything you do and say. People will draw conclusions about you and your organisation based on your behaviour towards them. You should always behave in a manner that shows you care.

For example, you might find that you hear customers saying similar things many times during the course of your work, and this could get boring. You might show your boredom by your voice becoming flat or by getting easily distracted and your listening skills might suffer. Your customer will pick up on your boredom, and may feel discouraged about your willingness to do the work needed. Remember that to your customer, the problem with their vehicle is not boring, and it is part of your job to be interested in finding the solutions that will help them.

Responding in a timely manner

Show your customer that their custom is important to you by answering their queries as quickly as you can. If you have to research information or seek help from others, do so. Never guess the answer. If the answer is taking longer to find than you anticipated, keep your customer informed of progress. If a customer has spent time queuing (either on the telephone or in person), acknowledge this before you answer their query, for example by saying 'I'm sorry to have kept you waiting.' This shows that you care and that you are treating the customer with respect.

Product and service knowledge

Make sure you know where to access information about all the products and services you deal with and that you keep this information up to date. This will help you to make sure that your responses to queries and requests are accurate. Know the limits to your authority. Do not make promises which cannot be kept. Know who to ask for help if you are unable to deal with your customers' requests.

6 Identifying & agreeing motor vehicle customer service needs

Using questions to check your understanding

To avoid misunderstandings and making assumptions, you will sometimes need to check that you have understood what the customer is telling you. You will ensure you have fully understood your customer by:

- actively listening
- asking the right questions
- repeating back information
- summarising.

Try using checking phrases such as:

- 'I didn't quite hear what you said … Did you mean?'
- 'To be sure I have understood correctly, let me repeat back what you need … Is that right?'

You can also use this type of question when you simply cannot hear what has been said. For example: 'Did you say you want to book your car in on the 5th of August?'

Sometimes you may feel confused by what the customer has said. Repeating back key words you think you have heard could help clarify things. For example: 'Did you say you want an MOT on the 15th of June at 10a.m. and need to wait while the car is being checked?'

It is also important to check that a customer has understood what you have said. If they haven't heard you, in their mind you haven't said it. For example: 'Let me clarify: the cost of your service will be £360.'

Figure 6.6 Asking questions will help you decide whether you have given the customer enough information

Positive and negative language

As with body language, the words you use can be positive or negative. Choosing the right words will be critical to your success. Customers need to trust you and know you are genuine in what you say and that you care. If you choose negative words, you could create an unwanted situation that could have been avoided with a more positive approach. Say what you can do rather than what you cannot do, but be honest.

Table 6.3 lists examples of positive and negative words and phrases.

Table 6.3 Positive and negative words and phrases

Positive words and phrases	Negative words and phrases
Yes	No
How may I help you?	What do you want?
I	They
Definitely	Unlikely
I will find out	I don't know
I will	I can't
Always	Never
I'll do it this morning	I'll do that when I can
I'll sort that out	It's not my fault

Tone of voice

Your customers will develop an impression of you based on your tone of voice. This is an important tool in customer service because it helps you to bring emotion into putting your message across. If your tone is clear and strong, you come across as confident. If it is halting and softly spoken, you may appear timid and lacking in knowledge. Someone who speaks with a flat tone (that is, with no rise and fall in their voice) may come across as boring and, again, possibly lacking in knowledge and confidence.

Some of the emotions people convey in their tone of voice include:

- boredom
- happiness
- sadness
- anger
- frustration
- worry.

You should always aim to speak clearly and with warmth and energy. Smiling while you speak will automatically bring warmth to your tone of voice.

> **Did you know?**
>
> Getting the words right is only one part of effective communication. Your tone of voice and the way you choose to emphasise certain words matter too. After all, 'it's not what you say, it's the way you say it'. This phrase has been in use for a long time, but it still holds true.

> **Skills for work**
>
> You are on a road test with your customer so that they can point out a strange noise that occurs when the car is driven. The customer is unable to recreate the noise during the road test and is becoming frustrated.
>
> 1. What questions would you ask to try to gather as much diagnostic information as possible from the customer?
> 2. This situation requires that you exhibit particular personal skills. Using the examples given in Table 6.1 on pages 304–305, choose one skill from each of the following categories that you think you need to demonstrate in this situation.
>   - General employment skills
>   - Self-reliance skills
>   - People skills
>   - Customer service skills
>   - Specialist skills
> 3. Now rank these skills in order of importance, starting with the one that it is most important for you to have in this situation.
> 4. Which of the skills chosen do you think you are good at?
> 5. Which of the skills chosen do you think you need to develop?
> 6. How can you develop these skills and what help might you need?

Being professional on the telephone

Why is dealing with customers on the telephone different from dealing with them face to face? The most obvious answer is that you cannot see each other, so you are both unable to observe each other's body language. Because customers cannot see what you're doing, never say, 'Hold on please', and leave the customer waiting. Explain exactly what you're going to do and how long it will take. You need to keep the customer informed of the actions you are taking.

Second, in today's world particular emphasis is placed on the importance of answering the telephone speedily, so when it rings you might be tempted to answer the call immediately. If you do this, you might speak far too quickly. Before picking up the phone, first take a deep breath and then answer the phone calmly. Put a smile in your voice and be polite and courteous.

Third, the customer will feel that time spent waiting (for example, for the call to be answered or for you to find out information) is time spent doing nothing. What might just be a few seconds to you, may feel like minutes to a customer who is literally hanging on the other end of the line. Customers get impatient and frustrated more quickly than they would do in a face-to-face situation. If finding information is going to take a while, note down the customer's number and say you will ring them back. Make sure you ring them back when you said you would with the required information.

Case study

The workshop receptionist at the garage where you work has phoned in sick today. Your boss asks you to cover the service reception for the day.

Having got changed and cleaned up, so that you present a good company image, you take your place in the reception area and wait for your first customer. After a couple of minutes, the phone rings.

Here's what you do:

- ✓ Answer the phone promptly, but don't rush.
- ✓ With a smile in your voice, identify yourself and the company, following your company's customer service procedures.
- ✓ Ask how you can help. (Mr Jenkins would like to find out the cost of a service on his car.)
- ✓ Gather and note down information, including: customer's name, vehicle type and the type of work required.
- ✓ Confirm the information to show that you have understood.
- ✓ Inform Mr Jenkins that you are just going to look up the costs, and tell him how long this will take you.
- ✓ Locate the costs using the computer-based estimating system, and store the estimate for future reference.
- ✓ Provide Mr Jenkins with the estimated cost of his service, and ask him if he would like to book the car in. (Offer days of the week that are available.)
- ✓ Confirm that you have understood the time, date and work to be completed by repeating the information back to Mr Jenkins.
- ✓ Book in the service using the company's workshop loading system and reference the estimate that you have provided.
- ✓ Thank Mr Jenkins for his call and tell him that you look forward to meeting him.

Throughout the call, remember to use positive language.

CHECK YOUR PROGRESS

1 List four organisational requirements that can promote good customer service
2 Explain the terms:
   - dress code
   - rapport
   - active listening.
3 Give three examples of positive body language and three examples of negative body language.

Being professional using the written word

Effective communication takes place when information is fully understood after it has been passed on and received. With the written word (letters, memos, emails, etc.), the customer cannot ask you questions on the spot as you will not be there. It is therefore very important to use words that your customer will understand and to get it right first time. Otherwise, costly errors could be made, your organisation's reputation will suffer and you will waste your customer's time.

When using the written word:

- Think carefully what you need to say before you write.
- Answer all questions and comments fully.
- Be concise and keep to the point; do not waffle.
- Summarise the key points and any actions to be taken.
- Avoid jargon and specialist technical terms.
- Check your communication for spelling and grammar.

Obtaining and providing relevant information to customers

The exchange of information between a garage and its customers is always necessary in order for a business to function correctly. The more information that is available, the more effective and efficient the customer service will be.

Customer needs

In order to provide your customer with the type and level of service they require, you will need to gather certain information. To assess their needs, the information you gather needs to be relevant and sufficient. The types of information needed can be broken down into three main areas, as described in Table 6.4.

Table 6.4 Types of information

Type	Information
Personal	Name Address Contact details Are they the owner/driver? Payment methods/terms
Vehicle	Make Model Registration number Body style Engine size/number Vehicle identification numbers (VIN) Mileometer reading Any pre-work damage
Work to be conducted	Service history Warranty information Symptoms Issues Faults Diagnostic data Technical data Legal requirements

Once you have gathered this information, you should be in a position to clarify both customer and vehicle needs. Always make sure that you refer to vehicle data and operating procedures when offering advice, and that any information gathered during this process is kept in accordance with the Data Protection Act (see pages 335–337).

Giving clear, non-technical explanations

It is important that you provide customers with accurate, current and relevant advice and information in a form that they will understand.

Level 3 Light Vehicle Technology

You may come across some situations where customers find it hard to understand what you're telling them. Vehicle operation and repairs is a very technical subject. You must be careful to avoid confusing your customer with technical information or **jargon**. If your customer does not understand fully what is to be done with their car, they may be cautious about having the work conducted or concerned by how much it is going to cost. Your customer may not always tell you that they do not understand, as they may not want to appear foolish, so listening and watching for clues is vital.

Clues could include:

- a customer who stays quiet when you're expecting a response
- a customer who gives a response which doesn't fit with the question you have asked
- a customer who fidgets and frowns and looks puzzled.

Key terms

Jargon – specialised technical terminology that others may find hard to understand.

Feedback – information given about the performance or quality of service of an individual or organisation.

Skills for work

You have been asked to mentor a young trainee while he deals with some customers in the service reception. You decide to observe him dealing with customers and give your **feedback** privately when he's finished.

Here's what he does:

- He comes into the office in his workshop overalls, sits down and puts his feet up on the desk.
- As a customer enters the office, the phone begins to ring at the same time.
- Ignoring the customer, he picks up the phone and says, 'Hello?'
- After listening to the customer's query on the phone, he says, 'I don't know. You'll have to phone back this afternoon when my manager's not busy', and puts the phone down.
- He turns to the customer in the office, and says, 'What can I do for you?'
- The customer asks to book their car in for an MOT on Tuesday.
- Your trainee looks at the booking diary and says, 'We can't do Tuesday, we're fully booked. Sorry but it's not my fault.'
- While he's waiting for the customer to check their diary for another day, he yawns and folds his arms.
- The customer asks if Friday has any space.
- The trainee says, 'Yeah, that's okay. What's your name?'
- He writes the booking in the diary, says 'thanks' and sits back down.

When the customer leaves, you give your trainee constructive feedback.

1. What things did he do wrong?
2. What could he have done differently?
3. How could you encourage the trainee to improve his skills?

6 Identifying & agreeing motor vehicle customer service needs

Giving clear explanations on the phone

You will not be able to see puzzled faces when talking to customers on the phone. For this reason, a key consideration is to quickly establish whether what you're saying makes sense. Listen out for verbal signs from customers that they do not understand you. Even silence may be a clue that they have not understood what you are saying.

It is very important to speak clearly and slowly, without being **patronising**. Don't be in a hurry! Rushing through what you have to say can come across as quite threatening and will only make your explanations even more difficult to understand. Use silence to give customers thinking time before moving on to your next point. Ways to help customers understand what you're saying include:

- Slow down your voice.
- Keep a smile in your voice.
- Avoid using technical terms or jargon.
- Use silence to give your customer time to understand and respond.
- Use language that is appropriate for the individual customer.
- Check with your customer that they have understood what is being said.
- Encourage the customer to ask questions and seek clarification during your conversation.

Fulfilling customer expectations

Providing a high-quality service involves fulfilling (ideally, exceeding) customer expectations within agreed time frames. Poor customer service is often the result of employees' attitudes and behaviour. For example, staff who are helpful, are willing to take responsibility and have the right level of knowledge to help customers will be on their way to providing the correct level of customer service. You, your colleagues and the systems and processes that support customer service in your organisation are what make or break the customer service experience. You have an enormous part to play in ensuring that customers are dealt with properly and to everyone's satisfaction.

Each customer will have their own ideas of what they expect from your organsation. These **customer expectations** are based on a number of factors, including:

- advertising and marketing
- word of mouth from friends or family
- past experiences with your organisation, products or services
- reputation of the organisation and its staff
- previous experiences with other garages.

Figure 6.7 overleaf gives examples of typical customer expectations.

> **Action**
> Think of a complex diagnostic and repair procedure. Try to explain the process to someone else in a non-technical manner.

> **Key terms**
> **Patronising** – treating someone as if you are superior.
>
> **Customer expectations** – what the customer thinks should happen and how they think they should be treated when asking for or receiving customer service.

> **Action**
> Conduct a small survey of family and friends to see if there is a local garage that they would recommend.
> Ask them why they take their car to that garage and their expectations of the service they will receive there.
> What is the most common reason given by the people that you asked?
> How can you make sure that you fulfil this customer expectation in your own work?

319

Typical customer expectations

Customer expectations:
- Friendliness of staff
- Quality of service or product
- Staff helpfulness
- Ease of doing business
- Staff appearance and behaviour
- Delivery times
- Staff product or service knowledge
- Price/cost
- Speed of service

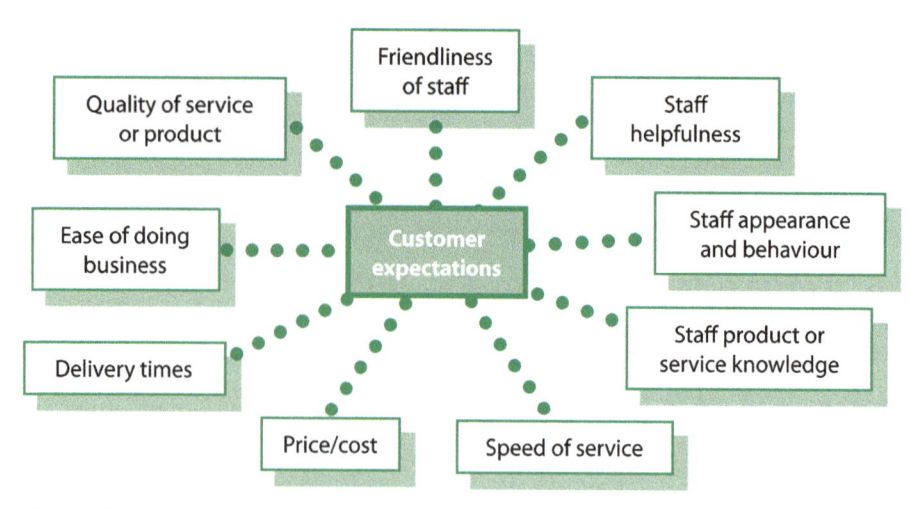

Figure 6.7 Typical customer expectations

Some examples of methods that will help you fulfil or exceed customer service expectations are shown in Table 6.5.

Table 6.5 Methods of fulfilling customer expectations

Expectation	Methods that can be used to fulfil expectation
Friendliness of staff	Greeting customers: Ensure that all staff are trained in the company's approved methods for greeting customers when they enter the garage. Answer your phone promptly and courteously: Make sure that someone is available to answer the phone when a customer calls your business. Having good staff morale: When employees are happy and are working well together, this will result in a friendly approach to customers.
Delivery times	Don't make promises unless you will keep them: Reliability is one of the keys to any good relationship, and good customer service is no exception. If you say, 'Your car will be ready on Tuesday', make sure it is ready on Tuesday. Otherwise, don't say it. The same rule applies to appointments, deadlines, etc. Think before you give any promise – because nothing annoys customers more than a broken one.
Ease of doing business	Listen to your customers: Is there anything more frustrating than telling someone what you want or what your problem is and then discovering that the person hasn't been paying attention and needs to have it explained again? Don't use sales pitches or product babble: Let your customer talk and show him or her that you are listening by making the appropriate responses, such as suggesting how to solve the problem.
Dealing with complaints	Listen attentively to the complaint: No one likes hearing complaints, and many people develop a reflex response to complaints, saying, 'You can't please all the people all the time'. Maybe not, but if you give the complaint your attention, you may be able to please this one person this one time – and your business will then reap the benefits of good customer service.
Staff helpfulness	Be helpful, even if there's no immediate profit in it: Occasionally it's good to give something away for free (especially if its second hand). Garages have often helped a customer out by fitting a small item that gets them back on the road – and charged them nothing! The result is that the customer comes back when they next need a service or repair, and they will tell all their friends and family about their positive experience.

CONTINUED ▶

Expectation	Methods that can be used to fulfil expectation
Staff product or service knowledge	Staff should be trained to be always helpful, courteous, and knowledgeable: This could be done 'in house' or by professional trainers. Talk to your colleagues about good customer service and what it is (and isn't) regularly. Most importantly, every member of staff should have enough information and power to make those small customer-pleasing decisions, so they never have to say, 'I don't know, but so-and-so will be back at …'
Staff behaviour	Take the extra step: For example, if someone walks into your workshop and asks you to help them find the reception or their car, don't just say, 'It's round the front'. Lead the customer to it. They may not say so to you, but customers will notice when you make an extra effort and will tell other people about it.
Quality of service price/cost	Throw in something extra: Whether it's a coupon for a future discount, additional information on how to get the most from their car, or a genuine smile, people love to get more than they anticipated. Also, don't think that a gesture has to be large to be effective: for example, a tax disc holder always comes in handy – it's a small thing, but will be appreciated.

Dealing with complaints

Your organisation may have a process in place for dealing with customer complaints; for example, when to say 'sorry', when to give refunds, what information to record and how to reach a satisfactory conclusion. Depending on your own organisation's practices, you may have full authority to give a refund or a sum of money as a gesture of **goodwill**. Alternatively, you may have to refer to someone else for permission to do so.

Many organisations actively encourage people to contact them with complaints and comments. This is because a complaint is a form of feedback; it tells an organisation about specific problems that need attention. Once the organisation knows that something has gone wrong, it can take action to protect its relationship with the customer. A complaint could be about a member of staff, a faulty product or poor communication – literally anything which has caused the customer to feel dissatisfied.

> **Key term**
>
> **Goodwill** – the positive reputation of a business that is not linked to income or money.

Complaints procedure

Many organisations have a complaints policy in place which describes in detail what a customer needs to do in order to make a complaint. This shows a willingness to help and to use the information to try to improve customer service in the future. For example, an organisation may describe the procedure in its literature or on a 'how to' page on their website.

Typically the complaints procedure will include answers to frequently asked questions such as:

- How does the complaints process work?
- Who will reply and when?
- What if I have not had a reply?
- What if I don't like the reply and what happens next?
- Can I complain to an independent body?

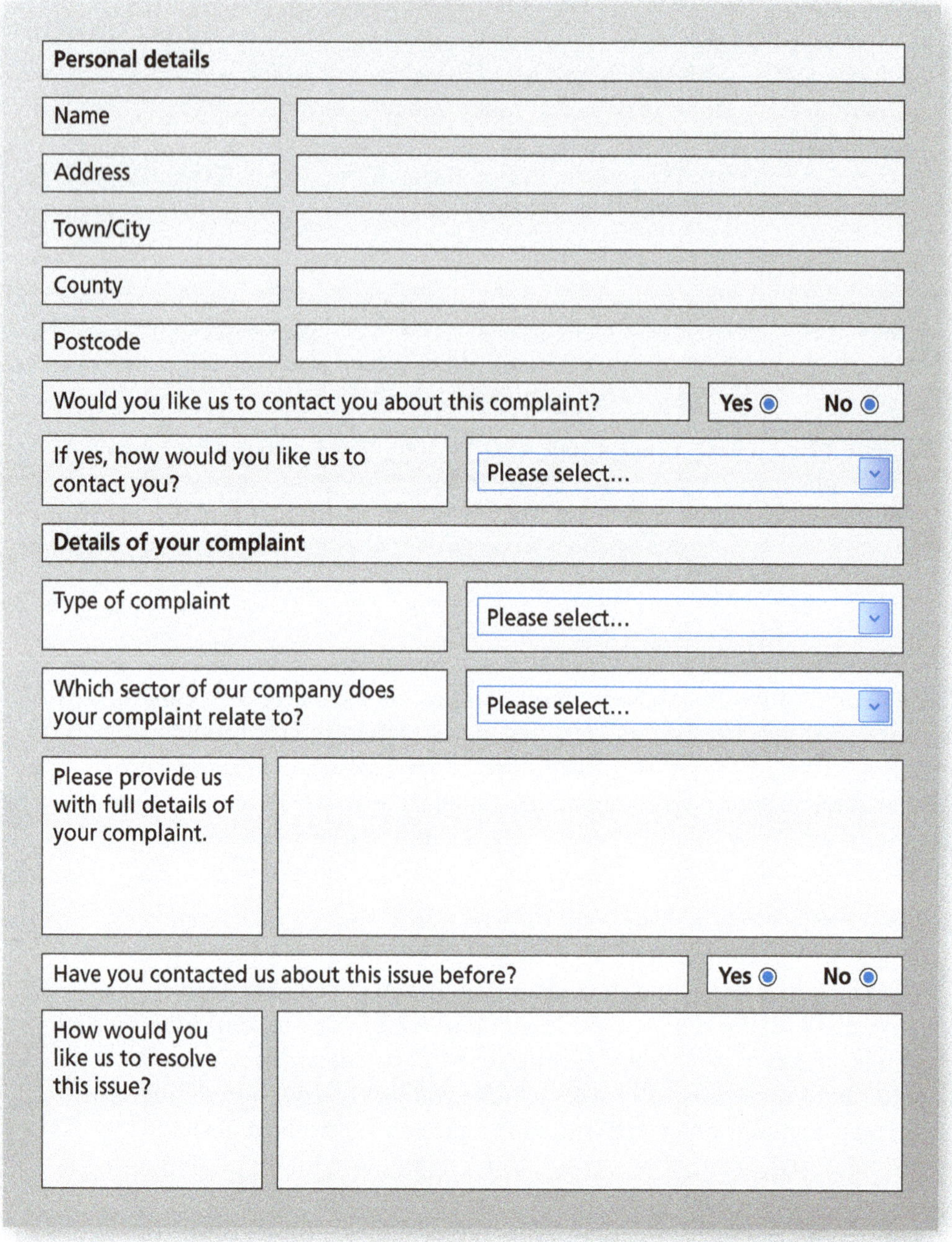

Level 3 Light Vehicle Technology

Figure 6.8 A sample online complaint/feedback form

Clearly, you need to know how your company's complaints procedures operate, because if a complaint is handled badly, a difficult situation will quickly escalate into something much worse. Make sure you know how to record details of any complaint made so that it can be followed up appropriately.

If there is no complaints procedure where you work, the flow chart in Figure 6.9 may help you to understand what role you should take. You may need to seek guidance from an appropriate person to help you with some of the answers.

The company's systems and procedures are there to protect you, your customer and the organisation. Try not to think of any system that helps you sort out problems as 'something else to worry about'. It is there to help you by providing you with a framework for dealing with customer problems.

There are websites that help people understand how to complain. Two examples of such websites are howtocomplain.com and the Directgov consumer rights website; please go to hotlinks and click on this chapter to view these websites.

Obtaining customer feedback

Your organisation may have a method or system in place for measuring the effectiveness of customer service. These methods and systems will vary, but they will involve obtaining feedback and then analysing and using it.

Feedback is information given about things that you or your organisation do. Sometimes customers will give it without prompting, for example sending a thank you letter or a letter of complaint. On other occasions, customers may give feedback as a result of a request from you or your organisation. You may also receive feedback from colleagues who have observed your work or from a line manager or supervisor as part of the organisation's performance and appraisal system.

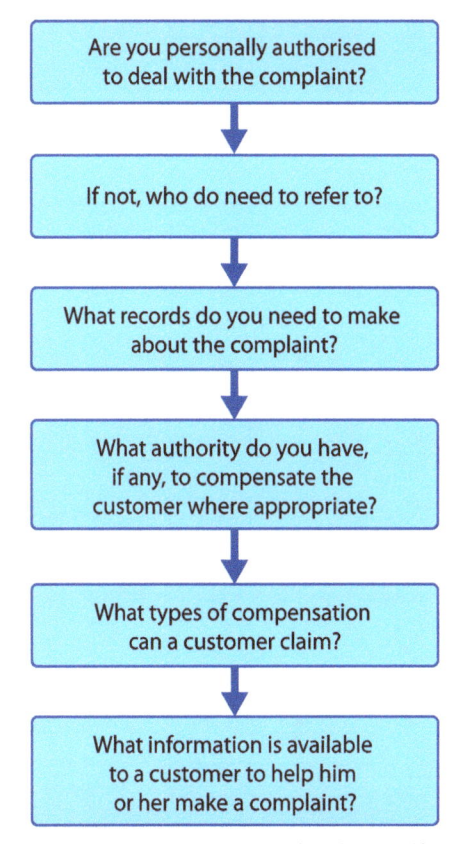

- Are you personally authorised to deal with the complaint?
- If not, who do need to refer to?
- What records do you need to make about the complaint?
- What authority do you have, if any, to compensate the customer where appropriate?
- What types of compensation can a customer claim?
- What information is available to a customer to help him or her make a complaint?

Figure 6.9 Questions you could ask yourself when dealing with a customer complaint

Moto Quick – top class repairs, top class service

How happy were you with your car service? Tick Excellent, Good, Fair or Poor for each statement

	Excellent	Good	Fair	Poor
The speed with which you were attended to when booking the appointment	☐	☐	☐	☐
The explanation of the work to be done	☐	☐	☐	☐
The availability of parts or accessories	☐	☐	☐	☐
How well you were kept informed of any changes to work agreed	☐	☐	☐	☐
The wait time when collecting your car	☐	☐	☐	☐
Overall value for money for the service or repair	☐	☐	☐	☐
How valued and respected you felt when dealing with our staff	☐	☐	☐	☐

Figure 6.10 An example questionnaire used by a garage to check customer satisfaction with car services

Every organisation will have its own way of obtaining feedback, depending on:

- the size of the organisation
- how sophisticated the organisation's systems and procedures are
- whether funds are available to undertake research
- whether the organisation wants to listen to customers
- the organisation's willingness and ability to implement (put into effect) any changes as a result of feedback.

While many people are comfortable making a complaint or mentioning when they would like something to be done differently, there is a huge silent majority who do not give any feedback. When the customer does not complain, it is very easy to assume that everything is fine and you are doing everything right. The organisation might believe its products and services are exactly what the customers want, but this might not be the case at all. If your organisation does not actively seek feedback, it runs the risk of making assumptions on behalf of its customers. Instead of giving feedback, customers might simply be walking away and finding what they want elsewhere.

Remember: it takes a long time to build a good reputation and a very short time to lose it.

There are many ways you and your organisation can set about obtaining customer feedback, including:

- questionnaires
- direct mailings
- telephone surveys
- comments/suggestion boxes.

Figure 6.11 A customer suggestion box

6 Identifying & agreeing motor vehicle customer service needs

Case study

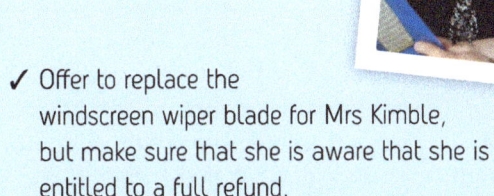

You are in the service reception of the garage where you work, when Mrs Kimble comes in and complains to you that the windscreen wiper blade that she bought from your garage the other day has broken.

No one else is around to deal with the issue, and because you have been given authority to deal with minor complaints, you decide to see if you can help.

Here's what you do:

- ✓ Listen to your customer's complaint and apologise for any inconvenience caused.
- ✓ Establish the facts, such as when the wiper blade was purchased and who fitted it.
- ✓ Go with the customer to check the component for yourself.
- ✓ Make sure that no further damage has been caused which might lead to extra expense.
- ✓ Decide that a replacement or refund is within your authority.
- ✓ Offer to replace the windscreen wiper blade for Mrs Kimble, but make sure that she is aware that she is entitled to a full refund.
- ✓ Replace the wiper blade and test for correct function and operation.
- ✓ Record all the details on the company's customer complaint form.
- ✓ Return the faulty wiper blade to your supplier for a refund.

Remember to follow up by getting feedback from your customer to ensure that the situation has been resolved satisfactorily.

Action

1. Look at the following customer complaints and find out which, if any, you would have the authority to deal with in your own workplace.
   - A customer complains about dirty marks on their car seat following a service.
   - A customer complains about the unexpected high cost of their invoice following repairs.
   - A customer complains that the headlamp bulb that they purchased for their car last week is the incorrect type.
2. How would you deal with these complaints?

CHECK YOUR PROGRESS

1. Give an example of the type of information that you might need to gather from a customer for each of the following areas:
   - personal
   - vehicle
   - work to be conducted.
2. Why should you avoid using jargon when communicating with customers?
3. List three customer expectations with regard to customer service.

> **Key term**
>
> **Third party** – someone who is not one of the people directly involved in a transaction or agreement.

Company products and services

No matter what size of company or organisation you work for, there will be standards in place that specify how work is conducted and how products are sold. These standards can be broken down into three different categories, as shown in Table 6.6.

Table 6.6 Service standard categories

Service standards	Example
National: standards set by nationally recognised organisations that seek to assure a high quality of service provided by a garage	Kitemark for Garage Services scheme: this is a voluntary, **third party** scheme that ensures that the standards of PAS 80 are met and maintained. The PAS 80 specification has been developed by the British Standards Institution (BSI) in conjunction with respected members of the automotive industry. It defines standards for customer service, ensures that technical and service standards are maintained, and provides a quality framework for garages that service and repair automotive vehicles, including fast-fit outlets. The specification includes a framework of performance measures that focus on aspects of service quality which have been identified as important to customers. Holding a garage services Kitemark licence shows your customers and competitors that you are serious about delivering a quality service. The reputation of the Kitemark ensures that your customers come back – reassured by your Kitemark status that you are committed to customer service, fair trading and safety. The Kitemark for Garage Services scheme covers the critical elements in delivering a quality service, including: • customer service • customer satisfaction • customer facilities • staff competencies • technical inspection.
Manufacturer: standards set by motor vehicle manufactures to ensure that work and products are of a quality expected by the original equipment manufacturer	If you work for a dealership, selling and repairing just one make of car, many manufacturers place requirements and stipulations on the garage to ensure that a certain level of service and corporate image is maintained. If you work for an independent garage, many manufacturers require that you use approved repair principles and parts that are of the same quality as those of the original equipment manufacturer (OEM). The use of inferior parts or working practices can jeopardise warranties and guarantees and may also make the vehicle unsafe.
Organisational: standards set by an individual garage to show the level of service expected from their staff	Many organisations have a set of guidelines for working practices within the garage. These are designed to promote a professional approach to the services provided. They are focused mainly on attracting and maintaining a solid customer base. They could include: • a dress code and corporate image • customer care standards • the level of training or qualification required to carry out certain repairs.

Figure 6.12 The British Standards Institution Kitemark

6 Identifying & agreeing motor vehicle customer service needs

In order for you to provide a good customer service experience, you need to be fully aware of the range and type of services offered by your organisation. Table 6.7 gives typical examples of services offered by many garages.

Did you know?

The British Standards Institution (BSI) is the national standards body of the UK. It has a globally recognised reputation for independence, integrity and innovation in the production of standards that promote best practice. It develops and sells standards in many different fields of business and industry.

Table 6.7 Types of services offered by garages

Service type	Description
Servicing	Servicing includes inspection and general maintenance designed to ensure that a customer's car is kept operating within legal requirements and with optimum performance and reliability.
Repair	Repair work normally includes the rectification of mechanical faults, whether in the form of a breakdown or as a result of wear and tear.
Warranty	Warranty work is normally conducted when a component has prematurely failed within a manufacturer's guarantee period. The cost of repair for both parts and labour is normally covered by the vehicle manufacturer, as long as the car has been maintained in accordance with the manufacturer's instructions. Warranty work is usually conducted by a garage at a reduced labour rate.
MOT testing	MOT testing is an annual inspection of a vehicle to ensure that testable items meet a minimum required legal standard. These standards are developed and regulated by the Vehicle and Operators Services Agency (VOSA). The MOT test is a visual inspection of safety critical items, and should not be considered a certificate of roadworthiness.
Fitment of accessories/ enhancements	Many garages will undertake the fitment of additional vehicle equipment and enhancements on behalf of a customer. Examples could include: • in-car entertainment • body styling • telecommunication devices • interior trim.
Diagnostic	Many garages are now offering a diagnostic service for advanced vehicle systems. This normally requires a large amount of investment on behalf of the garage in the way of equipment and knowledge. This specialist area is becoming a necessity for many garages in order to conduct normal maintenance and repair.

Figure 6.13 MOT logo

Action

Using sources of information available to you, research three local garages and list the products and services that they provide.

Resolving customer problems

When a customer comes to a garage to book in their vehicle, there are two main types of work that could be done:

- **Proactive** work generally involves scheduled maintenance, including servicing and annual MOT testing.
- **Reactive** maintenance is usually the result of a breakdown or mechanical failure.

> **Key terms**
>
> **Proactive** – taking action before something has happened.
>
> **Reactive** – taking action as a result of a problem.

> **Skills for work**
>
> A customer comes to collect their car, which has failed its MOT. As the car is not roadworthy, the vehicle cannot be driven legally on a public road. You need to explain this situation to your customer.
>
> This situation requires that you exhibit particular personal skills. Some examples of these skills are shown in Table 6.1 on pages 304–305.
>
> 1. Using the examples given in Table 6.1, choose one skill from each of the following categories that you think you need to demonstrate in this situation.
>   - General employment skills
>   - Self-reliance skills
>   - People skills
>   - Customer service skills
>   - Specialist skills
> 2. Now rank these skills in order of importance, starting with the one that it is most important for you to have in this situation.
> 3. Which of the skills chosen do you think you are good at?
> 4. Which of the skills chosen do you think you need to develop?
> 5. How can you develop these skills and what help might you need?

CHECK YOUR PROGRESS

1. List the three main types of standards used to ensure the quality of products and services offered by a garage.
2. List four services offered by most garages.
3. What is the difference between proactive and reactive maintenance and repair?

Agreeing and undertaking work for customers

It is very important that the type and scope of work to be conducted on the customer's car is agreed in advance, and that the customer understands the costs involved. When agreeing work with a customer, you are entering into a service contract which could be written or verbal. Any changes to the work agreed must be authorised by the customer before any further action is taken.

Extent and nature of the work to be undertaken

Garages usually have a set pricing scheme for scheduled maintenance and repairs, and the processes and timescales can normally be accurately predicted. Take care when you are describing service schedules and costs to a customer, as you must take into account any extra maintenance procedures that occur within a particular service due to time and mileage. These extra maintenance procedures, such as timing belt replacement, brake fluid change, gearbox oil change, airbag replacement, etc., are normally charged as extra costs.

It can be more difficult to predict costs and timescales for reactive maintenance and repairs. Because of the nature of breakdowns and mechanical failures, many tasks will require some form of diagnosis before an estimate can be given. If diagnosis is involved, many garages operate a policy of conducting some investigation work for a set price so that an estimate can be created to give the customer an idea of the costs and time involved. Some garages have a diagnostic labour rate that is different from the standard charge.

The difference between an estimate and a quote

When pricing a job for a customer, make sure you are clear whether you are giving the cost of repairs as a quotation or an estimate.

- A quotation is normally a fixed price, which can be legally binding.
- An estimate is normally an approximate price that may be subject to change if the situation develops further during the repair process.

It is vitally important to keep the customer informed of progress and costs throughout the entire diagnostic and repair procedures, so that they are able to give their authorisation for work to be conducted.

The terms and conditions of acceptance

When accepting a vehicle for repair, the garage is entering into a contract with the customer. This contract may be verbal and informal or it may be formalised, in which case the customer will have to sign to say that they agree that the work can be undertaken and that they will pay upon completion. At this point, the terms and conditions of payment are sometimes agreed, including the methods by which the customer may pay. This is often known as the **customer service agreement** and, no matter whether it is formal or informal, it will usually be legally

> **Key term**
>
> **Customer service agreement** – a contract between the organisation and the customer that agrees the level of service to be provided and what work will be undertaken.

binding. As work is conducted, it is common for situations to develop or change. Any changes will affect the contract that has been agreed with your customer. The contract will need to be updated so that the customer is aware of any changes to timescales and costs.

Factors affecting timescales and costs

A number of factors impact on the timescales and costs involved with the repair of the customer's car. Some of these will be fixed and you can explain them to the customer at the time when you estimate and book in a job. Other factors will change as work is being conducted, and the customer will need to be kept informed of these. Examples are shown in Table 6.8.

Table 6.8 Factors affecting timescales and costs

Factor	Reason and effect
Availability of equipment	During the diagnosis of a customer's car, you may discover that a specialist piece of equipment is required to conduct the repair. The equipment may need to be borrowed or hired from the manufacturer, adding extra cost and time to the job.
Availability of technicians	Staffing levels within a workshop will have an effect on how and when a repair can be conducted. Many workshops have technicians who specialise in certain types of diagnosis and repair. Depending on the task, there may only be one or two technicians with the skills required to do the work required. If technicians are on holiday or on sick leave, this will affect how long the job may take.
Workshop loading systems	Many garage workshops operate a booking system in order to make sure that all of the technicians, equipment and space are utilised in an efficient manner. This is known as a workshop loading system. If an unexpected problem occurs during the repair of a customer's car, this may mean that a technician or piece of equipment becomes unavailable. This will affect the workshop loading plan and the timescale of the job will also be affected.

How to access costing and work completion time information

No matter what job the customer books their car into your garage for, they will want to know roughly how much it will cost to repair. There are often set repair times and costs for scheduled maintenance, but other types of maintenance and repair will need to be estimated. Manufacturers supply given times for the inspection, replacement and repair of most vehicle components, which you can access using manuals or computer-based equipment. Repair times are normally shown to two decimal places, with 0.1 of an hour equalling six minutes. The table in Figure 6.14 gives examples of repair times.

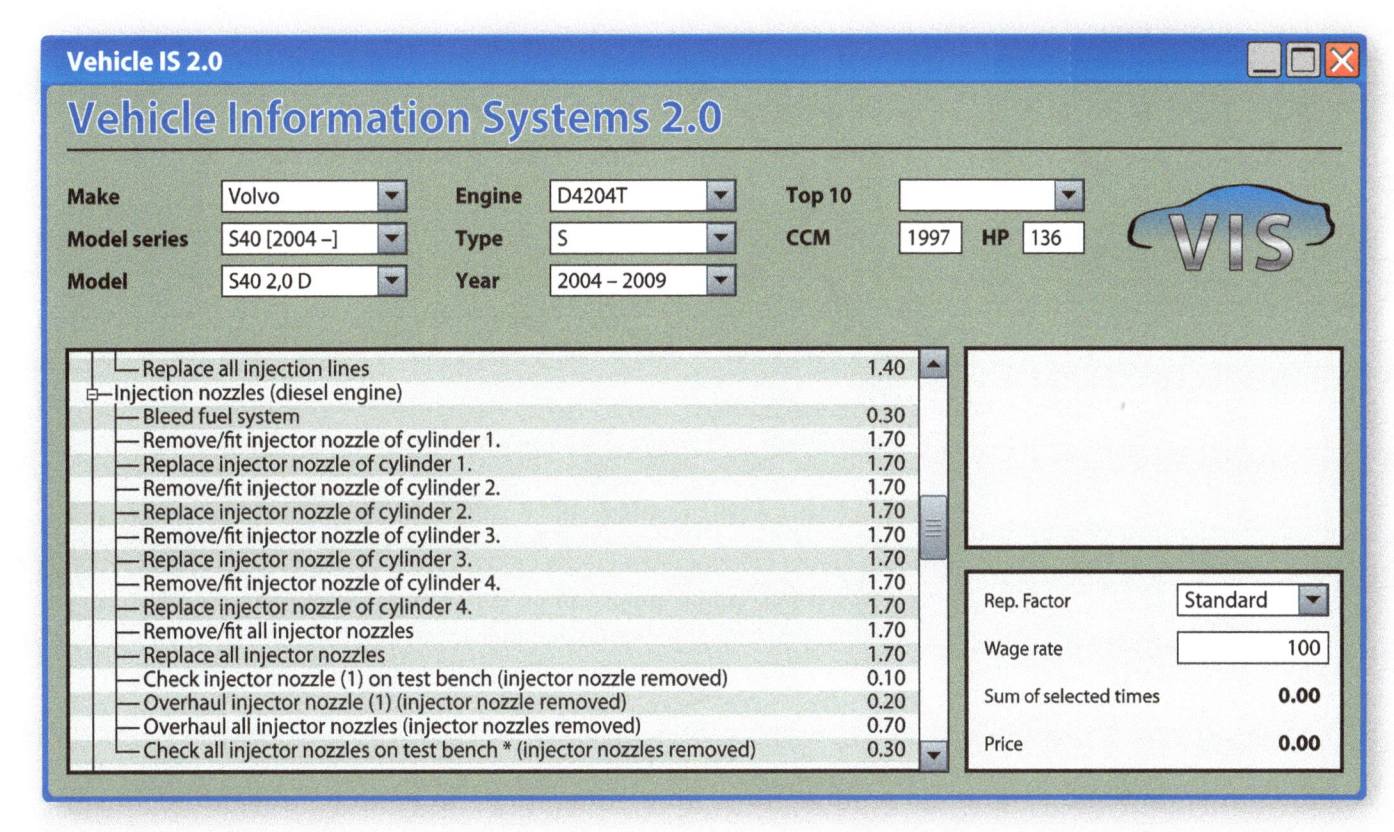

6 Identifying & agreeing motor vehicle customer service needs

Figure 6.14 Vehicle repair times

Vehicle information systems, servicing and repair requirements

During servicing and repair, you will need to have access to information. Table 6.9 gives some examples of information that you may need and where you can find it.

Table 6.9 Information needed during service and repair

Information type	Information source
Technical data, including diagnostics	Vehicle identification numbers (VIN)
	Technical data manuals
	Workshop manuals
	Wiring diagrams
	Safety recall sheets
	Technical help lines
	Advice from master technicians/colleagues
	Parts suppliers/catalogues
	Diagnostic trouble codes
	Oscilloscope wave forms

CONTINUED ▶

Information type	Information source
Servicing to manufacturer requirements/ standards	Manufacturer's literature Manufacturer's technical help line
Repair/operating procedures	Vehicle handbook Workshop manuals Manufacturer's technical help line
MOT standards/ requirements	The Vehicle and Operator Services Agency (VOSA) provides a range of licensing, testing and enforcement services with the aim of improving the roadworthiness standards of vehicles. VOSA oversees the MOT scheme for quality and standards.
Quality controls – interim and final	Company policy and standards Service or maintenance schedules Compliance with BSI Kitemark standards Involvement with Automotive Technician Accreditation (ATA) (To visit the ATA website, please go to hotlinks and click on this chapter.)
Requirements for cleanliness of vehicle on return to customer	Company policy and standards Post-work inspection sheets
Handover procedures	Company policy and standards Service history and handbook

Pre-work and post-work check sheets

It is good practice to perform pre-checks before you carry out any service. These checks usually include a visual inspection of the vehicle's exterior and interior for any damage or cosmetic faults. Many garages use pre-check sheets to identify and record any damage. By completing this inspection before you begin work, you can get the customer to sign the check sheet and confirm the condition of the vehicle prior to the repair. This will help to avoid any potential conflict after the repair if the customer claims they didn't know about any existing damage. Ideally, the customer should be present while the check is completed, so that any queries can be cleared up immediately. A typical process involved in carrying out pre-work checks is shown in Figure 6.15.

To meet the customer's expectations, you should carry out all service and repairs so that they conform to a high standard. You also need to return the vehicle in a clean and acceptable condition following any work done. You will need to complete a post-work check sheet by referring to the notes written on the original pre-work check sheet and the road test checklist. This will make sure that you haven't missed anything and that you send the vehicle back to the customer in a roadworthy condition.

6 Identifying & agreeing motor vehicle customer service needs

```
Collect keys from service reception
            ↓
      Are you inside the vehicle?
   YES ↙                    ↘ NO
```

YES branch:
- Check the seats for rips, tears and marks
- Check the trim and dashboard for rips, tears and marks
- Check to ensure no equipment or controls are missing, e.g. CD player
- Check floor mats are present and are not damaged or soiled

NO branch:
- Check the security and condition of bumpers
- Check the security and condition of light lenses
- Check the security and condition of door mirrors
- Check the condition of body panels for dents and scratches
- Check the security and condition of alloy wheels/wheel trims
- Check the condition of body windscreen and other glazing

↓

- Document any faults and gain customer confirmation of condition
- Request location of service book and wheel nut locking key if required

↓ ↓

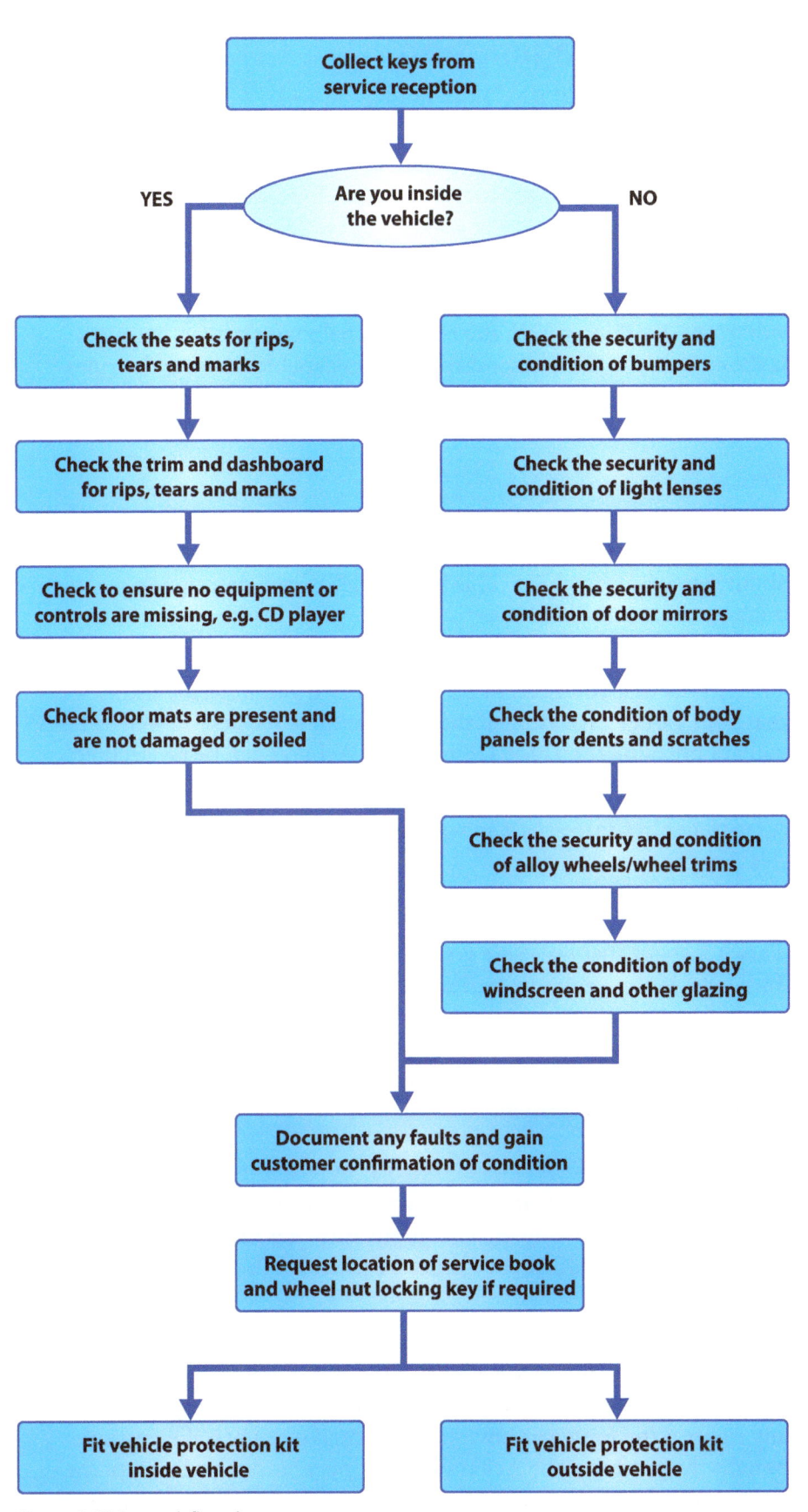

- Fit vehicle protection kit inside vehicle
- Fit vehicle protection kit outside vehicle

Figure 6.15 Pre-work flow chart

Action

Using a pre-work check sheet, examine a vehicle in your workshop and list all the faults and issues that you find. Compare your list with one made by someone else who has examined the same vehicle.

CHECK YOUR PROGRESS

1. What is the difference between an estimate and a quote?
2. What is a customer service agreement?
3. List eight information sources needed to undertake customer service and repair requirements.

333

Consumer legislation

Your organisation will have rules in place in order to comply with legislation. Some legislation applies to all workplaces, for example laws relating to data protection and health and safety. Other legislation is specifically designed to protect consumers, for example laws relating to the description and quality of goods that are sold.

Knowing about the legislation that is relevant to your job will help you to understand the reason for the rules you have to follow in your workplace. Working within the rules will help your career, because it shows that you are a responsible person capable of using the rules in order to deliver great service. There are advantages for both your customers and your employer of you doing so.

Trade Descriptions Act 1968

This Act states that traders must not falsely describe something that is on sale (it is a criminal offence to do so) and must not make false claims about services or facilities. This law applies to any description a garage might make, such as an advertisement in your service reception or a verbal description given by one of the members of staff.

This means that you need to be careful when you describe the services that your garage provides – if the description is reckless as well as false, you are breaking the Trade Descriptions Act and can be prosecuted under criminal consumer law.

A customer has three years in which to take any legal action under this Act.

Consumer Protection Act 1987

This Act states that only safe goods should be put on sale. Under the 28th rule, it also prohibits misleading price indications. When displaying goods that are being sold at a sale price, both the previous price and the sale price of the goods must be stated. The goods must have been on sale at the previous price for at least 28 consecutive days in the last six months at the same branch.

The Sale of Goods Act 1979 (as amended)

The Sale of Goods Act lays down several conditions that all goods sold by a trader must meet. The goods must be **merchantable** and of a satisfactory quality; they must be **as described** and **fit for purpose**.

To be of a satisfactory quality, goods must have nothing wrong with them (unless any defect was pointed out at the time of sale) and should last a reasonable time. This means that the Act does not give a customer any rights if the fault was obvious or pointed out when the customer bought the product. The Act covers the appearance and finish of the goods, as well as their safety and durability; this is known as **product liability**.

> **Key terms**
>
> **Merchantable** – suitable for purchase or sale.
>
> **As described** – this refers to any advertisement or verbal description made by the trader. For example, if an engine oil is described as fully synthetic then it must be so.
>
> **Fit for purpose** – good enough to do the job it was designed to do. This covers not only the obvious use or purpose of an item, but also anything you say the item will do when trying to sell the product.
>
> **Product liability** – the responsibility of the manufacturer or trader to compensate the customer for selling them a faulty product.

If any product bought by the customer does not meet any of the conditions set out in the Sale of Goods Act, the customer is entitled to a full refund. You cannot expect them to accept a repair, replacement or credit note instead. However, the customer does not have any rights to an automatic refund if he or she has had a change of mind, made a mistake and bought the wrong product, or been told about the fault before the purchase was made.

Supply of Goods and Services Act 1982

This Act covers work done and the products supplied by tradesmen and professionals. The Supply of Goods and Services Act makes the following recommendations:

- A tradesman or professional has a duty of care towards the customer and his or her property.
- Any price stated or agreed with the customer must be honoured. Where you and your customer have not agreed a price, the customer does not have to accept an outrageous bill. All the customer has to pay is what he or she considers is reasonable. A reasonable charge is considered to be the amount that other similar garages in the same geographical area would charge for the same job.
- The work must be done to a reasonable standard and at a reasonable cost (if not otherwise agreed in advance).

The Data Protection Act 1998

Most garages store information about their customers, either in written or electronic formats. Types of information stored can include:

- names
- addresses
- contact details
- vehicle details
- service history
- financial and payment history.

This information is known as data, and there are two types of personal data. Straightforward personal data includes information such as:

- names
- addresses
- banking details.

Sensitive personal data includes information such as:

- racial or ethnic origin
- political opinions
- religion
- membership of a trade union
- health
- criminal record.

People want to know that information about them is being kept private and that it will be used appropriately. Customers may have concerns about the information being misused or falling into the wrong hands.

For example, customers might have concerns about:
- Who can access the information?
- How accurate is it?
- Is it being copied?
- Is it being stored without their permission?

The Data Protection Act governs the use of personal data in the United Kingdom. The Act does not stop organisations storing information about people; it just ensures that they follow rules about how the data is stored. Some of these rules include:

- Data should only be used for the reasons given and not disclosed to unauthorised people.
- Data cannot be sold or given away without authority to do so.
- You must only hold enough detail to allow you to do the job for which the data is intended.
- The information you request from the customer must be obtained only for a genuine reason.
- You must keep the data away from people who are not authorised to access it. So, for example, leaving information out on a desk or storing it on a computer without password protection may mean it could be misused.
- Data should not be kept for longer than is necessary, although the Act does not state how long that should be.

Customers' rights

Under the Data Protection Act 1998, people have rights concerning the data kept about them. These include:

- Right of subject access: Anyone can request to see the personal data held about him or her.
- Right of correction: You are obliged to correct any mistakes in data held once these have been pointed out.
- Right to prevent distress: This right prevents the use of information if it will be likely to cause a person distress.
- Right to prevent direct marketing: People can prevent their data being used in attempts to sell them things, for example by direct mail or cold calling.
- Right to prevent automatic decisions: People can specify that they do not want a data user to make automated decisions about them, for example computerised credit scoring of a loan application.
- Right of complaint to the **Information Commissioner**: People can ask for the use of their personal data to be reviewed by the Information Commissioner, who can enforce a ruling using the Act. The Commissioner may inspect the computer on which data is held to help the investigation.
- Right to compensation: People can use the law to get compensation for damage caused if personal data about them is inaccurate, lost or disclosed without their permission.

> **Key term**
>
> **Information Commissioner** – an independent official responsible for ensuring that the Data Protection Act is enforced. The work of the Commissioner is supported by the Information Commissioner's Office.

Skills for work

A customer is booking their car in for some work to be done. When you take the customer's personal details, they express concern about how the information will be used. You need to reassure your customer that it will be held in compliance with the Data Protection Act.

This situation requires that you exhibit particular personal skills. Some examples of these skills are shown in Table 6.1 on pages 304–305.

1. Using the examples given in Table 6.1, choose one skill from each of the following categories that you think you need to demonstrate in order to reassure your customer.
   - General employment skills
   - Self-reliance skills
   - People skills
   - Customer service skills

2. Now rank these skills in order of importance, starting with the one that it is most important for you to have in this situation.
3. Which of the skills chosen do you think you are good at?
4. Which of the skills chosen do you think you need to develop?
5. How can you develop these skills and what help might you need?

Health and safety

In your working environment, there may well be a number of potential hazards to yourself, your customers and your colleagues. Some examples of hazards are:

- chemical substances
- dust and fumes
- excessive noise
- moving vehicles
- moving parts in machinery
- electricity
- extremes of heat and cold
- uneven floors
- exposed wiring
- cabling over the floor or ground.

The Health and Safety at Work Act 1974 (HASAWA) covers the responsibilities that employers have to their employees as well as to customers who are on their premises. An employer needs to take steps to put in place measures to control health and safety risks as far as is **reasonably practicable**. (For more information on the Health and Safety at Work Act, see Chapter 1, page 6.)

Key term

Reasonably practicable – can be carried out without incurring excessive effort or expense.

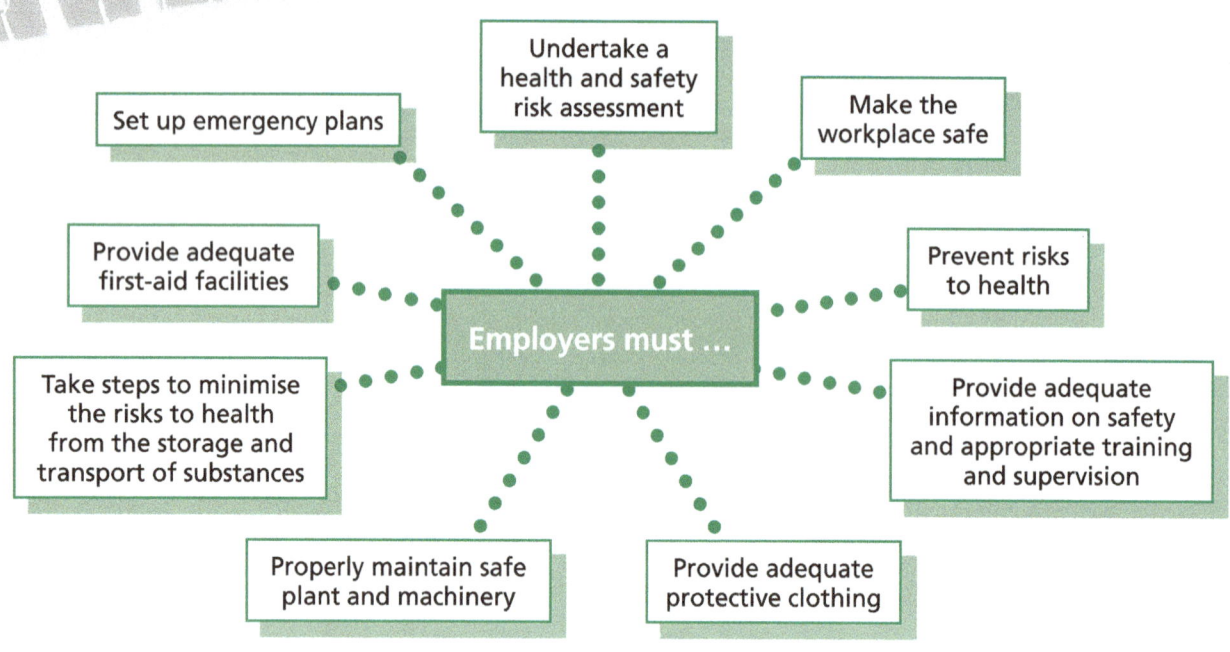

Employers must ...
- Set up emergency plans
- Undertake a health and safety risk assessment
- Make the workplace safe
- Provide adequate first-aid facilities
- Prevent risks to health
- Take steps to minimise the risks to health from the storage and transport of substances
- Provide adequate information on safety and appropriate training and supervision
- Properly maintain safe plant and machinery
- Provide adequate protective clothing

Figure 6.16 An employer's responsibilities to their employees with regard to health and safety

Discrimination

It is important to treat people fairly and equally regardless of who they are, where they live or how much you like or dislike them. In other words, while you are at work you should treat people as individuals – what makes one person different from another does not mean that he or she should have any advantage or disadvantage over anybody else in relation to customer service delivery. All your customers should be treated fairly, regardless of their age, gender, race, sexual orientation, disability, gender reassignment, religion or beliefs.

The Equality Act 2010

The Equality Act 2010 came into force on 1st October 2010 and replaces the Disability Discrimination Act 1995 (DDA). It is one of the most important pieces of legislation you need to know about regarding disability discrimination and equal rights. People who access your goods, facilities or services are protected from discrimination because of certain 'protected characteristics'. These are:

- age
- disability
- gender reassignment
- marital or civil partnership status
- pregnancy and maternity
- race – this includes ethnic or national origins, colour and nationality
- religion or belief
- sex
- sexual orientation.

6 Identifying & agreeing motor vehicle customer service needs

Definition of disability

Disability has a broad meaning. It is defined as a physical or mental **impairment** that has a **substantial** and long-term adverse effect on the ability to carry out normal day-to-day activities.

Some people are automatically protected as disabled people by the Act, including those with cancer, multiple sclerosis and HIV/AIDS.

Helping disabled people

Disabled people must not be treated less favourably than others because of their disability. Businesses have an obligation to make reasonable adjustments to help disabled people access their goods, facilities and services. Some organisations build ramps to ensure that wheelchair users can access their premises or provide information in Braille for visually impaired customers.

What is a reasonable adjustment? This will depend on a number of circumstances, including cost. The Equality Act 2010 requires that service providers must think ahead and take steps to put right things that may stop disabled people using their goods, facilities or services. This means that your organisation should not wait until a disabled person experiences difficulties using a service, as it may then be too late to make the necessary adjustment.

Reasonable adjustments a workplace might make to improve access for disabled people include:

- Changes in the way things are done: for example, changing practices, policies or procedures that could put disabled people at a substantial disadvantage, such as amending a 'no dogs' policy.
- Changes to premises: for example, altering the structure of the building to improve access, such as fitting handrails alongside steps.
- Providing auxiliary aids and services: for example, providing information in large print or an induction loop for customers with hearing aids.

To find out more information about the Equality Act 2010, please go to hotlinks and click on this chapter to visit the Home Office Equalities website and the Directgov website.

> **Key terms**
>
> **Impairment** – covers long-term medical conditions such as asthma or diabetes and fluctuating or progressive conditions such as rheumatoid arthritis and motor neurone disease. Mental impairment includes learning difficulties such as dyslexia and learning disabilities such as Down's syndrome and autism.
>
> **Substantial** – more than minor or trivial.

> **Did you know?**
>
> The Equalities and Human Rights Commission (EHRC) is the statutory body responsible for protecting, enforcing and promoting equality. It is recommended that you visit their website (by going to hotlinks and clicking on this chapter) for guidance and information, including equality matters within different industry sectors. You can also use this website to keep up to date with changes to equal opportunities legislation.

CHECK YOUR PROGRESS

1. Which piece of legislation is concerned with the storage of customer information?
2. Which piece of legislation is concerned with products and work done?
3. Which piece of legislation is concerned with descriptions of something on sale?

FINAL CHECK

1. What is the best way for you to help build your organisation's customer service reputation?
   a by attending training courses and reading books
   b by helping colleagues exceed their sales targets
   c by keeping to prices and offering a consistent quality of service
   d by participating in team meetings about sales targets

2. Customer satisfaction is all about:
   a the feeling a customer gets when he or she is happy with the service provided
   b making sure the customer is happy with the price paid for a service
   c making customers aware of additional products or services
   d inviting customers to events that allow them to get to know staff and the management team

3. Which statement contains examples of **unlawful** reasons to discriminate against people?
   a because of the time of day they require help
   b because of the number of staff at work
   c because of disability, age, religion or belief
   d because of budgets, cost, time or safety

4. Under the Data Protection Act 1998, personal information must be:
   a obtained in writing from customers and businesses
   b stored on an easily accessible computer system
   c provided in writing to anybody who requests it
   d stored in a way that prevents unauthorised access

5. Which of the following are designed to protect the interests of customers?
   a consumer legislation and data protection legislation
   b complaints procedures and feedback systems
   c advertising and marketing promotions
   d health and safety legislation and help desks

6. Why should records that include customers' personal information remain confidential?
   a competitors might try to obtain information about customers to win business
   b not all colleagues can be trusted to handle private information in an appropriate way
   c organisations need to keep records of customer types to develop their products and services
   d by law customers have the right to have their personal information treated with respect

7. What is the main difference between listening and hearing?
   a hearing involves asking questions to check which product is needed
   b listening involves ensuring a customer is comfortable when talking
   c listening means trying to understand the true meaning of the words
   d hearing means you need a quiet workplace

8. What do you need to do to actively listen to a customer?
   a listen to words and tone, observe body language and give feedback
   b listen to words and tone, offer a private place to talk and ask questions
   c listen to words and tone, use appropriate gestures and seek help if needed
   d listen to words and tone, offer advice, clarify jargon and seek agreement

9 Why is tone of voice important in effective communication?

   a because legislation requires people to be polite and courteous
   b because managers expect staff to exceed sales targets
   c because a customer always wants to get more than they asked for
   d because a customer will interpret emotions as well as words

10 Which of the following are examples of positive body language?

   a smiling and maintaining appropriate eye contact
   b keeping queues to a minimum
   c yawning and apologising for doing so
   d answering the telephone quickly

PREPARE FOR ASSESSMENT

The information contained in this chapter, as well as continued practical assignments, will help you to prepare for both the end-of-unit tests and diploma multiple-choice tests. This chapter will also help you to develop customer service routines that promote good practice.

You will need to be familiar with:

- Legal and organisational requirements and procedures
- Obtaining and providing relevant information to customers
- How to communicate and care for customers
- Agreeing and undertaking work for customers
- Company products and services
- Using systems for recording work undertaken

This chapter has given you a brief overview of how to identify and agree customer service needs within an automotive setting, and it will help you with both theory and practical assessments. It is possible that some of the evidence you generate when dealing with customers may contribute to more than one unit. You should ensure that you make best use of all your evidence to maximise the opportunities for cross-referencing between units.

You should choose the type of evidence that will be best suited to the type of assessment that you are undertaking (both theory and practical). These may include:

Assessment type	Evidence example
Workplace observation by a qualified assessor	A direct observation by your assessor of you working in the service reception and dealing with customers
Witness testimony	A signed statement or letter from a customer, stating that they were pleased with the service they received from you when they brought their car into your garage
Computer-based	A printout showing an estimate and work request that you produced for a customer during a routine enquiry
Audio recording	A timed and dated audio recording of you describing the company procedures for when a customer comes to collect their vehicle after repairs
Video recording	Short video clips showing you preparing an estimate for a customer during a repair enquiry
Photographic recording	Photographs showing you presenting a good corporate image while working in the service reception
Professional discussion	A recorded discussion with your assessor about how you dealt with a customer complaint
Oral questioning	Recorded answers to questions asked by your assessor, in which you explain how you gather customer information and ensure that you comply with the Data Protection Act
Personal statement	A written statement describing how you explained the work conducted on a customer's car in a non-technical manner

CONTINUED ▶

Assessment type	Evidence example
Competence/skills tests	A practical role-play task arranged by your training organisation, showing how you would deal with customers in a service reception
Written tests	A written answer to an end-of-unit test to check your knowledge and understanding of a customer service agreement
Multiple-choice tests	A multiple-choice test set by your awarding body to check your knowledge and understanding of customer service needs
Assignments/ projects	A written assignment arranged by your training organisation requiring you to show an in-depth knowledge and understanding of approved communication techniques

Before you attempt a theory end-of-unit or multiple-choice test, make sure you have reviewed and revised any key terms that relate to the topics in that unit. Ensure that you read all the questions carefully. Take time to digest the information so that you are confident about what each question is asking you. With multiple-choice tests, it is very important that you read all of the answers carefully, as it is common for two of the answers to be very similar, which may lead to confusion.

For practical assessments, it is important that you have had enough practice and that you feel that you are capable of passing. It is best to have a plan of action and work method that will help you.

Make sure that you have the correct technical information, in the way of vehicle data, and job times. Always check that your customer understands the work to be conducted.

When undertaking any practical assessment, always make sure that you are using approved techniques that comply with company and legal requirements.

Good luck!

COMMON ACRONYMS/ABBREVIATIONS

A – Amperes

A/C – Air Conditioning

A/F – Air/Fuel Ratio

A/T – Automatic Transmission

AAC – Auxiliary Air Control valve

AAT – Ambient Air Temperature

ABD – Automatic Brake Differential

ABS – Anti-lock Braking System

ABV – Air Bypass Valve

AC – Alternating Current

ACC – Air Conditioning Clutch

ACC – Automatic Climate Control

ACR – Air Conditioning Relay

ACR4 – Air Conditioning Refrigerant, Recovery, Recycling, Recharging

ACV – Air Control Valve

ADU – Analogue-Digital Unit

AFC – Air Flow Control

AFL – Advanced Front Lighting System

AFM – Air Flow Meter

AFR – Air Fuel Ratio

AFS – Air Flow Sensor

Ah – Amp hours

AIR – Secondary Air Injection System

AIS – Automatic Idle Speed

ALC – Automatic Level Control

AM – Amplitude Modulation

API – American Petroleum Institute

APS – Atmospheric Pressure Sensor

ARC – Automatic Ride Control

ARS – Automatic Restraint System

ASARC – Air Suspension Automatic Ride Control

ATC – Automatic Temperature Control

ATDC – After Top Dead Centre

ATF – Automatic Transmission Fluid

ATS – Air Temperature Sensor

AWD – All Wheel Drive

AWG – American Wire Gage

AYC – Active Yaw Control

B/MAP – Barometric/Manifold Absolute Pressure

BARO – Barometric Pressure

BCM – Body Control Module

BDC – Bottom Dead Centre

BEV – Battery Electric Vehicle

BHP – Brake Horsepower

BOB – Breakout Box

BP – Barometric Pressure

BPP – Brake Pedal Position Switch

BTDC – Before Top Dead Centre

BTS – Battery Temperature Sensor

Btu – British thermal unit

BUS N – Bus Negative

BUS P – Bus Positive

C – Celsius

CA – Cranking Amps

CAN – Controller Area Network

CANP – EVAP Canister Purge Solenoid

CAS – Crank Angle Sensor

CBW – Clutch By Wire

CC – Catalytic Converter

CC – Climate Control

CC – Cruise Control

CC – Cubic Centimetres

CCA – Cold Cranking Amps

CD – Compact Disc

CDI – Capacitor Discharge Ignition

CFC – Chlorofluorocarbons

CFI – Continuous Fuel Injection

CI – Compression Ignition

CKP – Crankshaft Position Sensor

Common Acronyms/Abbreviations

CL – Closed Loop

CLC – Converter Lockup Clutch

CLV – Calculated Load Value

CMP – Camshaft Position Sensor

CNG – Compressed Natural Gas

CO – Carbon Monoxide

CO₂ – Carbon Dioxide

COC – Conventional Oxidation Catalyst

COP – Coil On Plug Electronic Ignition

COSHH – Control Of Substances Hazardous to Health

CP – Canister Purge (GM)

CP – Crankshaft Position Sensor

CPP – Clutch Pedal Position

CPU – Central Processing Unit

CRC – Cyclic Redundancy Check

CRD – Common Rail Diesel

CRS – Common Rail System

CTP – Closed Throttle Position

CTS – Coolant Temperature Sensor

CV – Constant Velocity

CVT – Continuously Variable Transmission

DBW – Drive By Wire

DC – Direct Current

DC – Duty Cycle

DCS – Dual Clutch System

DI – Direct Ignition

DI – Distributor Ignition (System)

DIS – Direct Ignition (Waste Spark)

DIS – Distributorless Ignition System

DLC – Data Link Connector (OBD)

DMF – Dual Mass Flywheel

DOHC – Dual Overhead Cam

DPF – Diesel Particulate Filter

DRL – Daytime Running Lights

DTC – Diagnostic Trouble Code

DVD – Digitally Versitile Disc

EAIR – Electronic Secondary Air Injection

EBCM – Electronic Brake Control Module

EBD – Electronic Brake Force Distribution

EBP – Exhaust Back Pressure

ECC – Electronic Climate Control

ECM – Engine/Electronic Control Module

ECS – Emission Control System

ECT – Engine Coolant Temperature

ECU – Electronic Control Unit

EDC – Electronic Diesel Control

EECS – Evaporative Emission Control System

EEGR – Electronic EGR (Solenoid)

EEPROM – Electronically Erasable Programmable Read Only Memory

EFI – Electronic Fuel Injection

EFT – Engine Fuel Temperature

EGO – Exhaust Gas Oxygen Sensor

EGR – Exhaust Gas Recirculation

EGRT – Exhaust Gas Recirculation Temperature

EMF – Electromotive Force (voltage)

EMI – Electromagnetic Interference

EOBD – European On Board Diagnostics

EOP – Engine Oil Pressure

EOT – Engine Oil Temperature

EPA – Environmental Protection Act

EPB – Electronic Parking Brake

EPROM – Erasable Programmable Read Only Memory

EPS – Electronic Power-Assisted Steering

ESC – Electronic Stability Control

ESP – Electronic Stability Program

ESS – Engine Start-Stop

EVAP – Evaporative Emissions System

EVAP CP – Evaporative Canister Purge

FM – Frequency Modulation

FSD – Full Scale Deflection

FT – Fuel Trim

FWD – Front-Wheel Drive

GDI – Gasoline Direct Injection

GND – Electrical Ground Connection

GPS – Global Positioning System

H – Hydrogen

H$_2$O – Water

HC – Hydrocarbons

HCA – Hot Cranking Amps

HDI – High Pressure Direct Injection

HEGO – Heated Exhaust Gas Oxygen Sensor

HFC – Hydrogen Fuel Cell

HFC – Hydro-fluro Carbon

Hg – Mercury

HID – High Intensity Discharge (lighting)

HO$_2$S – Heated Oxygen Sensor

hp – Horsepower

HSE – Health and Safety Executive

HT – High Tension

HUD – Heads Up Display

Hz – Hertz

I/O – Input / Output

IA – Intake Air

IAC – Idle Air Control (motor or solenoid)

IAT – Intake Air Temperature

IC – Integrated Circuit

IC – Ignition Control

ICE – In Car Entertainment

ICM – Ignition Control Module

IFS – Inertia Fuel Switch

IGN – Ignition

IGN ADV – Ignition Advance

IGN GND – Ignition Ground

IPR – Injector Pressure Regulator

ISC – Idle Speed Control

ISO – International Standard of Organization

KAM – Keep Alive Memory

Kg/cm^2 – Kilograms/ Cubic Centimeters

kHz – Kilohertz

Km – Kilometers

KPA – Kilopascal

KPI – Kingpin Inclination

KS – Knock Sensor

KWP – Keyword Protocol

l – Liters

LCD – Liquid Crystal Display

LED – Light Emitting Diode

LHD – Left-Hand Drive

LOOP – Engine Operating Loop Status

LOS – Limited Operating Strategy

LPG – Liquid Petroleum Gas

LSD – Limited Slip Differential

LTFT – Long Term Fuel Trim

LWB – Long Wheel Base

M/T – Manual Transmission

MAF – Mass Air Flow Sensor

MAP – Manifold Absolute Pressure Sensor

MAT – Manifold Air Temperature

MFI – Multiport Fuel Injection

MIL – Malfunction Indicator Lamp

MPG – Miles Per Gallon

MPH – Miles Per Hour

mS or **ms** – Millisecond

mV or **mv** – Milivolt

N – Nitrogen

NCAPS – Non-Contact Angular Position Sensor

NCRPS – Non-Contact Rotary Position Sensor

NGV – Natural Gas Vehicles

Nm – Newton Meters

NOx – Oxides of Nitrogen

NTC – Negative Temperature Coefficient

O$_2$ – Oxygen

OBD I – On-Board Diagnostics Version I

OBD II – On-Board Diagnostics Version II

Common Acronyms/Abbreviations

OC – Oxidation Catalytic Converter

OD – Overdrive

OD – Outside Diameter

OE – Original Equipment

OEM – Original Equipment Manufacturer

OFN – Oxygen-Free Nitrogen

OHC – Overhead Cam Engine

OHV – Over Head Valve

OL – Open Loop

OS – Oxygen Sensor

P/N – Part Number

PAIR – Pulsed Secondary Air Injection

PAS – Passive Anti-Theft System

PAS – Power-Assisted Steering

PCB – Printed Circuit Board

PCM – Powertrain Control Module

PCV – Positive Crankcase Ventilation

Pd – Potential Difference (volts)

PFI – Port Fuel Injection

PGM-FI – Programmed Gas Management Fuel Injection (Honda)

PID – Parameter Identification Location

PKE – Passive Keyless Entry

POT – Potentiometer

PPE – Personal Protective Equipment

PPM – Parts Per Million

PPS – Accelerator Pedal Position Sensor

PROM – Programmable Read-Only Memory

PSI – Pounds per Square Inch

PTC – Positive Temperature Coefficient Resistor

PTO – Power Take Off (4WD Option)

PUWER – Provision and Use of Work Equipment Regulations

PWM – Pulse Width Modulation

RAM – Random Access Memory

RBS – Regenerative Braking system

RCM – Reserve Capacity Minutes

RDS – Radio Data System

REF – Reference

RFI – Radio Frequency Interference

RHD – Right-Hand Drive

RIDDOR – Reporting of Injuries Diseases and Dangerous Occurrences Regulations

RKE – Remote Keyless Entry

ROM – Read Only Memory

RON – Research Octane Number

RTV – Room Temperature Vulcanizing

RWD – Rear-Wheel Drive

SAE – Society of Automotive Engineers (Viscosity Grade)

SAI – Swivel Axis Inclination

SCR – Selective Catalytic Regeneration

SFI – Sequential Fuel Injection

SI – Spark Ignition

SIPS – Side Impact Protections System

SOHC – Single Overhead Cam

SPFI – Single Point Fuel Injection (throttle body)

SRI – Service Reminder Indicator

SRS – Supplementary Restraint System (airbag)

SRT – System Readiness Test

STFT – Short-Term Fuel Trim

SWB – Short Wheel Base

SWL – Safe Working Load

TAC – Throttle Actuator Control

TACH – Tachometer

TBI – Throttle Body Injection

TC – Turbocharger

TCC – Torque Converter Clutch

TCM – Transmission Control Module

TCS – Traction Control System

TD – Turbo Diesel

TDC – Top Dead Centre

TDI – Turbo Direct Injection

TOOT – Toe-Out On Turns

TP – Throttle Position

TPM – Tyre Pressure Monitor

TPP – Throttle Position Potentiometer

TPS – Throttle Position Sensor

TSB – Technical Service Bulletin

TV – Throttle Valve

TXV – Temperature-controlled Expansion Valve

TXV – Thermal Expansion Valve

UART – Universal Asynchronous Receiver-Transmitter

UJ – Universal Joint

USB – Universal Serial Bus

V – Volts

VAC – Vacuum

VAF – Vane Airflow Meter

VDP – Variable Diameter Pully

VDU – Visual Display Unit

VIN – Vehicle Identification Number

VPE – Vehicle Protective Equipment

VSS – Vehicle Speed Sensor

W/B – Wheelbase

WOT – Wide Open Throttle

WSS – Wheel Speed Sensor

YRS – Yaw Rate Sensor.

INDEX

Key terms are indicated by **bold type**.

0 per cent 35
10-minute rule 25
100 per cent 35

A

accelerometers 239
accidents, prevention of 10–11
accumulators 38
Ackerman steering principle 51–52
active air suspension systems 72
active hydro-pneumatic suspension systems 72–73
active listening 308–309
active yaw control (AYC) 49
actuators 25, 32, 82, 146, 176, 252
adaptions 26
advanced 90
advanced key 233
aerial systems 220–21
aftermarket suppliers 87
air-assisted suspension systems 70–71
air conditioning 165–67
air conditioning recovery stations 84
air induction
    air temperature sensors 143
    airflow meters 138–39
    auxiliary air valves 137
    barometric pressure sensors 140
    engine coolant temperature sensors (ECT) 142–43
    engine speed sensors 143
    hot wire mass airflow sensors (MAF) 140
    lambda sensors 143–45
    manifold absolute pressure sensors (MAP) 139
    plenum chambers 137–38
    quantity measurement 138–40
    throttle position sensors 137, 141
    throttle position switches 140–41
    wideband oxygen sensors 145
air temperature sensors 143
airbags 234–39
airflow meters 138–39
alarm systems 229
alternative fuel vehicles 158–63
alternator output 192
alternators 205, 206
ammonia 159
amplifiers 112, 221
amplitude 194
amps 179
analogue 37
anti-collision 219
anti-lock braking systems (ABS)
    brake-by-wire systems 39
    electronic control units (ECU) 38
    fault diagnosis 40–43
    mechanical systems 40–41
    modulator units 37–38
    operation and layout 35–36, 38–39
    wheel speed sensor alignment 43–45
    wheel speed sensors 36–37
appearance, personal 307–308, 311–12
arc 210
armatures 38
as described 334
atomised 146
attitude 49
audible continuity testing 189–90
automatic brake differential (ABD) 293
automatic transmission 274–783, 296
auxiliary 184
auxiliary air valves 137
auxiliary control systems
    batteries 203–208
    comfort and convenience systems 214–19
    communication systems 225–26
    diagnosis 201
    electrical faults 201–203
    electrical principles 179–81
    faults 214
    health and safety 174
    heating, cooling and ventilation 217–18
    in-car entertainment (ICE) system 219–23
    information sources 175
    lighting 208–14
    locking systems 232–33
    multiplexing 198–201
    navigation systems 223–25
    network systems 198–201
    record keeping 245
    security systems 227–30
    symptoms 191–93
    systems and components 176
    tooling 177–78, 182–97
    windscreen washers 232
    wiper systems 230–32

B

back-probe 183
barometric pressure sensors 140
barrelling 98
batteries 203–208
battery testers 193
baulk rings 270–71
bent connecting rods 96
bent valves 98
bevel 287
bioalcohol/ethanol 159
biodiesel 159
biogas 158
biogenic 158
bleeding 18
body language 309–310

bottom dead centre (BDC) 106
brake dynamometers 34
brake servos 35
braking systems
    active yaw control (AYC) 49
    anti-lock (ABS) 35–4536
    electronic brake force distribution 46
    electronic parking brakes (EPB) 47–48
    electronic stability control (ESC) 48–49
    electronic stability programs (ESP) 48–49
    emergency brake assistance 47
    emergency braking facility 47
    hybrid vehicle regenerative braking 50
    operation of 35
    symptoms of faults 41–43
    traction control systems (TCS) 48
    wheel speed sensors 36–37
brushes 133
bump 66
burns 19
bus systems 198–201

C

cadence braking 35
cam belt failure 97
cam lobe/follower wear 97
camber angle 54, 56
CAN bus 198–200
capacitor discharge ignition (CDI) 119–20
capacitors 119
case hardened 98
cassette tapes 222
caster angle 54–55, 56
catalytic convertors 127
cavitation 61
CD players 222–23
central locking 232
centrifugal force 264
charging batteries 205–207
chassis system faults. *see also* braking systems
    health and safety 30
    information sources 31
    power assisted steering (PAS) 57–65
    recommendations, making 75
    record keeping 75
    steering geometry 51–56
    suspension systems 66–74
    systems and components 32
    tools 33–34
checklists, diagnosis 26
chemical injuries 19
chemicals, safe use of 9, 10
CI fuel systems (diesel)
    common rail direct injection 149

349

CI fuel systems (diesel) – *contd.*
  electronic diesel control (EDC) 149
  faults and symptoms 153
  filters, fuel 150
  glow plugs 152–53
  injectors, fuel 151
  pumps, fuel 150, 151
  rail and pressure regulator 151
  sensors 152
  supply circuit 150–52
  tanks, fuel 150
cinching 239
climate control 167–68
clock springs 237
closed loop 144
clutches
  clutch by wire (CBW) systems 262
  coil and diaphragm spring 256–58
  dual mass flywheels (DMF) 262–63
  engagement/disengagement 258–61
  friction 255–56
  function of 255
  seamless shift dual clutch systems (DCS) 273–74
  symptoms and faults 296
code readers 34, 84, 178, 196–97, 254
collaboration 162
collaborative motor drive 163
combustion
  common terms 123–25
  control of emissions 121–22, 126–29
  diagnosis of emissions 122
  MOT requirements 125
  standards for emissions 126
comfort and convenience systems 214–19
common rail direct injection diesel 149
communication skills
  body language 309–310
  listening skills 308–309
  non-technical explanations 317–18
  telephone communication 319
  telephones 315–16
  written communication 316
communication systems 225–26
commutator 133
complaints 321–25
compliance 64
compressed natural gas (CNG) 158
compression ratio 87
compression testing 99–100
compromised 35
connecting rods 96
consciousness, loss of 19
constant velocity (CV) joints 285–86
consumer legislation 334–37
Consumer Protection Act 1987 334
contact breakers 110
continuity 184
continuously variable transmission (CVT) 280–83
Control of Substances Hazardous to Health Regulations 2002 (COSHH) 9, 10
control valves 58

controlled waste 12
controller area network (CAN) 198–200
converging 208
convex 209
coolant pressure testing 98
copyright 162
corporate manslaughter 5
corruption 200
coupling point 264
crank 230
crankshaft damage 97
creep 264
current, electrical, measuring 189
customer expectations 319
customer service
  agreeing and undertaking work 329–33
  body language 309–310
  checking your understanding 313
  complaints 321–25
  consumer legislation 334–37
  disability, people with 339
  discrimination 338–39
  estimates/quotes 329
  expectations, fulfilling 319–21
  feedback from customers 323–24
  health and safety 337–38
  image, personal 307–308
  information sources 331–32
  listening skills 308–309
  non-technical explanations 317–18
  obtaining/providing information 317–25
  organisational requirements 306
  personal skills 304–305, 307, 315, 318, 328, 337
  policy 306
  positive/negative language 314
  pre-/post-work checks 332–33
  products and services 326–28
  questions and comments, dealing with 311–12
  repair times 330–31
  skills for work 307, 315, 318, 328, 337
  standards 326
  telephone communication 315–16, 319
  tone of voice 314
  written communication 316
customer service agreements 329
cylinder balance testing 99–101
cylinder block testers 102
cylinder head cracks 97
cylinder liner damage 96

D
Data Protection Act 1998 335–37
dead reckoning 223–24
delta time 134
density 87
deployment 235
depression 85
diagnosis. *see also* chassis system faults
  10-minute rule 25

anti-lock braking systems (ABS) 40–43
  checklist 26
  gearboxes 278–79
  ignition systems 107
  multiplex and network 201
  Ohm's law for 182
  power assisted steering (PAS) 59–62
  pre-diagnostic questionnaire 24
  record keeping 25–26
  routes to 22–26
  supplementary restraint systems (SRS) 241–43
  suspension systems 74
  transmission and driveline 295–98
  turbochargers/superchargers 90
diesel fuel systems
  common rail direct injection 149
  electronic diesel control (EDC) 149
  filters, fuel 150
  glow plugs 152–53
  injectors, fuel 151
  pumps, fuel 150, 151
  rail and pressure regulator 151
  sensors 152
  supply circuit 150–52
  tanks, fuel 150
diesel particulate filters (DPF) 128
differentials 287–93
diffuse 208
digital 37
digital principles 32, 82, 176, 252
digital radios 221
diode testing 190
direct ignition systems 119
disability 339
discrimination 338–39
disengaging 167
disseminate 201
distributorless systems 117
dog teeth 270
drag 258, 291
dress code 307
drive shafts 285–86
dry clutch 256
dry test 100
dual mass flywheels (DMF) 262–63
dump valves 87
duty cycle 32, 82, 176, 252
DVDs 223

E
E-OBD 197
eccentric 94
ecotoxic 12
efficient 160
electric motors 160–61
electric windows 214–16
electrical injuries 18
electrical principles 179–81
electro-hydraulic power assisted steering (PAS) 62
electrodes 210
electrolysis 160
electromagnetic 167

Index

electromagnetic interference (EMI) 190
electronic brake force distribution 46
electronic control principles 154
electronic control unit (ECU) 32, 82, 176, 252
electronic diesel control (EDC) 149
electronic fuel injection (EFI) 130, 146–49
electronic parking brakes (EPB) 47–48
electronic power assisted steering (PAS) 62–64
electronic stability control (ESC) 48–49
electronic stability programs (ESP) 48–49
electrons 160
ellipsoidal 209
emergencies, first aid for 16–19
emergency brake assistance 47
emergency braking facility 47
emissions
   common terms 123–25
   control of 121–22, 126–29
   diagnosis of 122
   MOT requirements 125
   standards for 126
empathy 310
endoscopy 103
engaging 167
engine block structural failure 97
engine coolant temperature sensors (ECT) 142–43
engine speed sensors 143
engine systems
   air conditioning 165–67
   CI fuel systems 149–53
   climate control 167–68
   combustion 121–28
   health and safety 80
   heating 164
   ignition systems 106–21
   information sources 81
   management, engine 154–57
   measurement, engine 104
   mechanical 96–105
   pressure-charged induction systems 85–90
   recommendations, making 169
   record keeping 169
   restoration and repair 98–105
   restoration of components 104–105
   SI fuel systems (petrol) 129–48
   starting engines 207–208
   superchargers 88–90
   systems and components 82
   tooling 82–84
   turbocharging 85–88
   valve mechanisms 90–95
entertainment systems 219–23
environmental protection 12
epicyclic gear train 275–76
Equality Act 2010 338
estimates/quotes 329
ethanol 159
exhaust emissions. *see* emissions

exhaust gas analysers 84
exhaust gas recirculation (EGR) 128
expectations, customer's, fulfilling 319–21
extinguishers, fire 14–15

F
failsafe 38
faults 22
   air-conditioning 168
   climate control 168
   electrical 201–203
   fuel systems 153
   ignition systems 120
   infotainment systems 226
   lighting 214
   locking systems 233
   mechanical engine faults 96–105
   security systems 230
   supplementary restraint systems (SRS) 241–43
   transmission and driveline 295–98
   turbochargers/superchargers 89–90
   variable valve control 95
   windscreen systems 218
   wiper systems 232
feedback 323–24
fibre optic principles 32, 82, 176, 252
fibre optics 213–14
field winding 134
filament 208
filters, fuel 135, 150
final drive 286–87, 293–95, 298
fire safety 14–15
first aid 16–19
first time fix 20
fit for purpose 334
fluid couplings 263–65
fossil fuel 158
four-wheel drive vehicles 293–95
four-wheel steering 64–65
free play 40, 258
freewheel 264
frequency 37, 190
frequency testing 190
friction 35, 66, 255–56
fuel system control principles 154–55
fuel systems
   air induction 137–49
   alternative 158–63
   common rail direct injection 149
   electronic diesel control (EDC) 149
   electronic fuel injection (EFI) 130
   faults and symptoms 153
   filters, fuel 135, 150
   gasoline direct injection (GDI) 145–46
   glow plugs 152–53
   injectors, fuel 136–37, 151
   pressure and volume, testing 132–33
   pressure regulators, fuel 135–36
   pumps, fuel 131–35, 150, 151
   rail and pressure regulator 151
   sensors 152
   single point/throttle body injection 129–30

supply circuit 131–37, 150–52
   tanks, fuel 131, 150
fuelling common terms 123–25
fulcrum 258, 259

G
gasoline direct injection (GDI) 145–46
gearboxes
   automatic transmission 274–83, 296
   baulk rings 270–71
   continuously variable transmission (CVT) 280–83
   diagnosis and repair 278–79
   different engine types 266
   electronic control system 276–78
   epicyclic gear train 275–76
   gear types 267–68
   interlock mechanism 271–72
   manual 266–68, 272–73, 297
   need for 266
   ratios 268–69
   reverse gear 269
   seamless shift dual clutch systems (DCS) 273–74
   selector hubs 270
   sequential manual 272–73
   symptoms and faults 296
   synchromesh 270
generic 40
geometry 51
gestures 310
global positioning systems (GPS) 224–25
glow plugs 152–53, 192
goodwill 321
graduated container 133

H
Haldex coupling 294–95
Hall effect sensors 111–12
hazards 10
headlamps 209–210
health and safety
   auxiliary control systems 174
   car systems 234–44
   customer service 337–38
   diagnosis and rectification 30
   engine faults 80
   environmental protection 12
   fire 14–15
   first aid 16–19
   ignition systems 113
   legislation 5–10
   personal protective equipment (PPE) 2
   safety signs 12–14
   transmission and driveline faults 250
Health and Safety at Work Act 1974 (HASAWA) 6
Health and Safety Executive (HSE) 5–6
heating 164, 217–18
helical gears 267–68
helix 91
hertz 190
high intensity discharge (HID) 210–11

351

high resistance circuits 202, 218
high tension (HT) 106
hot wire mass airflow sensors (MAF)
   140
hubs 285
hybrid drive battery charging 207
hybrid vehicle regenerative braking 50
hybrid vehicles 161–63
hydraulic 57
hydraulic pistons 58
hydraulic power assistance 57–62
hydro-locked 96
hydrogen fuel cells 160–61
hypoid 287

I
ignition control principles 154
ignition module 112
ignition systems
   capacitor discharge ignition (CDI)
      119–20
   components of systems 106
   computer-controlled electronic 114
   diagnosis checklist 107
   direct 119
   distributorless systems 117
   electronic 110–16
   faults and symptoms 120
   Hall effect sensors 111–12
   health and safety 113
   ignition amplifiers 112
   ignition coils 108–109
   ignition control principles 154
   knock sensors 116–17
   primary circuit testing 109–10
   secondary circuits 114, 115–16
   spark plugs 107–10
   variable dwell angle 113
   voltage requirements 114
   wasted spark systems 117–19
image, personal 307–308, 311–12
immobilisers 229
impairment 339
impedance 114
improvement notices 5
in-car entertainment (ICE) system
   219–23
in series 60
inboard 37
incandescently 208
inch back 215
inductive amps clamp 178
inductive amps measurement 191–93
inductive magnetic sensors 37
inductive sensors 111
inert gas 208
inertia 66, 239
Information Commissioner 336
information sources
   auxiliary control systems 175
   chassis system faults 31
   engine faults 81
   transmission and driveline faults 251
injectors, fuel 136–37, 151
injuries 18–19

insulation 114
integrated circuits 112
integrated motor system 163
integrity 98
intelligent front lighting 213
intermediate 91
ionise 210

J
jargon 318
journal 98
judder 258

K
keyless entry 233
kinetic 35, 262
king pin inclination (KPI) 52–53
knock sensors 116–17

L
lambda sensors 143–45
lane change control 219
language, positive/negative 314
laser thermometers 34, 84, 178, 254
latent heat 87
launch control 263
lazy fuel pump 192
lead-acid batteries 203
leakdown testing 100–101
legislation 5
   alarm systems 229
   consumer 334–37
   Environmental Protection (Duty of
      Care) Regulations 1991 12
   Equality Act 2010 338
   health and safety 5–10
   radio signals 221
light emitting diodes (LED) 212–13
lighting 208–14
limited slip differentials (LSD) 289–92
liquefied petroleum gas (LPG) 158
liquid-cooled alternators 206
listening skills 308–309
locking systems 232–33
logic probes 33, 83, 177, 253
longitudinally 94
loss of consciousness 19

M
management, engine 154–57
manifold absolute pressure sensors
   (MAP) 139
manoeuvre 51
mass 66
measurement, engine 104
mechanical advantage 259
membrane 160
merchantable 334
metal hybride batteries 205
micro-switch 216
mirrors 216–17
mobile phone systems 225–26
molecules 160
momentum 266
MOT requirements for emissions 125

MP3 222
multimeters 33, 83, 177, 185–93, 253
multiplexing 198–201
mylar 237

N
naturally aspirated 85
navigation systems 223–25
network 198
networked systems 155–56, 198–201
new technology
   brake-by-wire systems 39
   collaborative motor drive 163
   digital radios 221
   emergency braking facility 47
   hybrid vehicle regenerative braking
      50
   integrated motor system 163
   launch control 263
   liquid-cooled alternators 206
   metal hybride batteries 205
   self-parking cars 63
   shut down period 88
   speed of network systems 201
   steer by wire 60
   wideband oxygen sensors
      145
Newton metres 268
nickel cadmium batteries 204
nodes 198

O
objects in the eye 19
obligatory 208
OEM 196–97
ohmmeters 43–44
ohms 179
Ohm's law 181–82
oil pressure testing 99
oil pump failure 98
one touch 215
opacity meters 125
open circuits 201–202, 218
open loop 144
organic matter 158
orifice 167
oscillation 71
oscilloscopes 33, 44–45, 83, 115–16,
   117, 138–39, 143, 147–48, 177,
   194–95, 253
ovality 98
overdrive 266
oxidisation 183

P
parallel 162
parallel circuits 180
parasites 88
parasitic drain 192, 202–203, 218
parking 231
particulate matter 125
patronising 319
perforations 96
performance dynamometer 84
personal protective equipment (PPE) 2